21

This report contains the colle
experts and does not necessai
of the United Nations Enviro
Organisation or the World H<

MW01532615

Environmental Health Criteria 213

CARBON MONOXIDE
(SECOND EDITION)

First draft prepared by Mr J. Raub, US Environmental Protection
Agency, Research Triangle Park, North Carolina, USA

Published under the joint sponsorship of the United Nations
Environment Programme, the International Labour
Organisation and the World Health Organization, and
produced within the framework of the Inter-Organization
Programme for the Sound Management of Chemicals.

World Health Organization
Geneva, 1999

The **International Programme on Chemical Safety (IPCS)**, established in 1980, is a joint venture of the United Nations Environment Programme (UNEP), the International Labour Organisation (ILO) and the World Health Organization (WHO). The overall objectives of the IPCS are to establish the scientific basis for assessment of the risk to human health and the environment from exposure to chemicals, through international peer review processes, as a prerequisite for the promotion of chemical safety, and to provide technical assistance in strengthening national capacities for the sound management of chemicals.

The **Inter-Organization Programme for the Sound Management of Chemicals (IOMC)** was established in 1995 by UNEP, ILO, the Food and Agriculture Organization of the United Nations, WHO, the United Nations Industrial Development Organization and the Organisation for Economic Co-operation and Development (Participating Organizations), following recommendations made by the 1992 UN Conference on Environment and Development to strengthen cooperation and increase coordination in the field of chemical safety. The purpose of the IOMC is to promote coordination of the policies and activities pursued by the Participating Organizations, jointly or separately, to achieve the sound management of chemicals in relation to human health and the environment.

WHO Library Cataloguing-in-Publication Data

Carbon monoxide.

(Environmental health criteria ; 213)

1.Carbon monoxide - adverse effects 2.Carbon monoxide - pharmacology
3.Environmental monitoring - methods 4.Environmental exposure 5.Risk factors
I.International Programme on Chemical Safety II.Series

ISBN 92 4 157213 2   (NLM classification: QV 662)
ISSN 0250-863X

Printed in Finland
99/12812 – Vammala – 5000

CONTENTS

ENVIRONMENTAL HEALTH CRITERIA FOR CARBON MONOXIDE

PREAMBLE

ABBREVIATIONS

NOTE TO READERS OF THE CRITERIA MONOGRAPHS

Every effort has been made to present information in the criteria monographs as accurately as possible without unduly delaying their publication. In the interest of all users of the Environmental Health Criteria monographs, readers are requested to communicate any errors that may have occurred to the Director of the International Programme on Chemical Safety, World Health Organization, Geneva, Switzerland, in order that they may be included in corrigenda.

* * *

A detailed data profile and a legal file can be obtained from the International Register of Potentially Toxic Chemicals, Case postale 356, 1219 Châtelaine, Geneva, Switzerland (telephone no. + 41 22 - 9799111, fax no. + 41 22 - 7973460, E-mail irptc@unep.ch).

Environmental Health Criteria

PREAMBLE

Objectives

In 1973, the WHO Environmental Health Criteria Programme was initiated with the following objectives:

(i) to assess information on the relationship between exposure to environmental pollutants and human health, and to provide guidelines for setting exposure limits;

(ii) to identify new or potential pollutants;

(iii) to identify gaps in knowledge concerning the health effects of pollutants;

(iv) to promote the harmonization of toxicological and epidemiological methods in order to have internationally comparable results.

The first Environmental Health Criteria (EHC) monograph, on mercury, was published in 1976, and since that time an ever-increasing number of assessments of chemicals and of physical effects have been produced. In addition, many EHC monographs have been devoted to evaluating toxicological methodology, e.g., for genetic, neurotoxic, teratogenic and nephrotoxic effects. Other publications have been concerned with epidemiological guidelines, evaluation of short-term tests for carcinogens, biomarkers, effects on the elderly and so forth.

Since its inauguration, the EHC Programme has widened its scope, and the importance of environmental effects, in addition to health effects, has been increasingly emphasized in the total evaluation of chemicals.

The original impetus for the Programme came from World Health Assembly resolutions and the recommendations of the 1972 UN Conference on the Human Environment. Subsequently, the work became an integral part of the International Programme on Chemical

Safety (IPCS), a cooperative programme of UNEP, ILO and WHO. In this manner, with the strong support of the new partners, the importance of occupational health and environmental effects was fully recognized. The EHC monographs have become widely established, used and recognized throughout the world.

The recommendations of the 1992 UN Conference on Environment and Development and the subsequent establishment of the Intergovernmental Forum on Chemical Safety with the priorities for action in the six programme areas of Chapter 19, Agenda 21, all lend further weight to the need for EHC assessments of the risks of chemicals.

Scope

The criteria monographs are intended to provide critical reviews on the effects on human health and the environment of chemicals and of combinations of chemicals and physical and biological agents. As such, they include and review studies that are of direct relevance for the evaluation. However, they do not describe *every* study carried out. Worldwide data are used and are quoted from original studies, not from abstracts or reviews. Both published and unpublished reports are considered, and it is incumbent on the authors to assess all the articles cited in the references. Preference is always given to published data. Unpublished data are used only when relevant published data are absent or when they are pivotal to the risk assessment. A detailed policy statement is available that describes the procedures used for unpublished proprietary data so that this information can be used in the evaluation without compromising its confidential nature (WHO (1990) Revised Guidelines for the Preparation of Environmental Health Criteria Monographs. PCS/90.69, Geneva, World Health Organization).

In the evaluation of human health risks, sound human data, whenever available, are preferred to animal data. Animal and *in vitro* studies provide support and are used mainly to supply evidence missing from human studies. It is mandatory that research on human subjects is conducted in full accord with ethical principles, including the provisions of the Helsinki Declaration.

The EHC monographs are intended to assist national and international authorities in making risk assessments and subsequent risk management decisions. They represent a thorough evaluation of risks and are not, in any sense, recommendations for regulation or standard setting. These latter are the exclusive purview of national and regional governments.

Content

The layout of EHC monographs for chemicals is outlined below.

*   Summary — a review of the salient facts and the risk evaluation of the chemical
*   Identity — physical and chemical properties, analytical methods
*   Sources of exposure
*   Environmental transport, distribution, and transformation
*   Environmental levels and human exposure
*   Kinetics and metabolism in laboratory animals and humans
*   Effects on laboratory mammals and *in vitro* test systems
*   Effects on humans
*   Effects on other organisms in the laboratory and field
*   Evaluation of human health risks and effects on the environment
*   Conclusions and recommendations for protection of human health and the environment
*   Further research
*   Previous evaluations by international bodies, e.g., IARC, JECFA, JMPR

Selection of chemicals

Since the inception of the EHC Programme, the IPCS has organized meetings of scientists to establish lists of priority chemicals for subsequent evaluation. Such meetings have been held in: Ispra, Italy, 1980; Oxford, United Kingdom, 1984; Berlin, Germany, 1987; and North Carolina, USA, 1995. The selection of chemicals has been based on the following criteria: the existence of scientific evidence that the substance presents a hazard to human health and/or the environment; the possible use, persistence, accumulation or degradation of the substance shows that there may be significant human or environmental exposure; the size and nature of populations at risk (both human and other species) and risks for the environment; international concern, i.e., the substance is of major interest to several countries; adequate data on the hazards are available.

If an EHC monograph is proposed for a chemical not on the priority list, the IPCS Secretariat consults with the cooperating organizations and all the Participating Institutions before embarking on the preparation of the monograph.

Procedures

The order of procedures that result in the publication of an EHC monograph is shown in the flow chart. A designated staff member of IPCS, responsible for the scientific quality of the document, serves as Responsible Officer (RO). The IPCS Editor is responsible for layout and language. The first draft, prepared by consultants or, more usually, staff from an IPCS Participating Institution, is based initially on data provided from the International Register of Potentially Toxic Chemicals and from reference databases such as Medline and Toxline.

The draft document, when received by the RO, may require an initial review by a small panel of experts to determine its scientific quality and objectivity. Once the RO finds the document acceptable as a first draft, it is distributed, in its unedited form, to well over 150 EHC contact points throughout the world who are asked to comment on its completeness and accuracy and, where necessary, provide additional material. The contact points, usually designated by governments, may be Participating Institutions, IPCS Focal Points or individual scientists known for their particular expertise. Generally, some four months are allowed before the comments are considered by the RO and author(s). A second draft incorporating comments received and approved by the Director, IPCS, is then distributed to Task Group members, who carry out the peer review, at least six weeks before their meeting.

The Task Group members serve as individual scientists, not as representatives of any organization, government or industry. Their function is to evaluate the accuracy, significance and relevance of the

EHC PREPARATION FLOW CHART

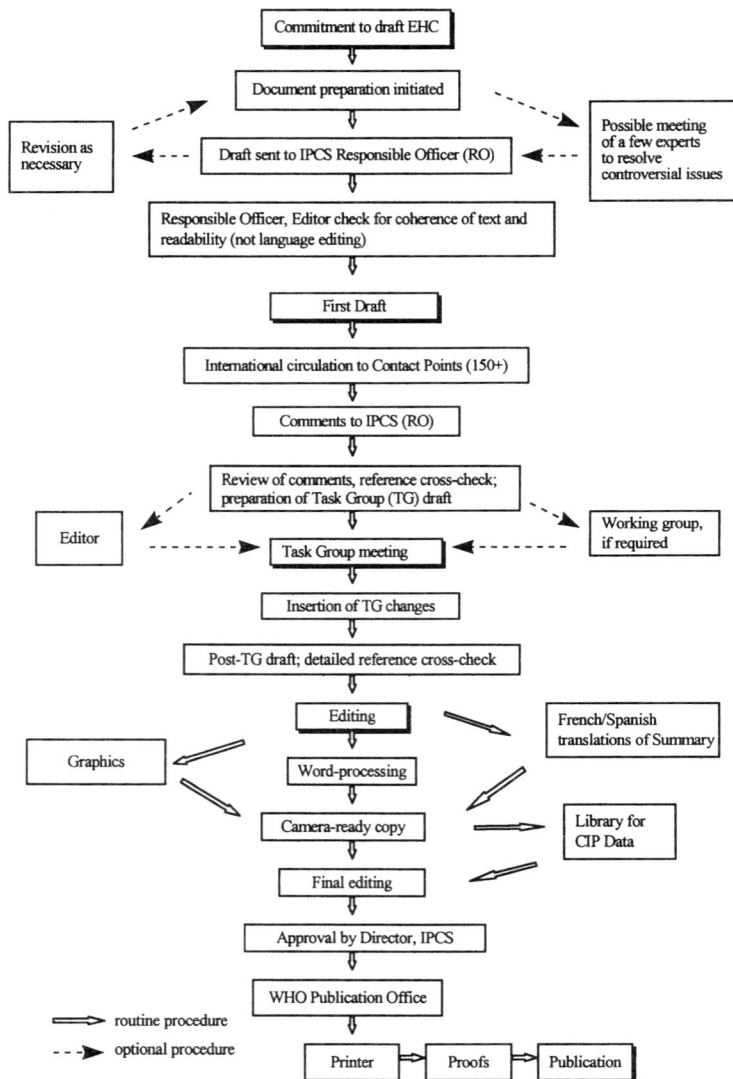

```
                        ┌──────────────────────────────┐
                        │     Commitment to draft EHC   │
                        └──────────────────────────────┘
                                     ⇩
                        ┌──────────────────────────────┐
                        │  Document preparation initiated│
                        └──────────────────────────────┘
           ┌───────────────┐    ⇩                    ┌──────────────────────┐
           │ Revision as   │ ←┄┄ ┌────────────────────────────────────┐ ←┄┄ │ Possible meeting     │
           │ necessary     │      │ Draft sent to IPCS Responsible Officer (RO) │     │ of a few experts     │
           └───────────────┘      └────────────────────────────────────┘      │ to resolve           │
                                     ⇩                                          │ controversial issues │
                        ┌──────────────────────────────────────────────┐       └──────────────────────┘
                        │ Responsible Officer, Editor check for coherence of text and │
                        │ readability (not language editing)           │
                        └──────────────────────────────────────────────┘
                                     ⇩
                        ┌──────────────────────────────┐
                        │          First Draft          │
                        └──────────────────────────────┘
                                     ⇩
                        ┌──────────────────────────────────────────────┐
                        │ International circulation to Contact Points (150+) │
                        └──────────────────────────────────────────────┘
                                     ⇩
                        ┌──────────────────────────────┐
                        │     Comments to IPCS (RO)     │
                        └──────────────────────────────┘
                                     ⇩
                        ┌──────────────────────────────────────────────┐
                        │ Review of comments, reference cross-check;     │
                        │ preparation of Task Group (TG) draft          │
                        └──────────────────────────────────────────────┘
           ┌──────────┐                                            ┌──────────────────┐
           │  Editor  │ ┄┄→ ┌──────────────────────┐ ←┄┄┄┄ │ Working group,    │
           └──────────┘      │   Task Group meeting  │              │ if required       │
                             └──────────────────────┘              └──────────────────┘
                                     ⇩
                        ┌──────────────────────────────┐
                        │    Insertion of TG changes    │
                        └──────────────────────────────┘
                                     ⇩
                        ┌──────────────────────────────────────────────┐
                        │ Post-TG draft; detailed reference cross-check │
                        └──────────────────────────────────────────────┘
                                     ⇩
                        ┌──────────────────────────────┐     ┌──────────────────────┐
                        │           Editing             │     │ French/Spanish         │
                        └──────────────────────────────┘     │ translations of Summary│
           ┌──────────┐        ⇩                              └──────────────────────┘
           │ Graphics │  ┌──────────────────────┐
           └──────────┘  │   Word-processing     │
                         └──────────────────────┘
                                     ⇩
                        ┌──────────────────────────────┐     ┌──────────────────┐
                        │      Camera-ready copy        │     │ Library for      │
                        └──────────────────────────────┘     │ CIP Data         │
                                     ⇩                         └──────────────────┘
                        ┌──────────────────────────────┐
                        │         Final editing         │
                        └──────────────────────────────┘
                                     ⇩
                        ┌──────────────────────────────┐
                        │  Approval by Director, IPCS   │
                        └──────────────────────────────┘
                                     ⇩
                        ┌──────────────────────────────┐
                        │    WHO Publication Office     │
                        └──────────────────────────────┘
   ⇒ routine procedure             ⇩
   ┄→ optional procedure   ┌─────────┐   ┌────────┐   ┌─────────────┐
                           │ Printer │ → │ Proofs │ → │ Publication │
                           └─────────┘   └────────┘   └─────────────┘
```

information in the document and to assess the health and environmental risks from exposure to the chemical. A summary and recommendations for further research and improved safety aspects are also required. The composition of the Task Group is dictated by the range of expertise required for the subject of the meeting and by the need for a balanced geographical distribution.

The three cooperating organizations of the IPCS recognize the important role played by nongovernmental organizations. Representatives from relevant national and international associations may be invited to join the Task Group as observers. While observers may provide a valuable contribution to the process, they can speak only at the invitation of the Chairperson. Observers do not participate in the final evaluation of the chemical; this is the sole responsibility of the Task Group members. When the Task Group considers it to be appropriate, it may meet *in camera*.

All individuals who as authors, consultants or advisers participate in the preparation of the EHC monograph must, in addition to serving in their personal capacity as scientists, inform the RO if at any time a conflict of interest, whether actual or potential, could be perceived in their work. They are required to sign a conflict of interest statement. Such a procedure ensures the transparency and probity of the process.

When the Task Group has completed its review and the RO is satisfied as to the scientific correctness and completeness of the document, the document then goes for language editing, reference checking and preparation of camera-ready copy. After approval by the Director, IPCS, the monograph is submitted to the WHO Office of Publications for printing. At this time, a copy of the final draft is sent to the Chairperson and Rapporteur of the Task Group to check for any errors.

It is accepted that the following criteria should initiate the updating of an EHC monograph: new data are available that would substantially change the evaluation; there is public concern for health or environmental effects of the agent because of greater exposure; an appreciable time period has elapsed since the last evaluation.

All Participating Institutions are informed, through the EHC progress report, of the authors and institutions proposed for the drafting of the documents. A comprehensive file of all comments received on drafts of each EHC monograph is maintained and is available on request. The Chairpersons of Task Groups are briefed before each meeting on their role and responsibility in ensuring that these rules are followed.

WHO TASK GROUP ON ENVIRONMENTAL HEALTH CRITERIA FOR CARBON MONOXIDE

Members

Dr M.R. Carratu, Institute of Pharmacology, Medical School, University of Bari, Policlinico, Piazza G. Cesare, I-70124 Bari, Italy

Dr Qing Chen, Beijing Medical University, School of Public Health, Beijing, People's Republic of China

Dr G. Cotti, Agenzia Regionale per la Prevenzione e l'Ambiente dell'Emilia-Romagna (ARPA), Sezione Provinciale di Bologna, via Triachini 17, I-40138 Bologna, Italy

Dr M.J. Hazucha, UNC Center for Environmental Medicine and Lung Biology, School of Medicine, The University of North Carolina, Chapel Hill, North Carolina 27599-7310, USA

Dr M. Jantunen, KTL Environmental Health, EXPOLIS, Mannerheiminitie 166, FIN-00300 Helsinki, Finland

Professor E. Lahmann, Schützallee 136, D-14169 Berlin 37 (Dahlem), Germany

Dr P. Lauriola, Direzione Tecnica, ARPA Emilia Romagna, Via Po 5, I-40139 Bologna, Italy

Dr M. Mathieu-Nolf, Centre Anti-Poisons, 5 avenue Oscar Lambret, F-59037 Lille Cédex, France

Dr D. Pankow, Institute of Environmental Toxicology, Martin Luther University, Franzosenweg 1a, D-06097 Halle (Saale), Germany

Professor D.G. Penney, Department of Physiology, Wayne State University, School of Medicine, Detroit, Michigan 48201, USA

Dr J.A. Raub, National Center for Environmental Assessment, US Environmental Protection Agency (MD-52), Research Triangle Park, North Carolina 27711, USA (*Rapporteur*)

Professor J.A. Sokal, Institute of Occupational Medicine, Koscielna 13 str., PL-41-200 Sosnowiec, Poland

Dr F.M. Sullivan, Harrington House, 8 Harrington Road, Brighton, E. Sussex, BN1 6RE, United Kingdom (*Chairman*)

Observers

Dr J.H. Duffus, International Union of Pure and Applied Chemistry (IUPAC), The Edinburgh Centre for Toxicology, Heriot-Watt University, Riccarton, Edinburgh, EH14 4AS, United Kingdom

Professor A. Mutti, International Commission on Occupational Health (ICOH), Laboratory of Industrial Toxicology, University of Parma Medical School, Via A. Gramsci 14, I-43100 Parma, Italy

Secretariat

Dr B.H. Chen, Medical Officer, Assessment of Risk and Methodologies, International Programme on Chemical Safety, World Health Organization, Geneva, Switzerland (*Secretary*)

Dr M. Mercier, Director, International Programme on Chemical Safety, World Health Organization, Geneva, Switzerland

IPCS TASK GROUP ON ENVIRONMENTAL HEALTH CRITERIA FOR CARBON MONOXIDE (SECOND EDITION)

The Environmental Health Criteria for Carbon Monoxide (First edition) was published in 1979. Since then, a lot of new health data on carbon monoxide have emerged, and the global exposure scenario has also changed. The information presented in this edition focuses primarily on the data that have become available since the publication of the first edition.

A WHO Task Group on Environmental Health Criteria for Carbon Monoxide met in Bologna, Italy, from 26 to 30 May 1997. The meeting was organized and supported financially by the Agenzia Regionale per la Prevenzione e l'Ambiente dell'Emilia-Romagna (ARPA), Sezione Provinciale di Bologna, Italy. Dr M. Mercier, Director of the International Programme on Chemical Safety (IPCS), opened the meeting and welcomed the participants on behalf of the three IPCS cooperating organizations (UNEP/ILO/WHO). From the Secretariat, Dr B.H. Chen, IPCS, served as Secretary of the meeting. The Task Group reviewed and revised the draft criteria monograph and made an evaluation of the risks for human health from exposure to carbon monoxide, and made a recommendation on the air quality guidelines for carbon monoxide.

The first draft of this monograph was prepared by Mr J. Raub of the US Environmental Protection Agency (US EPA) in Research Triangle Park. The second draft was also prepared by Mr J. Raub, incorporating comments received following the circulation of the first draft to the IPCS Contact Points for Environmental Health Criteria monographs. Dr P. Pankow contributed to the final text of the metabolism chapter, Drs Carratù, Hazucha, and Penney contributed to the final text of the animal study chapter, and Mr Raub contributed significantly to the final text of the document.

Dr B.H. Chen, member of the IPCS Central Unit, and Ms M. Sheffer, Scientific Editor, Ottawa, Canada, were responsible for the overall scientific content and linguistic editing, respectively.

The efforts of all who helped in the preparation and finalization of the document are gratefully acknowledged.

Financial support for this Task Group meeting was provided by the Agenzia Regionale per la Prevenzione e l'Ambiente dell'Emilia Romagna (Regional Agency for Prevention and Environment of Emilia-Romagna Region, Italy).

ABBREVIATIONS

ACGIH	American Conference of Governmental Industrial Hygienists
A/F	air to fuel
BAT	biological tolerance limit
BEI	biological exposure index
CFK	Coburn-Forster-Kane
CI	confidence interval
CO	carbon monoxide
COHb	carboxyhaemoglobin
DNA	deoxyribonucleic acid
EPA	Environmental Protection Agency (USA)
FEV_1	forced expiratory volume in 1 s
FVC	forced vital capacity
IQ	intelligence quotient
LC_{50}	median lethal concentratoin
LD_{50}	median lethal dose
LOEL	lowest-observed-effect level
NAAQS	National Ambient Air Quality Standard
NADPH	reduced nicotinamide adenine dinucleotide phosphate
NDIR	non-dispersive infrared
NIOSH	National Institute for Occupational Safety and Health (USA)
NOEL	no-observed-effect level
O_2Hb	oxyhaemoglobin
P_{CO}	partial pressure of carbon monoxide
P_{O_2}	partial pressure of oxygen
PEM	personal exposure monitor
PGI_2	prostacyclin
$PM_{2.5}$	particulate matter with mass median aerodynamic diameter less than 2.5 µm
PM_{10}	particulate matter with mass median aerodynamic diameter less than 10 µm
ppb	part per billion
ppbv	part per billion by volume
ppm	part per million
REL	recommended exposure limit
RNA	ribonucleic acid

RR	relative risk
SD	standard deviation
SMR	standardized mortality ratio
STEL	short-term exposure limit
TLV	threshold limit value
TWA	time-weighted average
UV	ultraviolet
$\dot{V}_E$	minute ventilation
$\dot{V}O_2$ max	maximal oxygen uptake
WHO	World Health Organization

1. SUMMARY AND CONCLUSIONS

Carbon monoxide (CO) is a colourless, odourless gas that can be poisonous to humans. It is a product of the incomplete combustion of carbon-containing fuels and is also produced by natural processes or by biotransformation of halomethanes within the human body. With external exposure to additional carbon monoxide, subtle effects can begin to occur, and exposure to higher levels can result in death. The health effects of carbon monoxide are largely the result of the formation of carboxyhaemoglobin (COHb), which impairs the oxygen carrying capacity of the blood.

1.1 Chemistry and analytical methods

Methods available for the measurement of carbon monoxide in ambient air range from fully automated methods using the non-dispersive infrared (NDIR) technique and gas chromatography to simple semiquantitative manual methods using detector tubes. Because the formation of carboxyhaemoglobin in humans is dependent on many factors, including the variability of ambient air concentrations of carbon monoxide, carboxyhaemoglobin concentration should be measured rather than calculated. Several relatively simple methods are available for determining carbon monoxide by analysis either of the blood or of alveolar air that is in equilibrium with the blood. Some of these methods have been validated by careful comparative studies.

1.2 Sources and environmental levels of carbon monoxide in the environment

Carbon monoxide is a trace constituent of the troposphere, produced by both natural processes and human activities. Because plants can both metabolize and produce carbon monoxide, trace levels are considered a normal constituent of the natural environment. Although ambient concentrations of carbon monoxide in the vicinity of urban and industrial areas can substantially exceed global background levels, there are no reports of these currently measured levels of carbon monoxide producing any adverse effects on plants or microorganisms. Ambient concentrations of carbon monoxide, however, can be detrimental to human health and welfare, depending on the levels that occur in areas where humans live and work and on the susceptibility of exposed individuals to potentially adverse effects.

Trends in air quality data from fixed-site monitoring stations show a general decline in carbon monoxide concentrations, which reflects the efficacy of emission control systems on newer vehicles. Highway vehicle emissions in the USA account for about 50% of total emissions; non-highway transportation sources contribute 13%. The other categories of carbon monoxide emissions are other fuel combustion sources, such as steam boilers (12%); industrial processes (8%); solid waste disposal (3%); and miscellaneous other sources (14%).

Indoor concentrations of carbon monoxide are a function of outdoor concentrations, indoor sources, infiltration, ventilation and air mixing between and within rooms. In residences without sources, average carbon monoxide concentrations are approximately equal to average outdoor levels. The highest indoor carbon monoxide concentrations are associated with combustion sources and are found in enclosed parking garages, service stations and restaurants, for example. The lowest indoor carbon monoxide concentrations are found in homes, churches and health care facilities. Exposure studies show that passive cigarette smoke is associated with increasing a non-smoker's exposure by an average of about 1.7 mg/m^3 (1.5 ppm) and that use of a gas cooking range at home is associated with an increase of about 2.9 mg/m^3 (2.5 ppm). Other sources that may contribute to carbon monoxide in the home include combustion space and water heaters and coal- or wood-burning stoves.

1.3 Environmental distribution and transformation

Recent data on global trends in tropospheric carbon monoxide concentrations indicate a decrease over the last decade. Global background concentrations fall in the range of 60–140 µg/m^3 (50–120 ppb). Levels are higher in the northern hemisphere than in the southern hemisphere. Average background concentrations also fluctuate seasonally. Higher levels occur in the winter months, and lower levels occur in the summer months. About 60% of the carbon monoxide found in the non-urban troposphere is attributed to human activities, both directly from combustion processes and indirectly through the oxidation of hydrocarbons and methane that, in turn, arise from agricultural activities, landfills and other similar sources. Atmospheric reactions involving carbon monoxide can produce ozone in the troposphere. Other reactions may deplete concentrations of the hydroxyl radical, a key participant in the global removal cycles of many other natural and anthropogenic trace gases, thus possibly contributing to changes in atmospheric chemistry and, ultimately, to global climate change.

1.4   Population exposure to carbon monoxide

During typical daily activities, people encounter carbon monoxide in a variety of microenvironments — while travelling in motor vehicles, working at their jobs, visiting urban locations associated with combustion sources, or cooking and heating with domestic gas, charcoal or wood fires — as well as in tobacco smoke. Overall, the most important carbon monoxide exposures for a majority of individuals occur in the vehicle and indoor microenvironments.

The development of small, portable electrochemical personal exposure monitors (PEMs) has made possible the measurement of carbon monoxide concentrations encountered by individuals as they move through numerous diverse indoor and outdoor microenvironments that cannot be monitored by fixed-site ambient stations. Results of both exposure monitoring in the field and modelling studies indicate that individual personal exposure determined by PEMs does not directly correlate with carbon monoxide concentrations determined by using fixed-site monitors alone. This observation is due to the mobility of people and to the spatial and temporal variability of carbon monoxide concentrations. Although they fail to show a correlation between individual personal monitor exposures and simultaneous nearest fixed-site monitor concentrations, large-scale carbon monoxide human exposure field studies do suggest that aggregate personal exposures are lower on days of lower ambient carbon monoxide levels as determined by the fixed-site monitors and higher on days of higher ambient levels. These studies point out the necessity of having personal carbon monoxide measurements to augment fixed-site ambient monitoring data when total human exposure is to be evaluated. Data from these field studies can be used to construct and test models of human exposure that account for time and activity patterns known to affect exposure to carbon monoxide.

Evaluation of human carbon monoxide exposure situations indicates that occupational exposures in some workplaces or exposures in homes with faulty or unvented combustion appliances can exceed 110 mg carbon monoxide/m^3 (100 ppm), often leading to carboxy-haemoglobin levels of 10% or more with continued exposure. In contrast, such high exposure levels are encountered much less commonly by the general public exposed under ambient conditions. More frequently, exposures to less than 29–57 mg carbon monoxide/ m^3 (25–50 ppm) for any extended period of time occur among the general population; at the low exercise levels usually engaged in under such circumstances, the resulting carboxyhaemoglobin levels most typically remain 1–2% among non-smokers. These levels can be compared with the physiological norm for non-smokers, which is estimated to be in the range of 0.3–0.7% carboxyhaemoglobin. In smokers, however, baseline carboxyhaemoglobin concentrations average 4%, with a usual range of 3–8%, reflecting absorption of carbon monoxide from inhaled smoke.

Studies of human exposure have shown that motor vehicle exhaust is the most important source for regularly encountered elevated carbon monoxide levels. Studies indicate that the motor vehicle interior has the highest average carbon monoxide concentration (averaging 10–29 mg/m^3 [9–25 ppm]) of all microenvironments. Furthermore, commuting exposures have been shown to be highly variable, with some commuters breathing carbon monoxide in excess of 40 mg/m^3 (35 ppm).

The workplace is another important setting for carbon monoxide exposures. In general, apart from commuting to and from work, exposures at work exceed carbon monoxide exposures during non-work periods. Occupational and non-occupational exposures may overlay one another and result in a higher concentration of carbon monoxide in the blood. Most importantly, the nature of certain occupations carries an increased risk of high carbon monoxide exposure (e.g., those occupations involved directly with vehicle driving, maintenance and parking). Occupational groups exposed to carbon monoxide from vehicle exhaust include auto mechanics; parking garage and gas station attendants; bus, truck or taxi drivers; police; and warehouse workers. Certain industrial processes can expose workers to carbon monoxide produced directly or as a by-product; they include steel production, coke ovens, carbon black production and petroleum refining. Firefighters, cooks and construction workers may also be exposed at work to high carbon monoxide levels. Occupational exposures in industries or settings with carbon monoxide production represent some of the highest individual exposures observed in field monitoring studies.

1.5 Toxicokinetics and mechanisms of action of carbon monoxide

Carbon monoxide is absorbed through the lungs, and the concentration of carboxyhaemoglobin in the blood at any time will depend on several factors. When in equilibrium with ambient air, the carboxyhaemoglobin content of the blood will depend mainly on the concentrations of inspired carbon monoxide and oxygen. However, if equilibrium has not been achieved, the carboxyhaemoglobin concentration will also depend on the duration of exposure, pulmonary ventilation and the carboxyhaemoglobin originally present before inhalation of the contaminated air. In addition to its reaction with haemoglobin, carbon monoxide combines with myoglobin, cytochromes and metalloenzymes such as cytochrome c oxidase and cytochrome P-450. The health significance of these reactions is not clearly understood but is likely to be of less importance at ambient exposure levels than that of the reaction of the gas with haemoglobin.

The exchange of carbon monoxide between the air we breathe and the human body is controlled by both physical (e.g., mass transport and diffusion) and physiological (e.g., alveolar ventilation and cardiac output) processes. Carbon monoxide is readily absorbed from the lungs into the bloodstream. The final step in this process involves competitive binding between carbon monoxide and oxygen to haemoglobin in the red blood cell, forming carboxyhaemoglobin and oxyhaemoglobin (O_2Hb), respectively. The binding of carbon monoxide to haemoglobin, producing carboxyhaemoglobin and decreasing the oxygen carrying capacity of blood, appears to be the principal mechanism of action underlying the induction of toxic effects of low-level carbon monoxide exposures. The precise mechanisms by which toxic effects are induced via carboxyhaemoglobin formation are not understood fully but likely include the induction of a hypoxic state in many tissues of diverse organ systems. Alternative or secondary mechanisms of carbon monoxide-induced toxicity (besides carboxyhaemoglobin) have been hypothesized, but none has been demonstrated to operate at relatively low (near-ambient) carbon monoxide exposure levels. Blood carboxyhaemoglobin levels, then, are currently accepted as representing a useful physiological marker by which to estimate internal carbon monoxide burdens due to the combined contribution of (1) endogenously derived carbon monoxide and (2) exogenously derived carbon monoxide resulting from exposure to external sources of carbon monoxide. Carboxyhaemoglobin levels likely to result from particular patterns (concentrations, durations, etc.) of external carbon monoxide exposure can be estimated reasonably well from the Coburn-Forster-Kane (CFK) equation.

A unique feature of carbon monoxide exposure, therefore, is that the blood carboxyhaemoglobin level represents a useful biological marker of the dose that the individual has received. The amount of carboxyhaemoglobin formed is dependent on the concentration and duration of carbon monoxide exposure, exercise (which increases the amount of air inhaled per unit time), ambient temperature, health status and the characteristic metabolism of the individual exposed. The formation of carboxyhaemoglobin is a reversible process; however, because of the tight binding of carbon monoxide to haemoglobin, the elimination half-time is quite long, ranging from 2 to 6.5 h, depending on the initial levels of carboxyhaemoglobin and the ventilation rate of the individuals. This might lead to accumulation of carboxyhaemoglobin, and even relatively low concentrations of carbon monoxide might produce substantial blood levels of carboxyhaemoglobin.

The level of carboxyhaemoglobin in the blood may be determined directly by blood analysis or indirectly by measuring carbon monoxide in exhaled breath. The measurement of exhaled breath has the advantages of ease, speed, precision and greater subject acceptance than measurement of blood carboxyhaemoglobin. However, the accuracy of the breath measurement procedure and the validity of the Haldane relationship between breath and blood remain in question for exposures at low environmental carbon monoxide concentrations.

Because carboxyhaemoglobin measurements are not readily available in the exposed population, mathematical models have been developed to predict carboxyhaemoglobin levels from known carbon monoxide exposures under a variety of circumstances. The best all-around model for carboxyhaemoglobin prediction is still the equation developed by Coburn, Forster and Kane. The linear solution is useful for examining air pollution data leading to relatively low carboxy-haemoglobin levels, whereas the non-linear solution shows good predictive power even for high carbon monoxide exposures. The two regression models might be useful only when the conditions of application closely approximate those under which the parameters were estimated.

Although the principal cause of carbon monoxide toxicity at low exposure levels is thought to be tissue hypoxia due to carbon monoxide binding to haemoglobin, certain physiological aspects of carbon monoxide exposure are not explained well by decreases in the intracellular oxygen partial pressure related to the presence of car-boxyhaemoglobin. Consequently, secondary mechanisms of carbon monoxide toxicity related to intracellular uptake of carbon monoxide have been the focus of a great deal of research interest. Carbon monoxide binding to many intracellular compounds has been well documented both *in vitro* and *in vivo*; however, it is still uncertain whether or not intracellular uptake of carbon monoxide in the presence of haemoglobin is sufficient to cause either acute organ system dys-function or long-term health effects. The virtual absence of sensitive techniques capable of assessing intracellular carbon monoxide binding under physiological conditions has resulted in a variety of indirect approaches to the problem, as well as many negative studies.

Current knowledge pertaining to intracellular carbon monoxide binding suggests that the proteins most likely to be inhibited function-ally at relevant levels of carboxyhaemoglobin are myoglobin, found predominantly in heart and skeletal muscle, and cytochrome oxidase. The physiological significance of carbon monoxide uptake by myo-globin is uncertain at this time, but sufficient concentrations of car-boxymyoglobin could potentially limit the maximal oxygen uptake of exercising muscle. Although there is suggestive evidence for signifi-cant binding of carbon monoxide to cytochrome oxidase in heart and brain tissue, it is unlikely that significant carbon monoxide binding would occur at low carboxyhaemoglobin levels.

1.6 Health effects of exposure to carbon monoxide

The health significance of carbon monoxide in ambient air is largely due to the fact that it forms a strong bond with the haemo-globin molecule, forming carboxyhaemoglobin, which impairs the oxygen carrying capacity of the blood. The dissociation of oxyhaemo-globin in the tissues is also altered by the presence of carboxyhaemo-globin, so that delivery of oxygen to tissues is reduced further. The affinity of human haemoglobin for carbon monoxide is roughly 240 times that for oxygen, and the proportions of carboxyhaemoglobin and oxyhaemoglobin formed in blood are dependent largely on the partial pressures of carbon monoxide and oxygen.

Concerns about the potential health effects of exposure to carbon monoxide have been addressed in extensive studies with both humans and various animal species. Under varied experimental protocols, considerable information has been obtained on the toxicity of carbon monoxide, its direct effects on the blood and other tissues, and the manifestations of these effects in the form of changes in organ function. Many of the animal studies, however, have been conducted at extremely high levels of carbon monoxide (i.e., levels not found in ambient air). Although severe effects from exposure to these high levels of carbon monoxide are not directly germane to the problems resulting from exposure to current ambient levels of carbon monoxide, they can provide valuable information about potential effects of accidental exposure to carbon monoxide, particularly those exposures occurring indoors.

1.6.1 *Cardiovascular effects*

Decreased oxygen uptake and the resultant decreased work capacity under maximal exercise conditions have clearly been shown to occur in healthy young adults starting at 5.0% carboxyhaemoglobin, and several studies have observed small decreases in work capacity at carboxyhaemoglobin levels as low as 2.3–4.3%. These effects may have health implications for the general population in terms of potential curtailment of certain physically demanding occupational or recreational activities under circumstances of sufficiently high carbon monoxide exposure.

However, of greater concern at more typical ambient carbon monoxide exposure levels are certain cardiovascular effects (i.e., aggravation of angina symptoms during exercise) likely to occur in a smaller, but sizeable, segment of the general population. This group, chronic angina patients, is currently viewed as the most sensitive risk group for carbon monoxide exposure effects, based on evidence for aggravation of angina occurring in patients at carboxyhaemoglobin levels of 2.9–4.5%. Dose–response relationships for cardiovascular effects in coronary artery disease patients remain to be defined more conclusively, and the possibility cannot be ruled out at this time that such effects may occur at levels below 2.9% carboxyhaemoglobin. Therefore, new published studies are evaluated in this document to determine the effects of carbon monoxide on aggravation of angina at levels in the range of 2–6% carboxyhaemoglobin.

Five key studies have investigated the potential for carbon monoxide exposure to enhance the development of myocardial ischaemia during exercise in patients with coronary artery disease. An early study found that exercise duration was significantly decreased by the onset of chest pain (angina) in patients with angina pectoris at post-exposure carboxyhaemoglobin levels as low as 2.9%, representing an increase of 1.6% carboxyhaemoglobin over the baseline. Results of a large multicentre study demonstrated effects in patients with reproducible exercise-induced angina at post-exposure carboxyhaemoglobin levels of 3.2%, corresponding to an increase of 2.0% carboxyhaemoglobin from the baseline. Others also found similar effects in patients with obstructive coronary artery disease and evidence of exercise-induced ischaemia at post-exposure carboxyhaemoglobin levels of 4.1% and 5.9%, respectively, representing increases of 2.2% and 4.2% carboxyhaemoglobin over the baseline. One study of subjects with angina found an effect at 3% carboxyhaemoglobin, representing an increase of 1.5% carboxyhaemoglobin from the baseline. Thus, the lowest-observed-adverse-effect level in patients with exercise-induced ischaemia is somewhere between 3% and 4% carboxyhaemoglobin, representing an increase of 1.5–2.2% carboxyhaemoglobin from the baseline. Effects on silent ischaemia episodes, which represent the majority of episodes in these patients, have not been studied.

The adverse health consequences of low-level carbon monoxide exposure in patients with ischaemic heart disease are very difficult to predict in the at-risk population of individuals with heart disease. Exposure to carbon monoxide that is sufficient to achieve 6% carboxyhaemoglobin, but not lower levels of carboxyhaemoglobin, has been shown to significantly increase the number and complexity of exercise-induced arrhythmias in patients with coronary artery disease and baseline ectopy. This finding, combined with the time-series studies of carbon monoxide-related morbidity and mortality and the epidemiological work of tunnel workers who are routinely exposed to automobile exhaust, is suggestive but not conclusive evidence that carbon monoxide exposure may provide an increased risk of sudden death from arrhythmia in patients with coronary artery disease.

Previous assessments of the cardiovascular effects of carbon monoxide have identified what appears to be a linear relationship between the level of carboxyhaemoglobin in the blood and decrements in human maximal exercise performance, measured as maximal oxygen uptake. Exercise performance consistently decreases at a blood level of about 5% carboxyhaemoglobin in young, healthy, non-smoking individuals. Some studies have even observed a decrease in short-term maximal exercise duration at levels as low as 2.3–4.3% carboxyhaemoglobin; however, this decrease is so small as to be of concern mainly for competing athletes rather than for ordinary people conducting the activities of daily life.

There is also evidence from both theoretical considerations and experimental studies in laboratory animals that carbon monoxide can adversely affect the cardiovascular system, depending on the exposure conditions utilized in these studies. Although disturbances in cardiac rhythm and conduction have been noted in healthy and cardiac-impaired animals, results from these studies are not conclusive. The lowest level at which effects have been observed varies, depending upon the exposure regime used and species tested. Results from animal studies also indicate that inhaled carbon monoxide can increase haemoglobin concentration and haematocrit ratio, which probably represents a compensation for the reduction in oxygen transport caused by carbon monoxide. At high carbon monoxide concentrations, excessive increases in haemoglobin and haematocrit may impose an additional workload on the heart and compromise blood flow to the tissues.

There is conflicting evidence that carbon monoxide exposure will enhance development of atherosclerosis in laboratory animals, and most studies show no measurable effect. Similarly, the possibility that carbon monoxide will promote significant changes in lipid metabolism that might accelerate atherosclerosis is suggested in only a few studies. Any such effect must be subtle, at most. Finally, carbon monoxide probably inhibits rather than promotes platelet aggregation. In general, there are few data to indicate that an atherogenic effect of exposure would be likely to occur in human populations at commonly encountered levels of ambient carbon monoxide.

.6.2   Acute pulmonary effects

It is unlikely that carbon monoxide has any direct effects on lung tissue except for extremely high concentrations associated with carbon monoxide poisoning. Human studies on the effects of carbon monoxide on pulmonary function are complicated by the lack of adequate exposure information, the small number of subjects studied and the short exposures explored. Occupational or accidental exposure to the products of combustion and pyrolysis, particularly indoors, may lead to acute decrements in lung function if the carboxyhaemoglobin levels are high. It is difficult, however, to separate the potential effects of carbon monoxide from those due to other respiratory irritants in the smoke and exhaust. Community population studies on carbon monoxide in ambient air have not found any significant relationship with pulmonary function, symptomatology and disease.

.6.3   Cerebrovascular and behavioural effects

No reliable evidence demonstrating decrements in neurobehavioural function in healthy, young adults has been reported at carboxyhaemoglobin levels below 5%. Results of studies conducted at or above 5% carboxyhaemoglobin are equivocal. Much of the research at 5% carboxyhaemoglobin did not show any effect even when behaviours similar to those affected in other studies at higher carboxyhaemoglobin levels were involved. However, investigators failing to find carbon monoxide-related neurobehavioural decrements at 5% or higher carboxyhaemoglobin levels may have utilized tests not sufficiently sensitive to reliably detect small effects of carbon monoxide. From the empirical evidence, then, it can be said that carboxyhaemoglobin levels greater than or equal to 5% may produce decrements in neurobehavioural function. It cannot be said confidently, however, that carboxyhaemoglobin levels lower than 5% would be without effect. However, only young, healthy adults have been studied using demonstrably sensitive tests and carboxyhaemoglobin levels of 5% or greater. The question of groups at special risk for neurobehavioural effects of carbon monoxide, therefore, has not been explored.

Of special note are those individuals who are taking drugs with primary or secondary depressant effects that would be expected to exacerbate carbon monoxide-related neurobehavioural decrements. Other groups at possibly increased risk for carbon monoxide-induced neurobehavioural effects are the aged and ill, but these groups have not been evaluated for such risk.

Under normal circumstances, the brain can increase blood flow or tissue oxygen extraction to compensate for the hypoxia caused by exposure to carbon monoxide. The overall responses of the cerebro-vasculature are similar in the fetus, newborn and adult animal; however, the mechanism of the increase in cerebral blood flow is still unclear. In fact, several mechanisms working simultaneously to increase blood flow appear likely, and these may involve metabolic and neural aspects as well as the oxyhaemoglobin dissociation curve, tissue oxygen levels and even a histotoxic effect of carbon monoxide. Whether these compensatory mechanisms will continue to operate successfully in a variety of conditions where the brain or its vasculature are compromised (i.e., stroke, head injury, atherosclerosis, hypertension) is also unknown. Aging increases the probability of such injury and disease. It is also possible that there exist individual differences with regard to carboxyhaemoglobin sensitivity and compensatory mechanisms.

Behaviours that require sustained attention or sustained performance are most sensitive to disruption by carboxyhaemoglobin. The group of human studies on hand–eye coordination (compensatory tracking), detection of infrequent events (vigilance) and continuous performance offers the most consistent and defensible evidence of carboxyhaemoglobin effects on behaviour at levels as low as 5%. These effects at low carbon monoxide exposure concentrations, however, have been very small and somewhat controversial. Nevertheless, the potential consequences of a lapse of coordination and vigilance on the continuous performance of critical tasks by operators of machinery such as public transportation vehicles could be serious.

1.6.4 Developmental toxicity

Studies in several laboratory animal species provide strong evidence that maternal carbon monoxide exposures of 170–230 mg/m^3 (150–200 ppm), leading to approximately 15–25% carboxyhaemoglobin, produce reductions in birth weight, cardiomegaly, delays in behavioural development and disruption in cognitive function. Isolated experiments suggest that some of these effects may be present at concentrations as low as 69–74 mg/m³ (60–65 ppm; approximately 6–11% carboxyhaemoglobin) maintained throughout gestation. Studies relating human carbon monoxide exposure from ambient sources or cigarette smoking to reduced birth weight are of concern because of the risk for developmental disorders; however, many of these studies have not considered all sources of carbon monoxide. The current data from children suggesting a link between environmental carbon monoxide exposures and sudden infant death syndrome are weak.

6.5 **Other systemic effects**

Laboratory animal studies suggest that enzyme metabolism of xenobiotic compounds may be affected by carbon monoxide exposure. Most of the authors of these studies have concluded, however, that effects on metabolism at low carboxyhaemoglobin levels (≤ 15%) are attributable entirely to tissue hypoxia produced by increased levels of carboxyhaemoglobin, because they are no greater than the effects produced by comparable levels of hypoxic hypoxia. At higher levels of exposure, where carboxyhaemoglobin concentrations exceed 15–20%, carbon monoxide may directly inhibit the activity of mixed-function oxidases. The decreases in xenobiotic metabolism shown with carbon monoxide exposure might be important to individuals receiving treatment with drugs.

Inhalation of high levels of carbon monoxide, leading to carboxyhaemoglobin concentrations greater than 10–15%, has been reported to cause a number of other systemic effects in laboratory animals, as well as effects in humans suffering from acute carbon monoxide poisoning. Tissues of highly active oxygen metabolism, such as heart, brain, liver, kidney and muscle, may be particularly sensitive to carbon monoxide poisoning. The effects of high levels of carbon monoxide on other tissues are not as well known and are, therefore, less certain. There are reports in the literature of effects on liver, kidney, bone and the immune capacity of the lung and spleen. It is generally agreed that the severe tissue damage occurring during acute carbon monoxide poisoning is due to one or more of the following: (1) ischaemia resulting from the formation of carboxyhaemoglobin, (2) inhibition of oxygen release from oxyhaemoglobin, (3) inhibition of cellular cytochrome function (e.g., cytochrome oxidases) and (4) metabolic acidosis.

Only relatively weak evidence points towards possible carbon monoxide effects on fibrinolytic activity, and then only at rather high carbon monoxide exposure levels. Similarly, whereas certain data also suggest that perinatal effects (e.g., reduced birth weight, slowed post-natal development, sudden infant death syndrome) are associated with carbon monoxide exposure, insufficient evidence exists by which to either qualitatively confirm such an association in humans or establish any pertinent exposure–effect relationships.

1.6.6 Adaptation

The only evidence for short- or long-term compensation for or adaptation to increased carboxyhaemoglobin levels in the blood is indirect. Experimental animal data indicate that increased carboxy-haemoglobin levels produce physiological responses that tend to offset other deleterious effects of carbon monoxide exposure. Such responses are (1) increased coronary blood flow, (2) increased cerebral blood flow, (3) increased haemoglobin through increased haematopoiesis and (4) increased oxygen consumption in muscle.

Short-term compensatory responses in blood flow or oxygen consumption may not be complete or might even be lacking in certain persons. For example, it is known from laboratory animal studies that coronary blood flow increases with increasing carboxyhaemoglobin, and it is known from human clinical studies that subjects with ischae-mic heart disease respond to the lowest levels of carboxyhaemoglobin (6% or less). The implication is that in some cases of cardiac impair-ment, the short-term compensatory mechanism is impaired.

From neurobehavioural studies, it is apparent that decrements due to carbon monoxide have not occurred consistently in all subjects, or even in the same studies, and have not demonstrated a dose–response relationship with increasing carboxyhaemoglobin levels. The implica-tion from these data is that there might be some threshold or time lag in a compensatory mechanism such as increased blood flow. Without direct physiological evidence in either laboratory animals or, prefer-ably, humans, this concept can only be hypothesized.

The mechanism by which long-term adaptation would occur, if it could be demonstrated in humans, is assumed to be an increased haemoglobin concentration via an increase in haematopoiesis. This alteration in haemoglobin production has been demonstrated repeat-edly in laboratory animal studies, but no recent studies have been conducted indicating or suggesting that some adaptational benefit has occurred or would occur. Furthermore, even if the haemoglobin increase is a signature of adaptation, it has not been demonstrated to occur at low ambient concentrations of carbon monoxide.

1.7 Combined exposure of carbon monoxide with altitude, drugs and other air pollutants and environmental factors

.7.1 High-altitude effects

Although there are many studies comparing and contrasting the effects of inhaling carbon monoxide with those produced by exposure to altitude, there are relatively few reports on the combined effects of inhaling carbon monoxide at altitude. There are data to support the possibility that the effects of these two hypoxia episodes are at least additive. These data were obtained at carbon monoxide concentrations that are too high to have much significance for regulatory concerns.

There are even fewer studies of the long-term effects of carbon monoxide at high altitude. These studies indicate few changes at carbon monoxide concentrations below 110 mg/m³ (100 ppm) and altitudes below 4570 m. The fetus, however, may be particularly sensitive to the effects of carbon monoxide at altitude; this is especially true with the high levels of carbon monoxide associated with maternal smoking.

.7.2 Carbon monoxide interaction with drugs

There remains little direct information on the possible enhancement of carbon monoxide toxicity by concomitant drug use or abuse; however, there are some data suggesting cause for concern. There is some evidence that interactions between drug effects and carbon monoxide exposure can occur in both directions; that is, carbon monoxide toxicity may be enhanced by drug use, and the toxic or other effects of drugs may be altered by carbon monoxide exposure. Nearly all the published data that are available on carbon monoxide combinations with drugs concern the use of alcohol.

The use and abuse of psychoactive drugs and alcohol are ubiquitous in society. Because of the effect of carbon monoxide on brain function, interactions between carbon monoxide and psychoactive drugs could be anticipated. Unfortunately, little systematic research has addressed this question. In addition, little of the research that has been done has utilized models for expected effects from treatment combinations. Thus, it is often not possible to assess whether the combined effects of drugs and carbon monoxide exposure are additive or differ from additivity. It is important to recognize that even additive effects of combinations can be of clinical significance, especially when the individual is unaware of the combined hazard. The greatest evidence for a potentially important interaction of carbon monoxide comes from studies with alcohol in both laboratory animals and humans, where at least additive effects have been obtained. The significance of this is augmented by the high probable incidence of combined alcohol use and carbon monoxide exposure.

1.7.3   Combined exposure of carbon monoxide with other air pollutants and environmental factors

Many of the data concerning the combined effects of carbon monoxide and other pollutants found in the ambient air are based on laboratory animal experiments. Only a few human studies are available. Early studies in healthy human subjects on common air pollutants such as carbon monoxide, nitrogen dioxide, ozone or peroxyacetyl nitrate failed to show any interaction from combined exposure. In laboratory studies, no interaction was observed following combined exposure to carbon monoxide and common ambient air pollutants such as nitrogen dioxide or sulfur dioxide. However, an additive effect was observed following combined exposure to high levels of carbon monoxide and nitric oxide, and a synergistic effect was observed after combined exposure to carbon monoxide and ozone.

Toxicological interactions of combustion products, primarily carbon monoxide, carbon dioxide and hydrogen cyanide, at levels typically produced by indoor and outdoor fires have shown a synergistic effect following carbon monoxide plus carbon dioxide exposure and an additive effect with hydrogen cyanide. Additive effects were also observed when carbon monoxide, hydrogen cyanide and low oxygen were combined; adding carbon dioxide to this combination was synergistic.

Finally, studies suggest that environmental factors such as heat stress and noise may be important determinants of health effects when combined with exposure to carbon monoxide. Of the effects described, the one potentially most relevant to typical human exposures is a greater decrement in the exercise performance seen when heat stress is combined with 57 mg carbon monoxide/m^3 (50 ppm).

.7.4 Tobacco smoke

Besides being a source of carbon monoxide for smokers as well as non-smokers, tobacco smoke is also a source of other chemicals with which environmental carbon monoxide could interact. Available data strongly suggest that acute and chronic carbon monoxide exposure attributed to tobacco smoke can affect the cardiopulmonary system, but the potential interaction of carbon monoxide with other products of tobacco smoke confounds the results. In addition, it is not clear if incremental increases in carboxyhaemoglobin caused by environmental exposure would actually be additive to chronically elevated carboxyhaemoglobin levels due to tobacco smoke, because some physiological adaptation may take place.

1.8 Evaluation of subpopulations potentially at risk from carbon monoxide exposure

Most information on the health effects of carbon monoxide involves two carefully defined population groups — young, healthy adults and patients with diagnosed coronary artery disease. On the basis of the known effects described, patients with reproducible exercise-induced ischaemia appear to be best established as a sensitive group within the general population that is at increased risk for experiencing health effects of concern (i.e., decreased exercise duration due to exacerbation of cardiovascular symptoms) at ambient or near-ambient carbon monoxide exposure concentrations that result in carboxyhaemoglobin levels down to 3%. A smaller sensitive group of healthy individuals experiences decreased exercise duration at similar levels of carbon monoxide exposure, but only during short-term maximal exercise. Decrements in exercise duration in the healthy population would therefore be of concern mainly to competing athletes, rather than to ordinary people carrying out the common activities of daily life.

It can be hypothesized, however, from both clinical and theoretical work and from experimental research on laboratory animals, that certain other groups in the population may be at probable risk from exposure to carbon monoxide. Identifiable probable risk groups can be categorized by gender differences; by age (e.g., fetuses, young infants and the elderly); by genetic variations (i.e., haemoglobin abnormalities); by pre-existing diseases, either known or unknown, that already decrease the availability of oxygen to critical tissues; or by the use of medications, recreational drugs or alterations in environment (e.g., exposure to other air pollutants or to high altitude). Unfortunately, little empirical evidence is currently available by which to specify health effects associated with ambient or near-ambient carbon monoxide exposures for most of these probable risk groups.

1.9 Carbon monoxide poisoning

Most of this document is concerned with the relatively low concentrations of carbon monoxide that induce effects in humans at, or near, the lower margin of carboxyhaemoglobin detection by current medical technology. Yet health effects associated with exposure to this pollutant range from the more subtle cardiovascular and neuro-behavioural effects at low ambient concentrations to unconsciousness and death after acute exposure to high concentrations of carbon monoxide. The morbidity and mortality resulting from the latter exposures can be a significant public health concern.

Carbon monoxide is responsible for a large percentage of the accidental poisonings and deaths reported throughout the world each year. Certain conditions exist in both the indoor and outdoor environments that cause a small percentage of the population to become exposed to dangerous levels of carbon monoxide. Outdoors, concentrations of carbon monoxide are highest near street intersections, in congested traffic, near exhaust gases from internal combustion engines and from industrial sources, and in poorly ventilated areas such as parking garages and tunnels. Indoors, carbon monoxide concentrations are highest in workplaces or in homes that have faulty or poorly vented combustion appliances or downdrafts or backdrafts.

The symptoms and signs of acute carbon monoxide poisoning correlate poorly with the level of carboxyhaemoglobin measured at the time of arrival at the hospital. Carboxyhaemoglobin levels below 10% are usually not associated with symptoms. At higher carboxyhaemoglobin saturations of 10–30%, neurological symptoms of carbon monoxide poisoning can occur, such as headache, dizziness, weakness, nausea, confusion, disorientation and visual disturbances. Exertional dyspnoea, increases in pulse and respiratory rates and syncope are observed with continuous exposure, producing carboxyhaemoglobin levels from 30% to 50%. When carboxyhaemoglobin levels are higher than 50%, coma, convulsions and cardiopulmonary arrest may occur.

Complications occur frequently in carbon monoxide poisoning (immediate death, myocardial impairment, hypotension, arrhythmias, pulmonary oedema). Perhaps the most insidious effect of carbon monoxide poisoning is the delayed development of neuropsychiatric impairment within 1–3 weeks and the neurobehavioural consequences, especially in children. Carbon monoxide poisoning during pregnancy results in high risk for the mother, by increasing the short-term complications rate, and for the fetus, by causing fetal death, developmental disorders and cerebral anoxic lesions. Furthermore, the severity of fatal intoxication cannot be assessed by the maternal rate.

Carbon monoxide poisoning occurs frequently, has severe consequences, including immediate death, involves complications and late sequelae and is often overlooked. Efforts in prevention and in public and medical education should be encouraged.

1.10 Recommended WHO guidelines

The following guideline values (ppm values rounded) and periods of time-weighted average exposures have been determined in such a way that the carboxyhaemoglobin level of 2.5% is not exceeded, even when a normal subject engages in light or moderate exercise:

100 mg/m^3 (87 ppm) for 15 min
60 mg/m^3 (52 ppm) for 30 min
30 mg/m^3 (26 ppm) for 1 h
10 mg/m^3 (9 ppm) for 8 h

2. CHEMISTRY AND ANALYTICAL METHODS

2.1 Physical and chemical properties

Carbon monoxide (CO) is a tasteless, odourless, colourless, non-corrosive and quite stable diatomic molecule that exists as a gas in the Earth's atmosphere. Radiation in the visible and near-ultraviolet (UV) regions of the electromagnetic spectrum is not absorbed by carbon monoxide, although the molecule does have weak absorption bands between 125 and 155 nm. Carbon monoxide absorbs radiation in the infrared region corresponding to the vibrational excitation of its electronic ground state. It has a low electric dipole moment (0.10 debye), short interatomic distance (0.123 nm) and high heat of formation from atoms or bond strength (2072 kJ/mol). These observations suggest that the molecule is a resonance hybrid of three structures (Perry et al., 1977), all of which contribute nearly equally to the normal ground state. General physical properties of carbon monoxide are given in Table 1.

2.2 Methods for measuring carbon monoxide in ambient air

2.2.1 Introduction

Because of the low levels of carbon monoxide in ambient air, methods for its measurement require skilled personnel and sophisticated analytical equipment. The principles of the methodology have been described by Smith & Nelson (1973). A sample introduction system is used, consisting of a sampling probe, an intake manifold, tubing and air movers. This system is needed to collect the air sample from the atmosphere and to transport it to the analyser without altering the original concentration. It may also be used to introduce known gas concentrations to periodically check the reliability of the analyser output. Construction materials for the sampling probe, intake manifold and tubing should be tested to demonstrate that the test atmosphere composition or concentration is not altered significantly. The sample introduction system should be constructed so that it presents no pressure drop to the analyser. At low flow and low concentrations, such operation may require validation.

Table 1. Physical properties of carbon monoxide[a]

Property	Value
Molecular weight	28.01
Critical point	-140 °C at 3495.7 kPa
Melting point	-199 °C
Boiling point	-191.5 °C
Density at 0 °C, 101.3 kPa at 25 °C, 101.3 kPa	 1.250 g/litre 1.145 g/litre
Specific gravity relative to air	0.967
Solubility in water[b] at 0 °C at 20 °C at 25 °C	 3.54 ml/100 ml (44.3 ppmm)[c] 2.32 ml/100 ml (29.0 ppmm)[c] 2.14 ml/100 ml (26.8 ppmm)[c]
Explosive limits in air	12.5–74.2%
Fundamental vibration transition	2143.3 cm^{-1}
Conversion factors at 0 °C, 101.3 kPa at 25 °C, 101.3 kPa	 1 mg/m^3 = 0.800 ppm[d] 1 ppm = 1.250 mg/m^3 1 mg/m^3 = 0.873 ppm[d] 1 ppm = 1.145 mg/m^3

[a] From NRC (1977).
[b] Volume of carbon monoxide is at 0 °C, 1 atm (atmospheric pressure at sea level = 101.3 kPa).
[c] Parts per million by mass (ppmm = μg/g).
[d] Parts per million by volume (ppm = mg/litre).

The analyser system consists of the analyser itself and any sample preconditioning components that may be necessary. Sample preconditioning might require a moisture control system to help minimize the false-positive response of the analyser (e.g., the non-dispersive infrared [NDIR] analyser) to water vapour and a particulate filter to help protect the analyser from clogging and possible chemical interference due to particulate buildup in the sample lines or analyser inlet. The sample preconditioning system may also include a flow metering and flow control device to control the sampling rate to the analyser.

A data recording system is needed to record the output of the analyser.

2.2.2   Methods

A reference method or equivalent method for air quality measurements is required for acceptance of measurement data. An equivalent method for monitoring carbon monoxide can be so designated when the method is shown to produce results equivalent to those from the approved reference monitoring method based on absorption of infrared radiation from a non-dispersed beam.

The designated reference methods are automated, continuous methods utilizing the NDIR technique, which is generally accepted as being the most reliable method for the measurement of carbon monoxide in ambient air. As of January 1988, no equivalent methods that use a principle other than NDIR have been designated for measuring carbon monoxide in ambient air.

There have been several excellent reviews on the measurement of carbon monoxide in the atmosphere (National Air Pollution Control Administration, 1970; Driscoll & Berger, 1971; Leithe, 1971; American Industrial Hygiene Association, 1972; NIOSH, 1972; Verdin, 1973; Stevens & Herget, 1974; Harrison, 1975; Schnakenberg, 1976; NRC, 1977; Repp, 1977; Lodge, 1989; OSHA, 1991a; ASTM, 1995; ISO, 1996).

2.2.2.1   Non-dispersive infrared photometry method

Currently, the most commonly used measurement technique is the type of NDIR method referred to as gas filter correlation (Acton et al., 1973; Burch & Gryvnak, 1974; Ward & Zwick, 1975; Burch et al., 1976; Goldstein et al., 1976; Gryvnak & Burch, 1976a,b; Herget et al., 1976; Bartle & Hall, 1977; Chaney & McClenny, 1977).

Carbon monoxide has a characteristic infrared absorption near 4.6 µm. The absorption of infrared radiation by the carbon monoxide molecule can therefore be used to measure the concentration of carbon monoxide in the presence of other gases. The NDIR method is based on this principle (Feldstein, 1967).

Most commercially available NDIR analysers incorporate a gas filter to minimize interferences from other gases. They operate at atmospheric pressure, and the most sensitive analysers are able to detect minimum carbon monoxide concentrations of about 0.05 mg/m^3

(0.044 ppm). Interferences from carbon dioxide and water vapour can be dealt with so as not to affect the data quality. NDIR analysers with detectors as designed by Luft (1962) are relatively insensitive to flow rate, require no wet chemicals, are sensitive over wide concentration ranges and have short response times. NDIR analysers of the newer gas filter correlation type have overcome zero and span problems and minor problems due to vibrations.

2.2.2   Gas chromatography method

A more sensitive method for measuring low background levels of carbon monoxide is gas chromatography (Porter & Volman, 1962; Feldstein, 1967; Swinnerton et al., 1968; Bruner et al., 1973; Dagnall et al., 1973; Tesarik & Krejci, 1974; Bergman et al., 1975; Smith et al., 1975; ISO, 1989). This technique is an automated, semicontinuous method in which carbon monoxide is separated from water, carbon dioxide and hydrocarbons other than methane by a stripper column. Carbon monoxide and methane are then separated on an analytical column, and the carbon monoxide is passed through a catalytic reduction tube, where it is converted to methane. The carbon monoxide (converted to methane) passes through a flame ionization detector, and the resulting signal is proportional to the concentration of carbon monoxide in the air. This method has been used throughout the world. It has no known interferences and can be used to measure levels from 0.03 to 50 mg/m^3 (0.026 to 43.7 ppm). These analysers are expensive and require continuous attendance by a highly trained operator to produce valid results. For high levels, a useful technique is catalytic oxidation of the carbon monoxide by Hopcalite or other catalysts (Stetter & Blurton, 1976), either with temperature-rise sensors (Naumann, 1975; Schnakenberg, 1976; Benzie et al., 1977) or with electrochemical sensors (Bay et al., 1972, 1974; Bergman et al., 1975; Dempsey et al., 1975; Schnakenberg, 1975; Repp, 1977). Numerous other analytical schemes have been used to measure carbon monoxide in air.

2.2.3   Other analysers

Other systems to measure carbon monoxide in ambient air include gas chromatography/flame ionization, in which carbon monoxide is separated from other trace gases by gas chromatography and catalytically converted to methane prior to detection; controlled-potential electrochemical analysis, in which carbon monoxide is measured by means of the current produced in aqueous solution by its electro-oxidation by an electro-catalytically active noble metal (the concentration of carbon monoxide reaching the electrode is controlled by its rate of diffusion through a membrane, which depends on its concentration in the sampled atmosphere; Bay et al., 1972, 1974); galvanic cells that can be used to measure atmospheric carbon monoxide continuously, in the manner described by Hersch (1964, 1966); coulometric analysis, which employs a modified Hersch-type cell; mercury replacement, in which mercury vapour formed by the reduction of mercuric oxide by carbon monoxide is detected photometrically by its absorption of UV light at 253.7 nm; dual-isotope fluorescence, which utilizes the slight difference in the infrared spectra of isotopes of carbon monoxide; catalytic combustion/thermal detection, which is based on measuring the temperature rise resulting from catalytic oxidation of the carbon monoxide in the sample air; second-derivative spectrometry, which utilizes a second-derivative spectrometer to process the transmission versus wavelength function of an ordinary spectrometer to produce an output signal proportional to the second derivative of this function; and Fourier-transform spectroscopy, which is an extremely powerful infrared spectroscopic technique.

Intermittent samples may be collected in the field and later analysed in the laboratory by the continuous analysing techniques described above. Sample containers may be rigid (glass cylinders or stainless steel tanks) or non-rigid (plastic bags). Because of location and cost, intermittent sampling may at times be the only practical method for air monitoring. Samples can be taken over a few minutes or accumulated intermittently to obtain, after analysis, either "spot" or "integrated" results.

Additional techniques for analysing intermittent samples include colorimetric analysis, in which carbon monoxide reacts in an alkaline solution with the silver salt of *p*-sulfamoyl-benzoate to form a coloured silver sol; a National Institute of Standards and Technology colorimetric indicating gel (incorporating palladium and molybdenum salts), which involves colorimetric comparison with freshly prepared indicating gels exposed to known concentrations of carbon monoxide; a length-of-stain indicator method, which uses an indicator tube containing potassium palladosulfite; and frontal analysis, in which air is passed over an adsorbent until equilibrium is established between the concentration of carbon monoxide in the air and the concentration of carbon monoxide on the adsorbent.

A simple and inexpensive measurement technique uses detector tubes (indicator tubes) (Leichnitz, 1993). This method is widely applied in industrial hygiene and is suitable for analysis of highly polluted atmospheric air. The measurement with Dräger tubes (Drägerwerk, 1994) is based on the reaction: $5CO + I_2O_5 \rightarrow I_2 + 5CO_2$. The iodine-coloured layer in the tube corresponds in length to the carbon monoxide concentration in the sample.

2.2.3 *Measurement using personal monitors*

Until the 1960s, most of the data available on ambient carbon monoxide concentrations came from fixed monitoring stations operated routinely in urban areas. The accepted measurement technique was NDIR spectrometry, but the instruments were large and cumbersome, often requiring vibration-free, air-conditioned enclosures. Without a portable, convenient monitor for carbon monoxide, it was extremely difficult to measure carbon monoxide concentrations accurately in the microenvironments that people usually visited.

Ultimately, small personal exposure monitors (PEMs) were developed that could measure carbon monoxide concentrations continuously over time and store the readings automatically on internal digital memories (Ott et al., 1986). These small PEMs made possible the large-scale field studies on human exposure to carbon monoxide in Denver, Colorado, and Washington, DC, USA, in the winter of 1982–83 (Akland et al., 1985).

2.2.4 *Carbon monoxide detectors/alarms*

Carbon monoxide detectors have been designed like residential smoke detectors — to be low cost, yet provide protection from a catastrophic event by sounding an audible alarm. The carbon monoxide detector industry is young, however, and is in a stage of rapid growth. In the USA, an estimated 7–8 million detectors have been purchased since the early 1990s, but the numbers used in homes will continue to rise as local municipalities change building codes to require the installation of carbon monoxide detectors in new residential structures containing combustion-source appliances, stoves or fireplaces.

Currently available carbon monoxide detectors are based on an interactive-type sensor (e.g., tin oxide, Figaro-type gel cell, fuel cell, artificial haemoglobin) that relies on direct interaction between carbon monoxide and the sensitive element in order to generate a response. They are battery-powered, alternating current-powered or both. The most popular alternating current-powered detectors have a heated metallic sensor that reacts with carbon monoxide; the battery-powered detectors have a chemically treated gel disk that darkens with exposure to carbon monoxide or a fuel cell. Small, inexpensive carbon monoxide detection cards or tablets that require frequent visual inspection of colour changes do not sound an alarm and are not recommended as a primary detector.

Carbon monoxide detectors are sensitive to location and environmental conditions, including temperature, relative humidity and the presence of other interfering gases. They may also become less stable with time. For example, they should not be installed in dead-space air (i.e., near ceilings), near windows or near doors where there is a lot of air movement, and they should not be exposed to temperature or humidity extremes. Excessive heat or cold will affect performance, and humidity extremes will affect the activation time. Utilization of non-interactive infrared technology (e.g., NDIR) in indoor carbon monoxide detection would overcome all of the shortcomings of the currently available carbon monoxide detectors.

In the USA, a new voluntary standard for carbon monoxide detectors was published in 1992 by the Underwriters Laboratories (UL Standard 2034) and revised in 1995. This standard provides alarm requirements for detectors that are based on both the carbon monoxide concentration and the exposure time. It is designed so that an alarm is activated within 90 min of exposure to 110 mg/m^3 (100 ppm), within 35 min of exposure to 230 mg/m^3 (200 ppm) or within 15 min of exposure to 460 mg/m^3 (400 ppm) (i.e., when exposures are equivalent to 10% carboxyhaemoglobin [COHb]; see section 2.3). Approximately 15 manufacturers produce detectors listed under UL Standard 2034.

Because UL Standard 2034 covers a wide range of exposure conditions, there has been some ambiguity about its interpretation. For example, it is not clear if a detector meets the standard if the alarm is activated anytime between 5 and 90 min in the presence of 110 mg carbon monoxide/mg^3 (100 ppm). In fact, alarm sensitivities are still a problem for the industry, and further discussion and direction are needed. Moreover, the 10% carboxyhaemoglobin level is protective of healthy individuals only (see chapter 8). It would be necessary to avoid exposures to 10 mg carbon monoxide/m^3 (9 ppm) for 8 h or 29 mg/m^3 (25 ppm) for 1 h in order to protect sensitive individuals with coronary heart disease at the 3% carboxyhaemoglobin level. Thus, current detectors provide warning against carbon monoxide levels that are protective of the healthy population only. Despite these limitations, carbon monoxide detectors are reliable and effective, continue to improve and should be recommended for use in homes in addition to smoke detectors and fire alarms.

2.3 Biological monitoring

A unique feature of carbon monoxide exposure is that there is a biological marker of the dose that the individual has received: the level of carbon monoxide in the blood. This level may be calculated by measuring carboxyhaemoglobin in blood or carbon monoxide in exhaled breath.

2.3.1 Blood carboxyhaemoglobin measurement

The level of dissolved carbon monoxide in blood is normally below the level of detection but may be of importance in the transportation of carbon monoxide between cells and tissues (see chapter 6). Thus, the blood level of carbon monoxide is conventionally represented as a percentage of the total haemoglobin available (i.e., the percentage of haemoglobin that is in the form of carboxyhaemoglobin, or simply percent carboxyhaemoglobin).

Any technique for the measurement of carboxyhaemoglobin in blood must be specific and must have sufficient sensitivity and accuracy for the purpose of the values obtained. The majority of technical methods that have been published on measurement of carbon monoxide in blood have been for forensic purposes. These methods are less accurate than generally required for the measurement of low levels of carboxyhaemoglobin (<5%). Blood levels of carbon monoxide resulting from exposure to existing ambient levels of carbon monoxide would not be expected to exceed 5% carboxyhaemoglobin in non-smoking subjects. The focus of the forensic methods has been the reliability of measurements over the entire range of possible values: from less than 1% to 100% carboxyhaemoglobin. These foren-sically oriented methods are adequate for the intended use of the values and the non-ideal storage conditions of the samples being analysed.

In the areas of exposure assessment and low-level health effects of carbon monoxide, it is more important to know the accuracy of any method in the low-level range of <5% carboxyhaemoglobin. There is little agreement upon acceptable reference methods in this range, nor are there accurate reference standards available in this range. The use of techniques that have unsubstantiated accuracy in the low range of carboxyhaemoglobin levels can lead to considerable differences in estimations of exposure conditions. Measurement of low levels of carboxyhaemoglobin demands careful evaluation because of the impli-cations, based upon these data, for the setting of air quality standards. Therefore, this section will focus on the methods that have been evaluated at levels below 10% carboxyhaemoglobin and the methods that have been extensively used in assessing exposure to carbon monoxide.

The measurement of carbon monoxide in blood can be accom-plished by a variety of techniques, both destructive and non-destructive. Carboxyhaemoglobin can be determined non-destructively by observing the change in the absorption spectrum in either the Soret or visible region brought about by the combination of carbon monox-ide with haemoglobin. With present optical sensing techniques, however, all optical methods are limited in sensitivity to approxi-mately 1% of the range of expected values. If attempts are made to expand the lower range of absorbances, sensitivity is lost on the upper end where, in the case of carboxyhaemoglobin, total haemoglobin is measured. For example, in the spectrophotometric method described by Small et al. (1971), a change in absorbance equal to the limit of resolution of 0.01 units can result in a difference in 0.6% carboxy-haemoglobin. Therefore, optical techniques cannot be expected to obtain the resolution that is possible with other means of detection of carbon monoxide (Table 2).

The more sensitive (higher-resolution) techniques require the release of the carbon monoxide from the haemoglobin into a gas phase; the carbon monoxide can then be detected directly by (1) infra-red absorption (Maas et al., 1970) following separation using gas

Table 2. Representative methods for the analysis of carbon monoxide in blood[a]

Source	Method	Resolution[b] (ml/dl)	CV (%)[c]	Reference method	r[d]
Gasometric detection					
Scholander & Roughton (1943)	Syringe capillary	0.02	2–4	Van Slyke	ND[e]
Horvath & Roughton (1942)	Van Slyke	0.03	6	Van Slyke– Neill	ND
Spectrophotometric detection					
Coburn et al. (1964)	Infrared	0.006	1.8	Van Slyke– Syringe	ND
Small et al. (1971)	Spectro- photometry	0.12	ND	Flame ionization	ND
Maas et al. (1970)	CO-Oximeter (IL-182)	0.21	5	Spectro- photometric	ND
Brown (1980)	CO-Oximeter (IL-282)	0.2	5	Flame ionization	0.999
Gas chromatography					
Ayres et al. (1966)	Thermal conductivity	0.001	2	ND	ND
Goldbaum et al. (1986)	Thermal conductivity	ND	1.35	Flame ionization	0.996
McCredie & Jose (1967)	Thermal conductivity	0.005	1.8	ND	ND
Dahms & Horvath (1974)	Thermal conductivity	0.006	1.7	Van Slyke	0.983
Collison et al. (1968)	Flame ionization	0.002	1.8	Van Slyke	ND
Kane (1985)	Flame ionization	ND	6.2	CO- Oximeter	1.00
Vreman et al. (1984)	Mercury vapour	0.002	2.2	ND	ND

[a] Modified from US EPA (1991d).
[b] The resolution is the smallest detectable amount of carbon monoxide or the smallest detectable difference between samples.
[c] Coefficient of variation (CV) was computed on samples containing less than 15% carboxyhaemoglobin, where possible.
[d] The r value is the correlation coefficient between the technique reported and the reference method used to verify its accuracy.
[e] ND indicates that no data were available.

chromatography, (2) the difference in thermal conductivity between carbon monoxide and the carrier gas (Ayres et al., 1966; McCredie & Jose, 1967; Dahms & Horvath, 1974; Goldbaum et al., 1986; Horvath et al., 1988b; Allred et al., 1989b), (3) the amount of ionization following quantitative conversion of carbon monoxide to methane and ionization of the methane (Collison et al., 1968; Dennis & Valeri, 1980; Guillot et al., 1981; Clerbaux et al., 1984; Kane, 1985; Katsumata et al., 1985; Costantino et al., 1986) or (4) the release of mercury vapour resulting from the combination of carbon monoxide with mercuric oxide (Vreman et al., 1984).

2.3.1.1 Sample handling

Carbon monoxide bound to haemoglobin is a relatively stable compound that can be dissociated by exposure to oxygen or UV radiation (Horvath & Roughton, 1942; Chace et al., 1986). If the blood sample is maintained in the dark under cool, sterile conditions, the carbon monoxide content will remain stable for a long period of time. Various investigators have reported no decrease in percent carboxyhaemoglobin over 10 days (Collison et al., 1968), 3 weeks (Dahms & Horvath, 1974), 4 months (Ocak et al., 1985) and 6 months (Vreman et al., 1984). The blood collection system used can influence the carbon monoxide level, because some ethylenediaminetetraacetic acid vacutainer tube stoppers contain carbon monoxide (Vreman et al., 1984). The stability of the carbon monoxide content in properly stored samples does not indicate that constant values will be obtained by all techniques of analysis. The spectrophotometric methods are particularly susceptible to changes in optical qualities of the sample, resulting in small changes in carboxyhaemoglobin with storage (Allred et al., 1989b).

Therefore, the care needed to make a carboxyhaemoglobin determination depends upon the technique that is being utilized. It appears as though measurement of low levels of carboxyhaemoglobin with optical techniques should be conducted as soon as possible following collection of the samples.

2.3.1.2 Potential reference methods

Exposure to carbon monoxide at equilibrium conditions results in carboxyhaemoglobin levels of between 0.1 and 0.2% for each milligram of carbon monoxide per cubic metre air (part per million).

A reference technique for the measurement of carboxyhaemoglobin should be able to discriminate between two blood samples with a difference of 0.1% carboxyhaemoglobin (approximately 0.02 ml/dl). To accomplish this task, the coefficient of variation (standard deviation of repeated measures on any given sample divided by the mean of the values times 100) of the method should be less than 5%, so that the two values that are different by 0.1 percentage points can be statistically proven to be distinct. In practical terms, a reference method should have the sensitivity to detect approximately 0.025% carboxyhaemoglobin to provide this level of confidence in the values obtained.

The accurate measurement of carboxyhaemoglobin requires the quantitation of the content of carbon monoxide released from haemoglobin in the blood. Optically based techniques have limitations of resolution and specificity due to the potential interference from many sources. The techniques that can be used as reference methods involve the quantitative release of carbon monoxide from the haemoglobin followed by the measurement of the amount of carbon monoxide released. Classically, this quantitation was measured manometrically with a Van Slyke apparatus (Horvath & Roughton, 1942) or a Roughton-Scholander syringe (Roughton & Root, 1945). These techniques have served as the "Gold Standard" in this field for almost 50 years. However, there are limitations of resolution with these techniques at the lower ranges of carboxyhaemoglobin. The gasometric standard methodology has been replaced with headspace extraction followed by the use of solid-phase gas chromatographic separation with several different types of detection: thermal conductivity, flame ionization and mercury vapour reduction. The carbon monoxide in the headspace can also be quantitated by infrared detection, which can be calibrated with gas standards. However, there is no general agreement that any of the more sensitive methods of carbon monoxide analysis are acceptable reference methods.

The following techniques all conform to all the requirements of a reference method:

(1) **Flame ionization detection:** This technique requires the separation of carbon monoxide from the other headspace gases and the reduction of the carbon monoxide to methane by catalytic reduction.

(2)   **Thermal conductivity detection:** This technique uses vacuum extraction of carbon monoxide from blood in a Van Slyke apparatus and gas chromatographic separation with thermal conductivity analysis of the carbon monoxide.

(3)   **Infrared detection:** This technique uses a method for extracting carbon monoxide from blood under normal atmospheric conditions and then injecting the headspace gas into an infrared analyser.

The conventional means of representing the quantity of carbon monoxide in a blood sample is the percent carboxyhaemoglobin: the percentage of the total carbon monoxide combining capacity that is in the form of carboxyhaemoglobin. This is conventionally determined by the use of the following formula:

$$\% \ COHb = [CO \ content/(Hb \times 1.389)] \times 100 \qquad (2\text{-}1)$$

where *CO content* is the carbon monoxide concentration, measured in millilitres per decilitre blood at standard temperature and pressure, dry; *Hb* is the haemoglobin concentration, measured in grams per decilitre blood; and 1.389 is the stoichiometric combining capacity of carbon monoxide for haemoglobin in units of millilitres of carbon monoxide per gram of haemoglobin at standard temperature and pressure, dry; however, in practice, a value of 1.36 ml carbon monoxide/g haemoglobin is used for the oxygen capacity of normal human blood because it is impossible to achieve 100% haemoglobin saturation.

The analytical methods that quantify the carbon monoxide content in blood require the conversion of these quantities to percent carboxyhaemoglobin. The product of the haemoglobin and the theoretical combining capacity (1.389, according to International Committee for Standardization in Haematology, 1978) yields the carbon monoxide capacity. With the use of capacity and the measured content, the percentage of carbon monoxide capacity (percent carboxyhaemoglobin) is calculated. To be absolutely certain of the accuracy of the haemoglobin measurement, the theoretical value should be routinely substantiated by direct measurement (internal validation) of the haemoglobin–carbon monoxide combining capacity. The total carbon monoxide–haemoglobin combining capacity should be determined as accurately as the content of carbon monoxide. The error of the techniques that measure carbon monoxide content is dependent on the error in haemoglobin analysis for the final form of the data, percent carboxyhaemoglobin. Therefore, the actual carbon monoxide–haemoglobin combining capacity should be measured and compared with the calculated value based upon the reference method for haemoglobin measurement. The measurement of carbon monoxide–haemoglobin combining capacity can be routinely performed by equilibration of a blood sample with carbon monoxide (Allred et al., 1989b).

The standard methods for haemoglobin determination involve the conversion of all species of haemoglobin to cyanomethaemoglobin with the use of a mixture of potassium ferricyanide, potassium cyanide and sodium bicarbonate.

3.1.3 Other methods of measurement

There is a wide variety of other techniques that have been described for the analysis of carbon monoxide in blood. These methods include UV-visible spectrophotometry (Small et al., 1971; Brown, 1980; Zwart et al., 1984, 1986), magnetic circular dichroism spectroscopy (Wigfield et al., 1981), photochemistry (Sawicki & Gibson, 1979), gasometric methods (Horvath & Roughton, 1942; Roughton & Root, 1945) and a calorimetric method (Sjostrand, 1948a). Not all of these methods have been as well characterized for the measurement of low levels of carboxyhaemoglobin as those listed above as potential reference methods.

1) Spectrophotometric methods

The majority of the techniques are based upon optical detection of carboxyhaemoglobin, which is more rapid than the reference techniques because it does not involve extraction of the carbon monoxide from the blood sample. These direct measurements also enable the simultaneous measurement of several species of haemoglobin, including reduced haemoglobin, oxyhaemoglobin (O_2Hb) and carboxyhaemoglobin. The limitations of the spectrophotometric techniques have been reviewed by Kane (1985). The optical methods utilizing UV wavelengths require dilution of the blood sample, which can lead to the loss of carbon monoxide as a result of competition with the dissolved oxygen in solution. Removing the dissolved oxygen with dithionite can lead to the formation of sulfhaemoglobin, which interferes with the measurement of carboxyhaemoglobin (Rai & Minty,

1987). Another limitation is that the absorption maxima (and spectral curves) are not precisely consistent between individuals. This may be due to slight variations in types of haemoglobin in subjects. For these reasons, the techniques using fixed-wavelength measurement points would not be expected to be as precise, accurate or specific as the proposed reference methods mentioned above.

2)   CO-Oximeter measurements of carboxyhaemoglobin

The speed of measurement and relative accuracy of spectrophoto-metric measurements over the entire range of expected values led to the development of CO-Oximeters. These instruments utilize from two to seven wavelengths in the visible region for the determination of proportions of oxyhaemoglobin, carboxyhaemoglobin, reduced hae-moglobin and methaemoglobin. The proportion of each species of haemoglobin is determined from the absorbance and molar extinction coefficients at present wavelengths. All of the commercially available instruments provide rapid results for all the species of haemoglobin being measured. In general, the manufacturers' listed limit of accuracy for all of the instruments is 1% carboxyhaemoglobin. However, this level of accuracy is not suitable for measurements associated with background carbon monoxide levels (<2% carboxyhaemoglobin) because it corresponds to errors exceeding 50%. The precision of measurement for these instruments is excellent and has misled users regarding the accuracy of the instruments. The relatively modest level of accuracy is adequate for the design purposes of the instruments; however, at low levels of carboxyhaemoglobin, the ability of the instruments to measure the percent carboxyhaemoglobin accurately is limited.

2.3.2   *Carbon monoxide in expired breath*

Carbon monoxide levels in expired breath can be used to estimate the levels of carbon monoxide in the subject's blood. The basic determinants of carbon monoxide levels in alveolar air have been described by Douglas et al. (1912), indicating that there are predictable equilibrium conditions that exist between carbon monoxide bound to haemoglobin and the partial pressure of the carbon monoxide in the blood. The equilibrium relationship for carbon monoxide between blood and the gas phase to which the blood is exposed can be described as follows:

$$P_{CO}/P_{O_2} = M \, (\% \, COHb/\% \, O_2Hb) \qquad (2\text{-}2)$$

where P_{CO} is the partial pressure of carbon monoxide in the blood, P_{O_2} is the partial pressure of oxygen in the blood, M is the Haldane coefficient (reflecting the relative affinity of haemoglobin for oxygen and carbon monoxide), $\%$ $COHb$ is the percentage of total haemoglobin combining capacity bound with carbon monoxide, and $\%$ O_2Hb is the percentage of total haemoglobin combining capacity bound with oxygen.

The partial pressure of carbon monoxide in the arterial blood will reach a steady-state value relative to the partial pressure of carbon monoxide in the alveolar gas. Therefore, by measuring the end-expired breath from a subject's lungs, one can measure the end-expired carbon monoxide partial pressure and, with the use of the Haldane relationship, estimate the blood level of carboxyhaemoglobin. This measurement will always be an estimate, because the Haldane relationship is based upon attainment of an equilibrium, which does not occur under physiological conditions.

The measurement of carbon monoxide levels in expired breath to estimate blood levels is based upon application of the Haldane relationship to gas transfer in the lung (Eq. 2-2). For example, when the oxygen partial pressure is increased in the alveolar gas, it is possible to predict the extent to which the partial pressure of carbon monoxide will increase in the alveolar gas. This approach is limited, however, because of the uncertainty associated with variables that are known to influence gas transfer in the lung and that mediate the direct relationship between liquid-phase gas partial pressures and air-phase partial pressures.

The basic mechanisms that are known to influence carbon monoxide transfer in the lung have been identified through the establishment of techniques to measure pulmonary diffusion capacity for carbon monoxide. Some of the factors that can result in decreased diffusion capacity for carbon monoxide (altering the relationship between expired carbon monoxide pressures and carboxyhaemoglobin levels) are increased membrane resistance, intravascular resistance, age, alveolar volume, pulmonary vascular blood volume, pulmonary blood flow and ventilation/perfusion inequality (Forster, 1964). The extent to which each of these variables actually contributes to the variability in the relationship has not been experimentally demonstrated. There are few experiments that focus on the factors leading to variability in the relationship between alveolar carbon monoxide and percent carboxyhaemoglobin at the levels of carboxyhaemoglobin currently deemed to be of regulatory importance. This may be due in part to the difficulties in working with analytical techniques, particularly blood techniques, that are very close to their limits of reproducibility.

The expired breath method for obtaining estimates of blood levels of carbon monoxide has a distinct advantage for monitoring large numbers of subjects, because of the non-invasive nature of the method. Other advantages include the ability to obtain an instantaneous reading and the ability to take an immediate replicate sample for internal standardization. The breath-holding technique for enhancing the normal carbon monoxide concentration in exhaled breath has been widely used; however, it should be noted that the absolute relationship between breath-hold carbon monoxide pressures and blood carbon monoxide pressures has not been thoroughly established for carboxyhaemoglobin levels below 5%. The breath-holding method allows time (20 s) for diffusion of carbon monoxide into the alveolar air so that carbon monoxide levels are higher than levels following normal tidal breathing.

Partial pressures of carbon monoxide in expired breath are highly correlated with percent carboxyhaemoglobin levels over a wide range of carboxyhaemoglobin levels. The accuracy of the breath-hold method is unknown owing to the lack of paired sample analyses of carbon monoxide partial pressures in exhaled breath and concurrent carboxyhaemoglobin levels in blood utilizing a sensitive reference method. No one has attempted to determine the error of estimate involved in applying group average regression relationships to the accurate determination of carboxyhaemoglobin. Therefore, the extrapolation of breath-hold carbon monoxide partial pressures to actual carboxyhaemoglobin levels must be made with reservation until the accuracy of this method is better understood.

2.3.2.1 Measurement methods

Ventilation in healthy individuals involves air movement through areas in the pulmonary system that are primarily involved in either conduction of gas or gas exchange in the alveoli. In a normal breath (tidal volume), the proportion of the volume in the non-gas exchanging area is termed the dead space. In the measurement of carbon monoxide in the exhaled air, the dead-space gas volume serves to dilute the alveolar carbon monoxide concentration. Several methods have been developed to account for the dead-space dilution. These include the mixed expired gas technique, which uses the Bohr equation to determine the physiological dead space; the breath-hold technique, a method of inspiration to total lung capacity followed by a breath-hold period of various durations (a breath-hold time of 20 s was found to provide near-maximal values for carbon monoxide pressures); and the rebreathing technique, in which 5 litres of oxygen are rebreathed for 2–3 min while the carbon dioxide is removed.

Kirkham et al. (1988) compared all three techniques for measuring expired carbon monoxide to predict percent carboxyhaemoglobin. The rebreathing and breath-hold methods both yield approximately 20% higher levels of "alveolar" carbon monoxide than does the Bohr computation from mixed expired gas. Both the mixed expired and breath-holding techniques show a significant decline in the alveolar carbon monoxide tension when the subject is standing. Therefore, measurements of expired carbon monoxide must be made in the same body position relative to control measurements or reference measurements.

2.2 Potential limitations

The measurement of exhaled breath has the advantages of ease, speed, precision and greater subject acceptance over measurement of blood carboxyhaemoglobin. However, the accuracy of the breath measurement procedure and the validity of the Haldane relationship between breath and blood at low environmental carbon monoxide concentrations remain in question.

3. SOURCES OF CARBON MONOXIDE IN THE ENVIRONMENT

3.1 Introduction

Carbon monoxide is produced by both natural and anthropogenic processes. About half of the carbon monoxide is released at the Earth's surface, and the rest is produced in the atmosphere. Many papers on the global sources of carbon monoxide have been published over the last 20 years; whether most of the carbon monoxide in the atmosphere is from human activities or from natural processes has been debated for nearly as long.

The recent budgets that take into account previously published data suggest that human activities are responsible for about 60% of the carbon monoxide in the non-urban troposphere, and natural processes account for the remaining 40%. It also appears that combustion processes directly produce about 40% of the annual emissions of carbon monoxide (Jaffe, 1968, 1973; Robinson & Robbins, 1969, 1970; Swinnerton et al., 1971), and oxidation of hydrocarbons makes up most of the remainder (about 50%) (Went, 1960, 1966; Rasmussen & Went, 1965; Zimmerman et al., 1978; Hanst et al., 1980; Greenberg et al., 1985), along with other sources such as the oceans (Swinnerton et al., 1969; Seiler & Junge, 1970; Lamontagne et al., 1971; Linnenbom et al., 1973; Liss & Slater, 1974; Seiler, 1974; Seiler & Schmidt, 1974; Swinnerton & Lamontagne, 1974; NRC, 1977; Bauer et al., 1980; Logan et al., 1981; DeMore et al., 1985) and vegetation (Krall & Tolbert, 1957; Wilks, 1959; Siegel et al., 1962; Seiler & Junge, 1970; Bidwell & Fraser, 1972; Seiler, 1974; NRC, 1977; Seiler & Giehl, 1977; Seiler et al., 1978; Bauer et al., 1980; Logan et al., 1981; DeMore et al., 1985). Some of the hydrocarbons that eventually end up as carbon monoxide are also produced by combustion processes, constituting an indirect source of carbon monoxide from combustion. These conclusions are summarized in Table 3, which is adapted from the 1981 budget of Logan et al., in which most of the previous work was incorporated (Logan et al., 1981; WMO, 1986). The total emissions of carbon monoxide are about 2600 million tonnes per year. Other budgets by Volz et al. (1981) and by Seiler & Conrad (1987) have been reviewed by Warneck (1988). Global emissions between 2000 and 3000 million tonnes per year are consistent with these budgets.

Table 3. Sources of carbon monoxide[a]

	Carbon monoxide production (million tonnes per year)[b]			
	Anthropogenic	Natural	Global	Range
Directly from combustion				
Fossil fuels	500	—	500	400–1000
Forest clearing	400	—	400	200–800
Savanna burning	200	—	200	100–400
Wood burning	50	—	50	25–150
Forest fires	—	30	30	10–50
Oxidation of hydrocarbons				
Methane[c]	300	300	600	400–1000
Non-methane hydrocarbons	90	600	690	300–1400
Other sources				
Plants	—	100	100	50–200
Oceans	—	40	40	20–80
Totals (rounded)	1500	1100	2600	2000–3000

[a] Adapted from Logan et al. (1981) and revisions reported by the WMO (1986).
[b] All estimates are expressed to one significant figure. The sums are rounded to two significant digits.
[c] Half the production of carbon monoxide from the oxidation of methane is attributed to anthropogenic sources and the other half to natural sources based on the budget of methane from Khalil & Rasmussen (1984c).

3.2 Principles of formation by source category

Carbon monoxide is produced in the atmosphere by reactions of hydroxyl radicals with methane and other hydrocarbons, both anthropogenic and natural, as well as by the reactions of alkenes with ozone and of isoprene and terpenes with hydroxyl radicals and ozone.

Carbon monoxide is also produced at the Earth's surface during the combustion of fuels. The burning of any carbonaceous fuel produces two primary products: carbon dioxide and carbon monoxide. The production of carbon dioxide predominates when the air or oxygen supply is in excess of the stoichiometric needs for complete combustion. If burning occurs under fuel-rich conditions, with less air or oxygen than is needed, carbon monoxide will be produced in abundance. In past years, most of the carbon monoxide and carbon dioxide formed were simply emitted into the atmosphere. In recent years, concerted efforts have been made to reduce ambient air concentrations of materials that are potentially harmful to humans. Much carbon monoxide, most notably from mobile sources, is converted to carbon dioxide, which is then emitted into the atmosphere.

Emission source categories in the USA (US EPA, 1991b) are divided into five individual categories: (1) transportation, (2) stationary source fuel combustion, (3) industrial processes, (4) solid waste disposal and (5) miscellaneous.

Transportation sources include emissions from all mobile sources, including highway and off-highway motor vehicles. Highway motor vehicles include passenger cars, trucks, buses and motorcycles. Off-highway vehicles include aircraft, locomotives, vessels and miscellaneous engines such as farm equipment, industrial and construction machinery, lawnmowers and snowmobiles.

Emission estimates from gasoline- and diesel-powered motor vehicles are based upon vehicle-mile (vehicle-kilometre) tabulations and emission factors. Eight vehicle categories are considered: (1) light-duty gasoline vehicles (mostly passenger cars), (2) light-duty diesel passenger cars, (3) light-duty gasoline trucks (weighing less than 6000 lb [2.7 tonnes]), (4) light-duty gasoline trucks (weighing 6000–8500 lb [2.7–3.9 tonnes]), (5) light-duty diesel trucks, (6) heavy-duty gasoline trucks and buses, (7) heavy-duty diesel trucks and buses and (8) motorcycles. The emission factors used are based on the US Environmental Protection Agency's (EPA) mobile source emission factor model, developed by the EPA Office of Mobile Sources, which uses the latest available data to estimate average in-use emissions from highway vehicles.

Aircraft emissions are based on emission factors and aircraft activity statistics reported by the Federal Aviation Administration (1988). Emissions are based on the number of landing–take-off cycles. Any emissions in cruise mode, which is defined to be above 3000 ft (1000 m), are ignored. Average emission factors for each year, which take into account the national mix of aircraft types for general aviation, military and commercial aircraft, are used to compute the emissions.

In the USA, the Department of Energy reports consumption of diesel fuel and residual fuel oil by locomotives (US Department of Energy, 1988a). Average emission factors applicable to diesel fuel consumption were used to calculate emissions. Vessel use of diesel fuel, residual oil and coal is also reported by the Department of Energy (US Department of Energy, 1988a,b). Gasoline use is based on national boat and motor registrations, coupled with a use factor (gallons [litres] per motor per year) (Hare & Springer, 1973) and marine gasoline sales (US Department of Transportation, 1988). Emission factors from EPA Report No. AP-42 are used to compute emissions (US EPA, 1985).

Gasoline and diesel fuel are consumed by off-highway vehicles in substantial quantities. The fuel consumption is divided into several categories (e.g., farm tractors, other farm machinery, construction equipment, industrial machinery, snowmobiles and small general utility engines such as lawnmowers and snowblowers). Fuel use is estimated for each category from estimated equipment population and an annual use factor of gallons (litres) per unit per year (Hare & Springer, 1973), together with reported off-highway diesel fuel deliveries (US Department of Energy, 1988a) and off-highway gasoline sales (US Department of Transportation, 1988).

Stationary combustion equipment, such as coal-, gas- or oil-fired heating or power generating plants, generates carbon monoxide as a result of improper or inefficient operating practices or inefficient combustion techniques. The specific emission factors for stationary fuel combustors vary according to the type and size of the installation and the fuel used, as well as the mode of operation. The US EPA's compilation of air pollutant emission factors provides emission data obtained from source tests, material balance studies, engineering estimates and so forth for the various common emission categories. For example, coal-fired electricity generating plants report coal use to the US Department of Energy (1988b,c). Distillate oil, residual oil, kerosene and natural gas consumed by stationary combustors are also reported by user category to the US Department of Energy (1988a). Average emission factors from EPA Report No. AP-42 (US EPA, 1985) were used to calculate the emission estimates. The consumption of wood in residential wood stoves has likewise been estimated by the US Department of Energy (1982, 1984).

In addition to fuel combustion, certain other industrial processes generate and emit varying quantities of carbon monoxide into the air. The lack of published national data on production, type of equipment and controls, as well as an absence of emission factors, makes it impossible to include estimates of emissions from all industrial process sources.

Solid waste carbon monoxide emissions result from the combustion of wastes in municipal and other incinerators, as well as from the open burning of domestic and municipal refuse.

Miscellaneous carbon monoxide emissions result from the burning of forest and agricultural materials, smouldering coal refuse materials and structural fires.

The Forest Service of the US Department of Agriculture publishes information on the number of forest fires and the acreage burned (US Forest Service, 1988). Estimates of the amount of material burned per acre are made to determine the total amount of material burned. Similar estimates are made to account for managed burning of forest areas. Average emission factors were applied to the quantities of materials burned to calculate emissions.

A study was conducted by the US EPA (Yamate, 1974) to obtain, from local agricultural and pollution control agencies, estimates of the number of acres and estimated quantity of material burned per acre in agricultural burning operations. These data have been updated and used to estimate agricultural burning emissions, based on average emission factors.

Estimates of the number of burning coal refuse piles existing in the USA are made in reports by the Bureau of Mines. McNay (1971) presents a detailed discussion of the nature, origin and extent of this source of pollution. Rough estimates of the quantity of emissions were obtained using this information by applying average emission factors for coal combustion. It was assumed that the number of burning refuse piles decreased to a negligible level by 1975.

The US Department of Commerce publishes, in its statistical abstracts, information on the number and types of structures damaged by fire (US Department of Commerce, 1987). Emissions were estimated by applying average emission factors for wood combustion to these totals.

The estimated total annual carbon monoxide emissions from the various source categories in the USA for 1970, 1975 and 1980–1990 (US EPA, 1991b) indicate that carbon monoxide emissions from all anthropogenic sources declined from 101.4 million tonnes (111.8 million short tons) in 1970 to 60.1 million tonnes (66.2 million short tons) in 1990. The majority, about 63%, of the carbon monoxide emissions total comes from transportation sources, 12% comes from stationary source fuel combustion, 8% comes from industrial processes, 3% comes from solid waste and 14% comes from miscellaneous sources.

The single largest contributing source of carbon monoxide emissions is highway vehicles, which emitted an estimated 50% of the national total in 1990. Because of the implementation of the Federal Motor Vehicle Control Program, carbon monoxide emissions from highway vehicles declined 54%, from 65.3 to 30.3 million tonnes, in the period 1970–1990. Fig. 1 displays the trend in estimated carbon monoxide emissions from the major highway vehicle categories from 1970 to 1990. Although the total annual vehicle-miles (vehicle-kilometres) travelled continue to increase in the USA (by 37% just in the period 1981–1990), total carbon monoxide emissions from highway vehicles have continued to decrease as a result of the Federal Motor Vehicle Control Program-mandated air pollution control devices on new vehicles.

Carbon monoxide emissions from other sources have also generally decreased. In 1970, emissions from burning of agricultural crop residues were greater than in more recent years. Solid waste disposal emissions have also decreased as the result of implementation of regulations limiting or prohibiting burning of solid waste in many areas. Emissions of carbon monoxide from stationary source fuel combustion occur mainly from the residential sector. These emissions were reduced somewhat through the mid-1970s as residential consumers converted to natural gas, oil or electric heating equipment. Recent growth in the use of residential wood stoves has reversed this trend, but increased carbon monoxide emissions from residential sources continue to be small compared with highway vehicle emissions. Nevertheless, in 1990, residential wood combustion accounted for

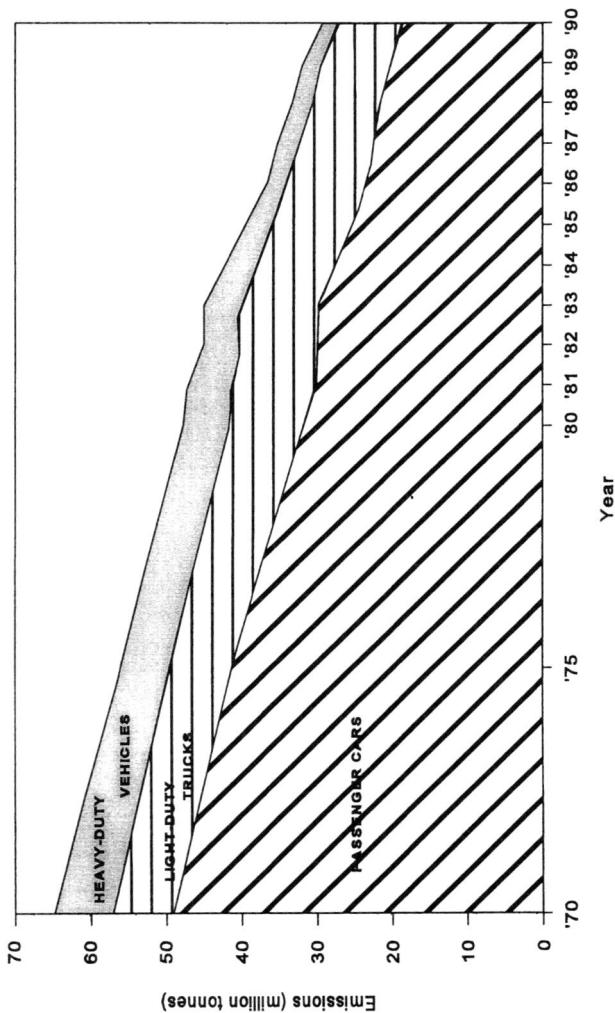

Fig. 1. Estimated emission of carbon monoxide from gasoline-fuelled highway vehicles in the USA (adapted from US EPA, 1991a).

44

about 10% of national carbon monoxide emissions, more than any source category except highway vehicles. Carbon monoxide emissions from industrial processes have generally been declining since 1970 as the result of the obsolescence of a few high-polluting processes such as manufacture of carbon black by the channel process and installation of controls on other processes.

Considerable effort has been made to reduce emissions of carbon monoxide and other pollutants to the atmosphere. Because the automobile engine is recognized to be the major source of carbon monoxide in most urban areas, special attention is given to the control of automotive emissions. Generally, the approach has been technological: reduction of carbon monoxide emissions to the atmosphere either by improving the efficiency of the combustion processes, thereby increasing the yield of carbon dioxide and decreasing the yield of carbon monoxide, or by applying secondary catalytic combustion reactors to the waste gas stream to convert carbon monoxide to carbon dioxide.

The development and application of control technology to reduce emissions of carbon monoxide from combustion processes generally have been successful and are continuing to receive deserved attention. The reduction of carbon monoxide emissions from 7.0 to 3.4 g/mi (from 4.4 to 2.1 g/km), scheduled for the 1981 model year, was delayed 2 years, reflecting in part the apparent difficulty encountered by the automobile industry in developing and supplying the required control technology. The carbon monoxide emission limit for light-duty vehicles at low altitude has been 3.4 g/mi (2.1 g/km) since 1983; since 1984, this limit has applied to light-duty vehicles at all altitudes.

Carbon monoxide emissions from 1990 to 1994 in Germany, as estimated by the Federal Environmental Agency (1997), are summarized in Table 4 and show a clear reduction with time. The estimated carbon monoxide emissions in Europe during 1990 are summarized in Table 5, including the sum of 28 European countries and individual data from 10 countries (European Environmental Agency, 1995). The Task Group noted that the original reference does not specify any details concerning the zero values listed in Table 5.

Table 4. Carbon monoxide national emission estimates: Germany[a]

Source category	1990		1991		1992		1993		1994	
	kt	%	kt	%	kt	%	kt	%	kt	%
Industrial processes	684	6.4	645	7.1	599	7.6	597	8.1	595	8.8
Production and distribution of fuel	27	0.3	23	0.3	16	0.2	14	0.2	13	0.2
Street traffic	6 487	60.4	5 593	61.8	4 962	62.6	4 457	60.4	3 953	58.7
Other transportation	252	2.3	208	2.3	184	2.3	183	2.5	183	2.7
Residential	2 085	19.4	1 512	16.7	1 165	14.7	1 169	15.8	1030	15.3
Small consumers	207	1.9	169	1.9	141	1.8	148	2	143	2.1
Industrial combustion	871	8.1	775	8.6	745	9.4	705	9.6	716	10.6
Power plants and heating plants	130	1.2	121	1.3	114	1.4	106	1.4	104	1.5
Total	10 743		9 046		7 926		7 379		6 737	

[a] From Federal Environmental Agency (1997).

Table 5. Carbon monoxide emission estimates in Europe in 1990[a]

| | Carbon monoxide emission estimates (kt)[b] | | | | | | | | | | | |
| | Europe (28 countries) | | Aus-tria | France | Ger-many | Italy | Neth-er-lands | Po-land | Roma-nia | Spain | Swe-den | United King-dom |
	kt	% of total										
Public power cogeneration and district heating	807	1	6	21	466	23	5	68	13	16	6	50
Commercial, institutional and residential combustion	9 947	14	776	1 892	2 053	260	101	1 343	373	890	72	294
Industrial combustion	8 200	12	27	599	1 174	620	12	3 389	782	406	24	71
Production processes	3 188	6	241	668	664	380	254	122	129	248	6	0
Extraction and distribution of fossil fuels	63	0	0	0	23	0	2	0	0	0	0	2
Solvent use	1	0	0	0	0	0	1	0	0	0	NE	0
Road transport	38 919	56	582	6 812	5 892	5 534	675	2 133	531	2 610	1 118	6 023

Table 5 (contd).

Carbon monoxide emission estimates (kt)[b]

	Europe (28 countries)		Aus-tria	France	Ger-many	Italy	Neth-er-lands	Po-land	Roma-nia	Spain	Swe-den	United King-dom
	kt	% of total										
Other mobile sources and machinery	2 223	3	NE	512	260	719	21	90	27	111	107	42
Waste treatment and disposal	4 427	6	0	232	0	1 705	2	225	1 332	527	14	220
Agriculture	579	1	60	NE	0	27	8	0	0	143	0	0
Nature	1 358	2	NE	194	0	1 079	26	18	0	26	2	0
Total	69 712	100	1 692	10 930	10 532	10 347	1 107	7 388	3 187	4 977	1 349	6 702

[a] From European Environmental Agency (1995).
[b] NE = no estimate.

1.2.1 General combustion processes

Incomplete combustion of carbon-containing compounds creates varying amounts of carbon monoxide. The chemical and physical processes that occur during combustion are complex because they depend not only on the type of carbon compound reacting with oxygen, but also on the conditions existing in the combustion chamber (Pauling, 1960; Mellor, 1972). Despite the complexity of the combustion process, certain general principles regarding the formation of carbon monoxide from the combustion of hydrocarbon fuels are accepted widely.

Gaseous or liquid hydrocarbon fuel reacts with oxygen in a chain of reactions that result in the formation of carbon monoxide. Carbon monoxide then reacts with hydroxyl radicals to form carbon dioxide. The second reaction is approximately 10 times slower than the first. In coal combustion, too, the reaction of carbon and oxygen to form carbon monoxide is one of the primary reactions, and a large fraction of carbon atoms go through the carbon monoxide form. Again, the conversion of carbon monoxide to carbon dioxide is much slower.

Four basic variables control the concentration of carbon monoxide produced in the combustion of all hydrocarbon fuels: (1) oxygen concentration, (2) flame temperature, (3) gas residence time at high temperatures and (4) combustion chamber turbulence. Oxygen concentration affects the formation of both carbon monoxide and carbon dioxide, because oxygen is required in the initial reactions with the fuel molecule and in the formation of the hydroxyl radical. As the availability of oxygen increases, more complete conversion of carbon monoxide to carbon dioxide results. Flame and gas temperatures affect both the formation of carbon monoxide and the conversion of carbon monoxide to carbon dioxide, because both reaction rates increase exponentially with increasing temperature. Also, the hydroxyl radical concentration in the combustion chamber is very temperature dependent. The conversion of carbon monoxide to carbon dioxide is also enhanced by longer residence time, because this is a relatively slow reaction in comparison with carbon monoxide formation. Increased gas turbulence in the combustion zones increases the actual reaction rates by increasing the mixing of the reactants and assisting the relatively slower gaseous diffusion process, thereby resulting in more complete combustion.

3.2.2   Combustion engines

3.2.2.1   Mobile combustion engines

Most mobile sources of carbon monoxide are internal combustion engines of two types: (1) gasoline-fuelled, spark ignition, reciprocating engines (carburetted or fuel-injected) and (2) diesel-fuelled reciprocating engines. The carbon monoxide emitted from any given engine is the product of the following factors: (1) the concentration of carbon monoxide in the exhaust gases, (2) the flow rate of exhaust gases and (3) the duration of operation.

1)   Internal combustion engines (gasoline-fuelled, spark ignition engines)

Exhaust concentrations of carbon dioxide increase with lower (richer) air-to-fuel (A/F) ratios and decrease with higher (leaner) A/F ratios, but they remain relatively constant with ratios above the stoichiometric ratio of about 15:1 (Hagen & Holiday, 1964). The behaviour of gasoline automobile engines before and after the installation of pollutant control devices differs considerably. Depending on the mode of driving, the average uncontrolled engine operates at A/F ratios ranging from about 11:1 to a point slightly above the stoichiometric ratio. During the idling mode, at low speeds with light load (such as low-speed cruise), during the full open throttle mode until speed picks up and during deceleration, the A/F ratio is low in uncontrolled cars, and carbon monoxide emissions are high. At higher-speed cruise and during moderate acceleration, the reverse is true. Cars with exhaust controls generally remain much closer to stoichiometric A/F ratios in all modes, and thus the carbon monoxide emissions are kept lower. The exhaust flow rate increases with increasing engine power output.

The decrease in available oxygen with increasing altitude has the effect of enriching the A/F mixture and increasing carbon monoxide emissions from carburetted engines. Fuel-injected gasoline engines, which predominate in the vehicle fleet today, have more closely controlled A/F ratios and are designed and certified to comply with applicable emission standards regardless of elevation (US EPA, 1983).

Correlations between total emissions of carbon monoxide in grams per vehicle-mile (vehicle-kilometre) and average route speed show a decrease in emissions with increasing average speed

(Simonaitis & Heicklen, 1972; Stuhl & Niki, 1972; US EPA, 1985). During low-speed conditions (below 32 km/h or 20 mi/h average route speed), the greater emissions per unit of distance travelled are attributable to (1) an increased frequency of acceleration, deceleration and idling encountered in heavy traffic and (2) the consequent increase in the operating time per mile (kilometre) driven.

The carbon monoxide and the unburned hydrocarbon exhaust emissions from an uncontrolled engine result from incomplete combustion of the fuel–air mixture. Emission control on new vehicles is being achieved by engine modifications, improvements in engine design and changes in engine operating conditions. Substantial reductions in carbon monoxide and other pollutant emissions result from consideration of design and operating factors such as leaner, uniform mixing of fuel and air during carburetion, controlled heating of intake air, increased idle speed, retarded spark timing, improved cylinder head design, exhaust thermal reactors, oxidizing and reducing catalysts, secondary air systems, exhaust recycle systems, electronic fuel injection, A/F ratio feedback controls and modified ignition systems (NAS, 1973).

2)   Internal combustion engines (diesel engines)

Diesel engines are in use throughout the world in heavy-duty vehicles, such as trucks and buses, and they are also extensively used in Western Europe in light-duty vans, taxis and some cars. Diesel engines allow more complete combustion and use less volatile fuels than do spark ignition engines. The operating principles are significantly different from those of the gasoline engine. In diesel combustion, carbon monoxide concentrations in the exhaust are relatively low because high temperature and large excesses of oxygen are involved in normal operation.

2.2.2   *Stationary combustion sources (steam boilers)*

This section refers to fuel-burning installations such as coal-, gas- or oil-fired heating or power generating plants (external combustion boilers).

In these combustion systems, the formation of carbon monoxide is lowest at a ratio near or slightly above the stoichiometric A/F ratio. At lower than stoichiometric A/F ratios, high carbon monoxide concentrations reflect the relatively low oxygen concentration and the possibility of poor reactant mixing from low turbulence. These two factors can increase emissions even though flame temperatures and residence time are high. At higher than stoichiometric A/F ratios, increased carbon monoxide emissions result from decreased flame temperatures and shorter residence time. These two factors remain predominant even when oxygen concentrations and turbulence increase. Minimal carbon monoxide emissions and maximum thermal efficiency therefore require combustor designs that provide high turbulence, sufficient residence time, high temperatures and near-stoichiometric A/F ratios. Combustor design dictates the actual approach to that minimum.

3.2.3   Other sources

There are numerous industrial activities that result in the emission of carbon monoxide at one or more stages of the process (Walsh & Nussbaum, 1978; US EPA, 1979a, 1985). Manufacturing pig iron can produce as much as 700–1050 kg carbon monoxide/tonne of pig iron. Other methods of producing iron and steel can produce carbon monoxide at a rate of 9–118.5 kg/tonne. However, most of the carbon monoxide generated is normally recovered and used as fuel. Conditions such as "slips," abrupt collapses of cavities in the coke–ore mixture, can cause instantaneous emissions of carbon monoxide that temporarily exceed the capacity of the control equipment. Grey iron foundries can produce 72.5 kg carbon monoxide/tonne of product, but an efficient afterburner can reduce the carbon monoxide emissions to 4.5 kg/tonne. Nevertheless, industrial carbon monoxide emissions may constitute an important part of total emissions in industrial cities — for example, in the Ruhr area in Germany.

Charcoal production results in average carbon monoxide emissions of 172 kg/tonne. Emissions from batch kilns are difficult to control, although some may have afterburners. Afterburners can more easily reduce, by an estimated 80% or more, the relatively constant carbon monoxide emissions from continuous charcoal production. Emissions from carbon black manufacture can range from 5 to 3200 kg carbon monoxide/tonne depending on the efficiency and quality of the emission control systems.

Some chemical processes, such as phthalic anhydride production, give off as little as 6 kg carbon monoxide/tonne with proper controls or as much as 200 kg carbon monoxide/tonne if no controls are installed. There are numerous other chemical processes that produce relatively low carbon monoxide emissions per tonne of product: sulfate pulping for paper produces 1–30 kg carbon monoxide/tonne, lime manufacturing normally produces 1–4 kg carbon monoxide/ tonne, and carbon monoxide emissions from adipic acid production are zero or slight with proper controls. Other industrial chemical processes that cause carbon monoxide emissions are the manufacture of terephthalic acid and the synthesis of methanol and higher alcohols. As a rule, most industries find it economically desirable to install suitable controls to reduce carbon monoxide emissions.

Even though some of these carbon monoxide emission rates seem excessively high, they are, in fact, only a small part of the total pollutant load. Mention of these industries is made to emphasize the concern for localized pollution problems when accidents occur or proper controls are not used.

In some neighbourhoods, wintertime carbon monoxide emissions include a significant component from residential fireplaces and wood stoves. Emissions of carbon monoxide can range from 18 to 140 g/kg, depending on design, fuel type and skill of operation.

Although the estimated carbon monoxide emissions resulting from forest wildfires in the USA have fluctuated between about 4 and 9 million tonnes per year since 1970 and were 6.2 million tonnes in 1989, the estimated total carbon monoxide emissions from industrial processes in the USA declined from 8.9 million tonnes in 1970 to 4.6 million tonnes in 1989 (US EPA, 1991c).

3.3 Indoor carbon monoxide

3.3.1 Introduction

Carbon monoxide is introduced to indoor environments through emissions from a variety of combustion sources and in the infiltration or ventilation air from outdoors. The resulting indoor concentration, both average and peak, is dependent on a complex interaction of several interrelated factors affecting the introduction, dispersion and removal of carbon monoxide. These factors include, for example, such variables as (1) the type, nature (factors affecting the generation rate of carbon monoxide) and number of sources, (2) source use characteristics, (3) building characteristics, (4) infiltration or ventilation rates, (5) air mixing between and within compartments in an indoor space, (6) removal rates and potential remission or generation by indoor surfaces and chemical transformations, (7) existence and effectiveness of air contaminant removal systems and (8) outdoor concentrations.

Source emissions from indoor combustion are usually characterized in terms of emission rates, defined as the mass of pollutant emitted per unit of fuel input (micrograms per kilojoule). They provide source strength data as input for indoor modelling, promote an understanding of the fundamental processes influencing emissions, guide field study designs assessing indoor concentrations, identify and rank important sources and aid in developing effective mitigation measures. Unfortunately, source emissions can vary widely. Although it would be most useful to assess the impact of each of the sources on indoor air concentrations of carbon monoxide by using models, the high variability in the source emissions and in other factors affecting the indoor levels does not make such an effort very useful. Such an estimate will result in predicted indoor concentrations ranging over several orders of magnitude, making them of no practical use, and may be misleading.

3.3.2 Emissions from indoor sources

Carbon monoxide emitted directly into the indoor environment is one of several air contaminants resulting from combustion sources. Such emissions into occupied spaces can be unintentional or the result of accepted use of unvented or partially vented combustion sources. Faulty or leaky flue pipes, backdrafting and spillage from combustion appliances that draw their air from indoors (e.g., Moffatt, 1986), improper use of combustion sources (e.g., use of a poorly maintained kerosene heater) and air intake into a building from attached parking garages are all examples of unintentional or accidental indoor sources of carbon monoxide. In the USA, the National Center for Health Statistics (1986) estimates that between 700 and 1000 deaths per year are due to accidental carbon monoxide poisoning. Mortality statistics are similar for other developed countries as well. The number of individuals experiencing severe adverse health effects at sublethal carbon monoxide concentrations from accidental indoor sources is no doubt many times the number of estimated deaths. Although the unintentional or accidental indoor sources of carbon monoxide represent a serious health hazard, little is known about the extent of the problem throughout the world. Such sources cannot be characterized for carbon monoxide emissions in any standard way that would make the results extendable to the general population.

The major indoor sources of carbon monoxide emissions that result from the accepted use of unvented or partially vented combustion sources include gas cooking ranges and ovens, gas appliances, unvented gas space heaters, unvented kerosene space heaters, coal- or wood-burning stoves and cigarette combustion.

3.2.1 Gas cooking ranges, gas ovens and gas appliances

Estimates indicate that gas (natural gas and liquid propane) is used for cooking, heating water and drying clothes in approximately 45.1% of all homes in the USA (US Bureau of the Census, 1982) and in nearly 100% of the homes in some other countries (e.g., the Netherlands). Unvented, partially vented and improperly vented gas appliances, particularly the gas cooking range and oven, represent an important source category of carbon monoxide emissions into the indoor residential environment. Emissions of carbon monoxide from these gas appliances are a function of a number of variables relating to the source type (range top or oven, water heater, dryer, number of pilot lights, burner design, etc.), source condition (age, maintenance, combustion efficiency, etc.), source use (number of burners used, frequency of use, fuel consumption rate, length of use, improper use, etc.) and venting of emissions (existence and use of outside vents over ranges, efficiency of vents, venting of gas dryers, etc.).

The source emission studies typically have been conducted in the laboratory setting and have involved relatively few gas ranges and gas appliances. The reported studies indicate that carbon monoxide emissions are highly variable among burners on a single gas cooking range and between gas cooking ranges and ovens, varying by as much as an order of magnitude. Operating a gas cooking range or oven under improperly adjusted flame conditions (yellow-tipped flame) can result in greater than a fivefold increase in emissions compared with properly operating flame conditions (blue flame). Use of a rich or lean fuel appeared to have little effect on carbon monoxide emissions. In general, carbon monoxide emissions were roughly, on average, comparable for top burners, ovens, pilot lights and unvented gas dryers when corrected for fuel consumption rate. The emission rates gathered by either the direct or mass balance method were comparable. Only one study attempted to evaluate gas stove emissions in the field for a small number (10) of residences. This study found carbon monoxide emissions to be as much as a factor of 4 higher than in chamber studies. Given the prevalence of the source, limited field measurements and poor agreement between existing laboratory- and field-derived carbon monoxide emission data, there is a need to establish a better carbon monoxide emission database for gas cooking ranges in residential settings.

3.3.2.2 Unvented space heaters

Unvented kerosene and gas space heaters are used in the colder climates to supplement central heating systems or in more moderate climates as the primary source of heat. During the heating season, space heaters generally will be used for a number of hours during the day, resulting in emissions over relatively long periods of time.

Over the last several years, there has been a dramatic increase in the use of unvented or poorly vented kerosene space heaters in residential and commercial establishments, primarily as a supplemental heat source. For example, in the USA, an estimated 16.1 million such heaters had been sold through 1986 (S.E. Womble, personal communication, US Consumer Product Safety Commission, 1988). An additional 3 million residences use unvented gas space heaters (fuelled by natural gas or propane). The potentially large number of unvented space heaters used throughout the world, particularly during periods when energy costs rise quickly, makes them an important source of carbon monoxide indoors.

Carbon monoxide emissions from unvented kerosene and gas space heaters can vary considerably and are a function of heater design (convective, radiant, combination, etc.), condition of heater and manner of operation (e.g., flame setting).

Carbon monoxide emissions from unvented gas space heaters were found to be variable from heater to heater, but were roughly comparable to those for gas cooking ranges. Infrared gas space heaters produced higher emissions than the convective or catalytic heaters. Emissions of carbon monoxide for these heaters were higher for maltuned heaters and for the mass balance versus direct method of testing. No differences for rich or lean fuel were found, but use of natural gas resulted in higher emissions than did use of propane. Lower fuel consumption settings resulted in lower carbon monoxide emissions. Emissions were observed to vary in time during a heater run and increase when room or chamber oxygen levels decreased.

Among the three principal unvented kerosene space heater designs (radiant, convective and two-stage burners), radiant heaters produced the highest carbon monoxide emissions and convective heaters produced the lowest emissions. Wick setting (low, normal or high) had a major impact on emissions, with the low-wick setting resulting in the highest carbon monoxide emissions. Data from different laboratories are in good agreement for this source.

3.2.3   Coal or wood stoves

Use of coal- or wood-burning stoves has been a popular cost savings alternative to conventional cooking and heating systems using gas or oil. Carbon monoxide and other combustion by-products enter the indoor environment during fire start-up, during fire-tending functions or through leaks in the stove or venting system. Hence, it is difficult to evaluate indoor carbon monoxide emission rates for these sources. Traynor et al. (1987) evaluated indoor carbon monoxide levels from four wood-burning stoves (three airtight stoves and one non-airtight stove) in a residence. The non-airtight stove emitted substantial amounts of carbon monoxide to the residence, particularly when operated with a large fire. The airtight stoves contributed considerably less. The average carbon monoxide source strengths during stove operation ranged from 10 to 140 cm^3/h for the airtight stoves and from 220 to 1800 cm^3/h for the non-airtight stoves.

3.2.4   Tobacco combustion

The combustion of tobacco represents an important source of indoor air contaminants. Carbon monoxide is emitted indoors from tobacco combustion through the exhaled mainstream smoke and from the smouldering end of the cigarette (sidestream smoke). Carbon monoxide emission rates in mainstream and sidestream smoke have been evaluated extensively in small chambers (less than a litre in volume) using a standardized smoking machine protocol. The results of these studies have been summarized and evaluated in several reports (e.g., NRC, 1986a; Surgeon General of the United States,

1986). These results indicate considerable variability in total (mainstream plus sidestream smoke) carbon monoxide emissions, with a typical range of 40–67 mg per cigarette. A small chamber study of 15 brands of Canadian cigarettes (Rickert et al., 1984) found the average carbon monoxide emission rate (mainstream plus sidestream smoke) to be 65 mg per cigarette. A more limited number of studies have been done using large chambers with the occupants smoking or using smoking machines. Girman et al. (1982) reported a carbon monoxide emission rate of 94.6 mg per cigarette for a large chamber study in which one cigarette brand was evaluated. A carbon monoxide emission factor of 88.3 mg per cigarette was reported by Moschandreas et al. (1985) for a large chamber study of one reference cigarette.

On average, a smoker smokes approximately two cigarettes per hour, with an average smoking time of approximately 10 min per cigarette. Using the above range of reported carbon monoxide emission rates for environmental tobacco smoke, this would roughly result in the emission of 80–190 mg of carbon monoxide per smoker per hour into indoor spaces where smoking occurs. This value compares with an approximate average carbon monoxide emission rate of 260–545 mg/h for one range-top burner (without pilot light) operating with a blue flame. Two smokers in a house would produce hourly carbon monoxide emissions comparable to the hourly production rate of a single gas burner. Tobacco combustion therefore represents an important indoor source of carbon monoxide, particularly in locations where many people are smoking.

4. ENVIRONMENTAL DISTRIBUTION AND TRANSFORMATION

4.1 Introduction

Carbon monoxide was first discovered to be a minor constituent of the Earth's atmosphere in 1948 by Migeotte (1949). While taking measurements of the solar spectrum, he observed a strong absorption band in the infrared region at 4.6 μm, which he attributed to carbon monoxide (Lagemann et al., 1947). On the twin bases of the belief that the solar contribution to that band was negligible and his observation of a strong day-to-day variability in absorption, Migeotte (1949) concluded that an appreciable amount of carbon monoxide was present in the terrestrial atmosphere of Columbus, Ohio, USA. In the 1950s, many more observations of carbon monoxide were made, with measured concentrations ranging from 0.09 to 110 mg/m^3 (0.08 to 100 ppm) (Migeotte & Neven, 1952; Benesch et al., 1953; Locke & Herzberg, 1953; Faith et al., 1959; Robbins et al., 1968; Sie et al., 1976). On the basis of these and other measurements available in 1963, Junge (1963) stated that carbon monoxide appeared to be the most abundant trace gas, other than carbon dioxide, in the atmosphere. The studies of Sie et al. (1976) indicated higher mixing ratios near the ground than in the upper atmosphere, implying a source in the biosphere, but Junge (1963) emphasized that knowledge of the sources and sinks of atmospheric carbon monoxide was extremely poor. It was not until the late 1960s that concerted efforts were made to determine the various production and destruction mechanisms for carbon monoxide in the atmosphere.

Far from human habitation in remote areas of the southern hemisphere, natural background carbon monoxide concentrations average around 0.05 mg/m^3 (0.04 ppm), primarily as a result of natural processes such as forest fires and the oxidation of methane. In the northern hemisphere, background concentrations are 2–3 times higher because of more extensive human activities. Much higher concentrations occur in cities, arising from technological sources such as automobiles and the production of heat and power. Carbon monoxide emissions are increased when fuel is burned in an incomplete or inefficient way.

The physical and chemical properties of carbon monoxide suggest that its atmospheric removal occurs primarily by reaction with hydroxyl radicals. Almost all the carbon monoxide emitted into the atmosphere each year is removed by reactions with hydroxyl radicals (85%), by soils (10%) and by diffusion into the stratosphere. There is a small imbalance between annual emissions and removal, causing an increase of about 1% per year. It is very likely that the imbalance is due to increasing emissions from anthropogenic activities. The average concentration of carbon monoxide is about 100 µg/m^3 (90 ppbv), which amounts to about 400 million tonnes in the atmosphere, and the average lifetime is about 2 months. This view of the global cycle of carbon monoxide is consistent with the present estimates of average hydroxyl radical concentrations and the budgets of other trace gases, including methane and methyl chloroform.

4.2   Global sources, sinks and lifetime

The largest sources of carbon monoxide in the global atmosphere are combustion processes and the oxidation of hydrocarbons (see chapter 3). The mass balance of a trace gas in the atmosphere can be described as a balance between the rate of change of the global burden added to the annual rate of loss on the one side and global emissions on the other side (dC/dt + loss rate = source emissions, where C = concentration). In steady state, the atmospheric lifetime (τ) is the ratio of the global burden to the loss rate. The global burden is the total number of molecules of a trace gas in the atmosphere or its total mass. The concentration of a trace gas can vary (dC/dt is not 0) when either the loss rate or the emissions vary cyclically in time, representing seasonal variations, or vary over a long time, often representing trends in human industrial activities or population. For carbon monoxide, both types of trends exist. There are large seasonal cycles driven mostly by seasonal variations in the loss rate but also affected by seasonal variations in emissions, and there are also indications of long-term trends probably caused by increasing anthropogenic emissions.

4.2.1   Sinks

It is believed that reaction with hydroxyl radicals is the major sink for removing carbon monoxide from the atmosphere. The cycle of the hydroxyl radical itself cannot be uncoupled from the cycles of carbon monoxide, methane, water and ozone. In the troposphere, hydroxyl radicals (OH·) are produced by the photolysis of ozone ($h\upsilon$ + $O_3 \rightarrow O(^1D) + O_2$) followed by the reaction of the excited oxygen atoms with water vapour to produce two hydroxyl radicals ($O(^1D)$ + $H_2O \rightarrow OH· + OH·$). The production of hydroxyl radicals is balanced by their removal principally by reactions with carbon monoxide and methane. On a global scale, carbon monoxide may remove more hydroxyl radicals than methane; however, methane is more important in the southern hemisphere, where there is much less carbon monoxide, than in the northern hemisphere, but the amount of methane is only slightly less in the northern hemisphere.

The amount of carbon monoxide that is removed by reactions with hydroxyl radicals can be estimated by calculating the loss as $loss = K_{\it eff} [OH·]_{ave} [CO]_{ave}$, where $K_{\it eff}$ is the effective reaction rate constant, $[OH·]_{ave}$ is the average hydroxyl radical concentration and $[CO]_{ave}$ is the average concentration of carbon monoxide. The reaction rate constant of CO + OH· is $K = (1.5 \times 10^{-13}) (1 + 0.6\ P_{atm})$ cm³/molecule per second (DeMore et al., 1987), where P_{atm} is the atmospheric pressure. The constant $K_{\it eff}$ describes the effective reaction rate, taking into account the decreasing atmospheric pressure and decreasing carbon monoxide concentrations with height. Estimating $K_{\it eff}$ to be 2×10^{-13} cm³/molecule per second and taking $[OH·]_{ave}$ to be 8×10^5 molecules/cm³ and $[CO]_{ave}$ to be 90 ppbv (equivalent to about 100 μg/m³), the annual loss of carbon monoxide from reactions with hydroxyl radicals is about 2200 million tonnes per year. The values adopted for $[OH·]_{ave}$ and $[CO]_{ave}$ are discussed in more detail later in this chapter.

Uptake of carbon monoxide by soils has been documented and may amount to about 250 million tonnes per year, or about 10% of the total emitted into the atmosphere (Inman et al., 1971; Ingersoll et al., 1974; Seiler & Schmidt, 1974; Bartholomew & Alexander, 1981), although arid soils may release carbon monoxide into the atmosphere (Conrad & Seiler, 1982). Another 100 million tonnes (5%) or so are probably removed annually in the stratosphere (Seiler, 1974).

4.2.2 Atmospheric lifetime

Based on the global sources and sinks described above, the average atmospheric lifetime of carbon monoxide can be calculated to be about 2 months, with a range between 1 and 4 months, which reflects the uncertainty in the annual emissions of carbon monoxide

($\tau = C/S$, where C is the tropospheric mixing ratio and S is the total annual emissions). The lifetime, however, can vary enormously with latitude and season compared with its global average value. During winters at high and middle latitudes, carbon monoxide has a lifetime of more than a year, but during summers at middle latitudes, the lifetime may be closer to the average global lifetime of about 2 months. Moreover, in the tropics, the average lifetime of carbon monoxide is probably about 1 month. These calculated variations reflect the seasonal cycles of hydroxyl radicals at various latitudes.

4.2.3   Latitudinal distribution of sources

When the sources, sinks, transport and observed concentrations of carbon monoxide are combined into a mass balance model, it is possible to calculate any one of these four components if the others are known. In the case of carbon monoxide, the sources can be estimated assuming that the sinks (hydroxyl radical reaction and soils), transport and concentrations are known. The latitudinal distribution of sources can be described in a one-dimensional model (Khalil & Rasmussen, 1990b). This model is similar to that described by Czeplak & Junge (1974) and Fink & Klais (1978). A time-averaged version was applied to the carbon monoxide budget by Hameed & Stewart (1979), and a somewhat modified and time-dependent version, mentioned above, was applied by Khalil & Rasmussen (1990b) to derive the latitudinal distribution of carbon monoxide shown in Fig. 2. Calculations by Khalil & Rasmussen (1990b) also suggest that emissions are higher in spring and summer than in the other seasons, particularly in the middle northern latitudes. This is expected for three reasons: (1) oxidation of methane and other hydrocarbons is faster during the summer because of the seasonal variation of hydroxyl radicals; (2) other direct emissions are also greater during spring and summer; and (3) at middle and higher latitudes, methane and non-methane hydrocarbons build up during the winter, and this reservoir is oxidized when hydroxyl radical concentrations rise during the spring.

From Fig. 2, the emissions from the northern and southern tropical latitudes sum up to 480 million tonnes per year and 330 million tonnes per year, respectively; the emissions from the northern and southern middle latitudes are 960 million tonnes per year and 210 million tonnes per year, respectively; some 50 million tonnes are emitted each year from the Arctic; and some 10 million tonnes per year come from the Antarctic. The largest fluxes of carbon monoxide

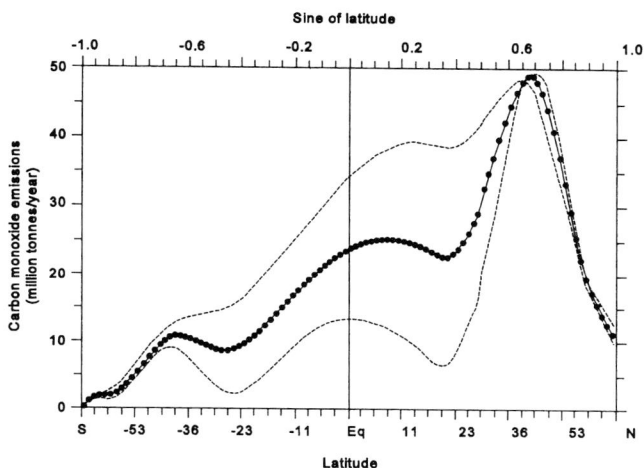

Fig. 2. The estimated emissions of carbon monoxide as a function of latitude. The emissions are in million tonnes/year (Mt/year) in each latitude band 0.02 units in sine of latitude. The dashed lines are estimates of uncertainties as hydroxyl radical concentrations and the rate of dispersion are varied simultaneously so that the maximum values of each of these parameters are twice the minimum values (from Khalil & Rasmussen, 1990b).

are from the industrial band of latitudes between 30 and 50 °N. From this region, some 620 million tonnes per year are emitted, representing about 30% of the total emissions of 2050 million tonnes per year. The model does not distinguish between anthropogenic and natural sources, nor does it distinguish between direct emissions and photo-chemical production of carbon monoxide from the oxidation of hydro-carbons. A large part of the estimated fluxes from the mid-northern latitudes and from tropical regions is likely to be of anthropogenic origin. The latitudinal distribution in Fig. 2 is compatible with the estimate (from Table 3 in chapter 3) that about 60% of the total carbon monoxide emissions are from anthropogenic activities.

4.2.4 Uncertainties and consistencies

The first consistency one notes is that the total emissions of carbon monoxide estimated from the various sources are balanced by the estimated removal of carbon monoxide. The approximate balance between sources and sinks is expected because the trends are showing an increase of only about 4–8 million tonnes per year compared with the total global emission rate of more than 2000 million tonnes per year.

On the other hand, there are large uncertainties in the sources and sinks that in the future may upset the apparently cohesive present budget of carbon monoxide. Although the patterns of the global distribution are becoming established, there are still uncertainties about the absolute concentrations. Estimates of emissions from individual sources are very uncertain (see Table 3 in chapter 3). In most cases, the stated uncertainty is a qualitative expression of the likely range of emissions, and it cannot be interpreted statistically. Therefore, the resulting uncertainty in the total emissions, obtained by adding up the uncertainties in individual sources, appears to be large.

There are two difficulties encountered in improving the estimates of carbon monoxide emissions from individual sources. First, although many critical experiments to determine the production and emissions of carbon monoxide from individual sources are yet to be done, there is a limit to the accuracy with which laboratory data can be extrapolated to the global scale. Second, the cycle of carbon monoxide may be so intimately tied up with the cycles of hydrocarbons that accurate global estimates of carbon monoxide emissions may not be possible until the hydrocarbon cycles are better understood.

Whereas the global distribution and seasonal variations in the global distribution of hydroxyl radicals can be calculated, there are no direct measurements of hydroxyl radicals that can be used to estimate the removal of carbon monoxide. The effective average concentration of hydroxyl radicals that acts on trace gases can be estimated indirectly from the cycles of other trace gases with known global emissions. Therefore, the total emissions of carbon monoxide are constrained by the budgets of other trace gases, even though the estimates of emissions from individual sources may remain uncertain. The most notable constraint may be the budget of methyl chloroform. Methyl chloroform is a degreasing solvent that has been emitted into the atmosphere in substantial quantities for more than 20 years. It is thought to be removed principally by reacting with hydroxyl radicals and to a lesser extent by photodissociation in the stratosphere. Because industry records on methyl chloroform production and sales have been kept for a long time, methyl chloroform can be used to estimate the average amount of hydroxyl radicals needed to explain the observed concentrations compared with the emissions. The accuracy of the source estimates of methyl chloroform is improved by the patterns of its uses; most methyl chloroform tends to be released shortly after purchase, so large unknown or unquantified reservoirs probably do not exist. The recent budgets of methyl chloroform suggest that, on average, there are about 8×10^5 molecules of hydroxyl radical per cubic centimetre, although significant uncertainties remain (see, for example, Khalil & Rasmussen, 1984c). This is the value used above in estimating the loss of carbon monoxide from reaction with hydroxyl radicals. The same average value of hydroxyl radicals also explains the methane concentrations compared with estimated sources, lending more support to the accuracy of the estimated hydroxyl radical concentrations. Neither of these constraints is very stringent; however, if the total global emissions of carbon monoxide from all sources are much different from the estimated 2600 million tonnes per year, then revisions of the budgets of both methane and methyl chloroform may be required.

Although there are other sources and sinks of carbon monoxide, these are believed to be of lesser importance on a global scale (Swinnerton et al., 1971; Chan et al., 1977).

4.3 Global distributions

Atmospheric concentrations, and thus the global distribution, are generally the most accurately known components of a global mass balance of a trace gas, because direct atmospheric measurements can be taken (Wilkniss et al., 1973; Seiler, 1974; Ehhalt & Schmidt, 1978; Pratt & Falconer, 1979; Heidt et al., 1980; Dianov-Klokov & Yurganov, 1981; Seiler & Fishman, 1981; Rasmussen & Khalil, 1982; Reichle et al., 1982; Hoell et al., 1984; Fraser et al., 1986; Khalil & Rasmussen, 1988, 1990a). Much has been learned about the global distribution of carbon monoxide over the last decade. The experiments leading to the present understanding range from systematic global observations at ground level for the last 8–10 years, reported by Khalil & Rasmussen (1988, 1990a) and Seiler (Seiler & Junge, 1970; Seiler, 1974), to finding the instantaneous global distribution of carbon monoxide from remote-sensing instruments on board the US National Aeronautics and Space Administration's space shuttle, as reported by Reichle et al. (1982, 1990).

4.3.1   Seasonal variations

The seasonal variations in carbon monoxide are well established (Dianov-Klokov & Yurganov, 1981; Seiler et al., 1984; Fraser et al., 1986; Khalil & Rasmussen, 1990a). High concentrations are observed during the winters in each hemisphere, and the lowest concentrations are seen in late summer. The amplitude of the cycle is largest at high northern latitudes and diminishes as one moves towards the equator until it is reversed in the southern hemisphere, reflecting the reversal of the seasons. The seasonal variations are small in the equatorial region. These patterns are expected from the seasonal variations in hydroxyl radical concentrations and carbon monoxide emissions. At mid and high latitudes, diminished solar radiation, water vapour and ozone during winters cause the concentrations of hydroxyl radicals to be much lower than during summer. The removal of carbon monoxide is slowed down, and its concentrations build up. In summer, the opposite pattern exists, causing the large seasonal variations in carbon monoxide.

On the hemispheric scale, the seasonal variation in carbon monoxide is approximately proportional to the concentration. Therefore, because there is much more carbon monoxide in the northern hemisphere than in the southern hemisphere, the decline of concentrations in the northern hemisphere during the summer is not balanced by the rise of concentrations in the southern hemisphere. This causes a global seasonal variation. The total amount of carbon monoxide in the Earth's atmosphere undergoes a remarkably large seasonal variation; the global burden is highest during northern winters and lowest during northern summers.

4.3.2   Latitudinal variation

The global seasonal variation in carbon monoxide content in the Earth's atmosphere also creates a seasonal variation in the latitudinal distribution (Newell et al., 1974; Seiler, 1974; Reichle et al., 1982, 1986; Khalil & Rasmussen, 1988, 1990a). During northern winters, carbon monoxide levels are at their highest in the northern hemisphere, whereas concentrations in the southern hemisphere are at a minimum. The interhemispheric gradient, defined as the ratio of the amounts of carbon monoxide in the northern and southern hemispheres, is at its maximum of about 3.2 during northern hemisphere winters and falls to about 1.8 during northern hemisphere summers, which is about half the winter value. The average latitudinal gradient is about 2.5, which means that, on average, there is about 2.5 times as much carbon monoxide in the northern hemisphere as in the southern hemisphere. Early data on latitudinal variations did not account for the seasonal variations.

4.3.3 Variations with altitude

In the northern hemisphere troposphere, the concentrations of carbon monoxide generally decline with altitude, but in the southern hemisphere, the vertical gradient may be reversed as a result of the transport of carbon monoxide from the northern hemisphere into the southern hemisphere. Above the tropopause, concentrations decline rapidly, so that there is very little carbon monoxide between 20 and 40 km; at still higher altitudes, the mixing ratio may again increase (Seiler & Junge, 1969; Seiler & Warneck, 1972; Fabian et al., 1981).

4.3.4 Other variations

The concentration of carbon monoxide is generally higher in the air over populated continental areas than in the air over oceans, even though oceans release carbon monoxide into the atmosphere. Other regions, such as tropical forests, may also be a source of isoprene and other hydrocarbons that may form carbon monoxide in the atmosphere. Such sources produce shifting patterns of high carbon monoxide concentrations over regional and perhaps even larger spatial scales. Variations in carbon monoxide concentrations were measured during the 1984 flights of the space shuttle, as reported by Reichle et al. (1990).

Occasionally, significant diurnal variations in carbon monoxide concentrations may also occur in some locations. For instance, diurnal variations have been observed over some parts of the oceans, with high concentrations during the day and low concentrations at night. Because similar patterns also exist in the surface seawater, the diurnal variations in carbon monoxide concentrations in the air can be explained by emissions from the oceans.

Finally, after the repeating cycles and other trends are subtracted, considerable random fluctuations still remain in time series of measurements. These fluctuations reflect the short lifetime of carbon monoxide and the vicinity of the sources, and they complicate the detection of long-term trends.

4.4   Global trends

Global concentrations of carbon monoxide were reported to be increasing during the late 1970s and early 1980s, because some 60% of the global emissions of carbon monoxide come from anthropogenic sources, which had increasing emissions over this period. Direct atmospheric observations reported by Khalil & Rasmussen (1984a) showed a detectable increasing trend at Cape Meares in Oregon, USA, between 1979 and 1982, when the rate of increase was about 5% per year. Subsequent data from the same site showed that the rate was not sustained for long, and a much smaller increasing trend of somewhat less than 2% per year emerged over the longer period of 1970–1987 (Khalil & Rasmussen, 1988). Similar data from other sites distributed worldwide also showed a global increase of about 1% per year (Khalil & Rasmussen, 1988). The trends were strongest in the mid-northern latitudes where most of the sources were located and became smaller and weaker in the southern hemisphere. At the mid-southern latitude site, the trends persisted but were not statistically significant (Khalil & Rasmussen, 1988). Rinsland & Levine (1985) reported estimates of carbon monoxide concentrations from spectroscopic plates from Europe showing that between 1950 and 1984, carbon monoxide concentrations increased at about 2% per year. Spectroscopic measurements of carbon monoxide taken by Dvoryashina et al. (1982, 1984) and Dianov-Klokov and colleagues (Dianov-Klokov et al., 1978; Dianov-Klokov & Yurganov, 1981) in the Soviet Union also suggested an increase of about 2% per year between 1974 and 1982 (Khalil & Rasmussen, 1984b, 1988).

More recent reports of carbon monoxide measurements in air samples show that from 1988 to 1993, global carbon monoxide concentrations started to decline rapidly. Novelli et al. (1994) collected air samples from 27 locations between 71 °N and 41 °S about once every 3 weeks from a ship during the period June 1990 to June 1993. In the northern latitudes, carbon monoxide concentrations decreased at a spatially and temporally average rate of 8.4 ± 1.0 μg/m^3 (7.3 ± 0.9 ppb) per year (6.1% per year). In the southern latitudes, carbon monoxide concentrations decreased at a rate of 4.8 ± 0.6 μg/m^3 (4.2 ± 0.5 ppb) per year (7.0% per year). Khalil & Rasmussen (1994) reported a slightly smaller decline in global carbon monoxide concentrations of $2.6 \pm 0.8\%$ per year during the period from 1988 to 1992. The rate of decrease reported by Khalil & Rasmussen (1994) was particularly rapid in the southern hemisphere. The authors hypothesize that this decline may reflect a reduction in tropical biomass burning.

Since 1993, the downward trend in global carbon monoxide concentrations has levelled off, and it is not clear if carbon monoxide will continue to decline or increase. These reported trends in global carbon monoxide concentrations are relatively small, and the random variability is large. Nevertheless, they are extremely important towards an understanding of global atmospheric chemistry and possible effects on global climate. Such changes in tropospheric carbon monoxide concentrations can cause shifts in hydroxyl radical concentrations and affect the oxidizing capacity of the atmosphere, thereby influencing the concentration of other trace gases, including methane. This change could also be a significant factor contributing to levels of ozone in the non-urban troposphere.

The likely future global-scale concentrations of carbon monoxide are completely unknown at present. It is possible that in the next decade, carbon monoxide concentrations will remain stable or even decline further. Emissions from automobiles are probably on the decline worldwide, emissions from biomass burning may be stabilizing or even declining, as speculated above, and the contribution from methane oxidation may no longer be increasing as rapidly as before. Because the atmospheric lifetime of carbon monoxide is short compared with those of other contributors to global change, the ambient concentrations adjust rapidly to existing emissions of carbon monoxide or its precursors.

5. ENVIRONMENTAL LEVELS AND PERSONAL EXPOSURES

5.1 Introduction

Air quality guidelines for carbon monoxide are designed to protect against actual and potential human exposures in ambient air that would cause adverse health effects. The World Health Organization's guidelines for carbon monoxide exposure (WHO, 1987) are expressed at four averaging times, as follows:

100 mg/m³ for 15 min
60 mg/m³ for 30 min
30 mg/m³ for 1 h
10 mg/m³ for 8 h

The guideline values and periods of time-weighted average exposures have been determined so that the carboxyhaemoglobin level of 2.5% is not exceeded, even when a normal subject engages in relatively heavy work.

Cigarette consumption represents a special case of carbon monoxide exposure; for the smoker, it almost always dominates over personal exposure from other sources. Studies by Radford & Drizd (1982) show that carboxyhaemoglobin levels of cigarette smokers average 4%, whereas those of non-smokers average 1%. Therefore, this summary focuses on environmental exposure of non-smokers to carbon monoxide.

People encounter carbon monoxide in a variety of environments — while travelling in motor vehicles, working at their jobs, visiting urban locations associated with combustion sources, or cooking and heating with domestic gas, charcoal or wood fires — as well as in tobacco smoke. Studies of human exposure have shown that among these settings, the motor vehicle is the most important for regularly encountered elevations of carbon monoxide. Studies conducted by the US EPA in Denver, Colorado, and Washington, DC, for example, have demonstrated that the motor vehicle interior has the highest average carbon monoxide concentrations (averaging 8–11 mg/m³ [7–10 ppm]) of all microenvironments (Johnson, 1984).

Another important setting for carbon monoxide exposure is the workplace. In general, carbon monoxide exposures at work exceed exposures during non-work periods, apart from commuting to and from work. Average concentrations may be elevated during this period because workplaces are often located in congested areas that have higher background carbon monoxide concentrations than do many residential neighbourhoods. Occupational and non-occupational exposures may overlay one another and result in a higher concentration of carbon monoxide in the blood. Certain occupations also increase the risk of high carbon monoxide exposure. These include those occupations involved directly with vehicle driving, maintenance and parking, such as auto mechanics; parking garage and gas station attendants; bus, truck or taxi drivers; traffic police; and warehouse workers. Some industrial processes produce carbon monoxide directly or as a by-product, including steel production, nickel refining, coke ovens, carbon black production and petroleum refining. Firefighters, cooks and construction workers may also be exposed to higher carbon monoxide levels at work. Occupational exposures in industries or settings with carbon monoxide production also represent some of the highest individual exposures observed in field monitoring studies.

The highest indoor non-occupational carbon monoxide exposures are associated with combustion sources and include enclosed parking garages, service stations and restaurants. The lowest indoor carbon monoxide concentrations are found in homes, churches and health care facilities. The US EPA's Denver Personal Monitoring Study showed that passive cigarette smoke is associated with increasing a non-smoker's exposure by an average of about 1.7 mg/m^3 (1.5 ppm) and that use of a gas range is associated with an increase of about 2.9 mg/m^3 (2.5 ppm) at home. Other sources that may contribute to higher carbon monoxide levels in the home include combustion space heaters and wood- and coal-burning stoves.

5.2 Population exposure to carbon monoxide

2.1 Ambient air monitoring

Many early attempts to estimate exposure of human populations used data on ambient air quality from fixed monitoring stations. An example of such an analysis can be found in the 1980 annual report of the President's Council on Environmental Quality (1980). In this analysis, a county's exposure to an air pollutant was estimated as the product of the number of days on which violations of the primary standard were observed at county monitoring sites multiplied by the county's population. Exposure was expressed in units of person-days. National exposure to an air pollutant was estimated by the sum of all county exposures.

The methodology employed by the Council on Environmental Quality provides a relatively crude estimate of exposure and is limited by four assumptions:

(1) The exposed populations do not travel outside areas represented by fixed-site monitors.

(2) The air pollutant concentrations measured with the network of fixed-site monitors are representative of the concentrations breathed by the population throughout the area.

(3) The air quality in any one area is only as good as that at the location that has the worst air quality.

(4) There are no violations in areas of the county that are not monitored.

Many studies cast doubt on the validity of these assumptions for carbon monoxide. These studies are reviewed in Ott (1982) and in Spengler & Soczek (1984). Doubts over the ability of fixed-site monitors alone to accurately depict air pollutant exposures are based on two major findings on fixed-site monitor representativeness:

(1) Indoor and in-transit concentrations of carbon monoxide may be significantly different from ambient carbon monoxide concentrations.

(2) Ambient outdoor concentrations of carbon monoxide with which people come in contact may vary significantly from carbon monoxide concentrations measured at fixed-site monitors.

In estimating exposure, the Council on Environmental Quality also assumed that each person in the population spends 24 h at home.

This assumption permitted the use of readily available demographic data from the US Bureau of the Census. Data collected 20 years ago indicate that people spend a substantial portion of their time away from home. In a study of metropolitan Washington, DC, residents during 1968, Chapin (1974) found that people spent an average 6.3 h away from home on Sunday and 10.6 h away from home on Friday. This translates to between 26.3% and 44.2% of the day spent away from home. More recent personal exposure and time budget studies (e.g., Johnson, 1987; Schwab et al., 1990) also indicate that a substantial portion of time is spent away from home.

Fixed-site monitors measure concentrations of pollutants in ambient air. Ambient air has been defined by the US EPA in the Code of Federal Regulations (OSHA, 1991b) as air that is "external to buildings, to which the general public has access." But the nature of modern urban lifestyles in many countries, including the USA, indicates that people spend an average of over 20 h per day indoors (Meyer, 1983). Reviews of studies on this subject by Yocom (1982), Meyer (1983) and Spengler & Soczek (1984) show that measurements of indoor carbon monoxide concentrations vary significantly from simultaneous measurements in ambient air. The difference between indoor and outdoor air quality and the amount of time people spend indoors reinforce the conclusion that using ambient air quality measurements alone will not provide accurate estimates of population exposure.

2.2 Approaches for estimating population exposure

In recent years, researchers have focused on the problem of determining actual population exposures to carbon monoxide. There are three alternative approaches for estimating the exposures of a population to air pollution: the "direct approach," using field measurement of a representative population carrying PEMs; the "indirect approach," involving computation from field data of activity patterns and measured concentration levels within microenvironments (Ott, 1982); and a hybrid approach that combines the direct and indirect approaches (Mage, 1991).

In the direct approach, as study participants engage in regular daily activities, they are responsible for recording their exposures to the pollutant of interest using a personal monitor. Subjects can record their exposures in a diary, the method used in a US pilot study in

Los Angeles, California (Ziskind et al., 1982), or they can auto-matically store exposure data in a data logger, the method used in studies in Denver, Colorado (Johnson, 1984), and Washington, DC (Hartwell et al., 1984), which are summarized by Akland et al. (1985). In all of these studies, subjects also recorded the time and nature of their activities while they monitored personal exposures to carbon monoxide.

The direct approach can be used to obtain an exposure inventory of a representative sample from either the general population or a specific subpopulation, which can be defined by many demographic, occupational and health factors. The inventory can cover a range of microenvironments encountered over a period of interest (e.g., a day), or it can focus on one particular microenvironment. With this flexi-bility, policy analysts can assess the problem that emission sources pose to a particular subgroup (e.g., commuters) active in a specific microenvironment (e.g., automobiles).

The indirect approach to estimating personal exposure is to use PEMs or microenvironmental monitors to monitor microenvironments rather than individuals. Combined with the ambient data and addition-al data on human activities that occur in these microenvironments, data from the indirect approach can be used to estimate the percentage of a subpopulation that is at risk for exposure to pollutant concentra-tions that exceed national or regional air quality standards. Flachsbart & Brown (1989) conducted this type of study to estimate merchant exposure to carbon monoxide from motor vehicle exhaust at the Ala Moana Shopping Center in Honolulu, Hawaii, USA.

5.2.3 *Personal monitoring field studies*

The development of small PEMs made possible the large-scale carbon monoxide human exposure field studies in Denver, Colorado, and Washington, DC, in the winter of 1982–83 (Akland et al., 1985). These monitors proved effective in generating 24-h carbon monoxide exposure profiles on 450 persons in Denver and 800 persons in Washington, DC. The Denver–Washington, DC, study is the only large-scale field study on population exposure to carbon monoxide that has been undertaken to date.

Results from the Denver–Washington, DC, study (Akland et al., 1985) show that over 10% of the Denver residents and 4% of the

Washington, DC, residents were exposed to 8-h average carbon monoxide levels above 10 mg/m³ (9 ppm) during the winter study period. This degree of population exposure could not be accurately deduced from simultaneous data collected by the fixed-site monitors without taking into account other factors, such as contributions from indoor sources, elevated levels within vehicles and individuals' activity patterns. In Denver, for example, the fixed-site monitors exceeded the 10 mg/m³ (9 ppm) level only 3.1% of the time. These results indicate that the effects of personal activity, indoor sources and, especially, time spent commuting all greatly contribute to a person's carbon monoxide exposure.

This study emphasizes that additional strategies are required to augment data from fixed-site monitoring networks in order to evaluate actual human carbon monoxide exposures and health risks within a community. The cumulative carbon monoxide data for both Denver and Washington, DC, show that personal monitors often measure higher concentrations than do fixed stations. As part of this study, 1-h exposures to carbon monoxide concentrations as determined by personal monitors were compared with measured ambient concentrations at fixed monitor sites. Correlations between personal monitor data and fixed-site data were consistently poor; the fixed-site data usually explained less than 10% of the observed variation in personal exposure. For example, 1-h carbon monoxide measurements taken at the nearest fixed stations were only weakly correlated ($0.14 \leq r \leq 0.27$) with office or residential measurements taken with personal monitors (Akland et al., 1985).

The conclusion that exposure of persons to ambient carbon monoxide and other pollutants does not directly correlate with concentrations determined at fixed-site monitors is supported by the work of others (Ott & Eliassen, 1973; Cortese & Spengler, 1976; Dockery & Spengler, 1981; Wallace & Ott, 1982; Wallace & Ziegenfus, 1985). Results from the Finnish Liila study in Helsinki, in which personal carbon monoxide and nitrogen dioxide exposures of preschool children were monitored, showed that their short-term personal carbon monoxide exposures did not correlate with carbon monoxide levels in ambient air and that gas stove use at home was the dominant determinant of carbon monoxide exposure (Alm et al., 1994).

In view of the high degree of variability of ambient carbon monoxide concentrations over both space and time and the presence of indoor sources of carbon monoxide, the reported results are not surprising. A given fixed monitor is unable to track the exposure of individuals to ambient carbon monoxide as they go about their daily activities, moving from one location to another, all of which are seldom in the immediate vicinity of the monitor. This does not necessarily mean, however, that fixed monitors do not give some useful general information on the overall level of exposure of a population to carbon monoxide. The Denver and Washington, DC, data, although failing to show a correlation between exposures measured by individual personal monitors and simultaneous concentrations measured by the nearest fixed-site monitors, did suggest that, in Denver, aggregate personal exposures were lower on days of lower ambient carbon monoxide levels as determined by fixed-site monitors and higher on days of higher ambient levels. Also, both fixed-site and personal exposures were higher in Denver than in Washington, DC. For example, the median ambient daily 1-h maximum carbon monoxide concentration was measured by fixed monitors to be 3.7 mg/m^3 (3.2 ppm) higher in Denver than in Washington, DC, and the personal median daily 1-h maximum carbon monoxide exposure was measured by PEMs to be 4.5 mg/m^3 (3.9 ppm) higher in Denver. Likewise, the median ambient daily 8-h maximum carbon monoxide concentration measured by fixed monitors was found to be 3.3 mg/m^3 (2.9 ppm) higher in Denver, whereas the personal median daily 8-h maximum carbon monoxide exposure was 3.9 mg/m^3 (3.4 ppm) higher in Denver.

The in-transit microenvironment with the highest estimated carbon monoxide concentration was the motorcycle, whereas walking and bicycling had the lowest carbon monoxide concentrations. Outdoor microenvironments can also be ranked for these data. Outdoor public garages and outdoor residential garages and carports had the highest carbon monoxide concentrations; outdoor service stations, vehicle repair facilities and parking lots had intermediate concentrations. In contrast, school grounds and residential grounds had relatively low concentrations, whereas extremely low carbon monoxide concentrations were found in outdoor sports arenas, amphitheatres, parks and golf courses. Finally, a wide range of concentrations was found in Denver within indoor microenvironments. The highest indoor carbon monoxide concentrations occurred in service stations, vehicle repair facilities and public parking garages; intermediate concentrations were found in shopping malls, residential garages, restaurants, offices, auditoriums, sports arenas, concert halls and stores; and the lowest concentrations were found in health care facilities, public buildings, manufacturing facilities, homes, schools and churches.

One activity that influences personal exposure is commuting. An estimated 1% of the non-commuters in Washington, DC, were exposed to concentrations above 10 mg/m^3 (9 ppm) for 8 h. By comparison, an estimated 8% of persons reporting that they commuted more than 16 h per week had 8-h carbon monoxide exposures above the 10 mg/m^3 (9 ppm) level. Finally, certain occupational groups whose work brings them in close proximity to the internal combustion engine had a potential for elevated carbon monoxide exposures. These include automobile mechanics; parking garage or gas station attendants; crane deck operators; cooks; taxi, bus and truck drivers; firemen; policemen; and warehouse and construction workers. Of the 712 carbon monoxide exposure profiles obtained in Washington, DC, 29 persons fell into this "high-exposure" category. Of these, 25% had 8-h carbon monoxide exposures above the 10 mg/m^3 (9 ppm) level.

Several field studies have also been conducted by the US EPA to determine the feasibility and effectiveness of monitoring selected microenvironments for use in estimating exposure profiles indirectly. One study (Flachsbart et al., 1987), conducted in Washington, DC, in 1982 and 1983, concentrated on the commuting microenvironment, because earlier studies identified this microenvironment type as the single most important non-occupational microenvironment relative to total carbon monoxide population exposure. It was observed that for the typical automobile commuter, the time-weighted average carbon monoxide exposure while commuting ranged from 10 to 16 mg/m^3 (9 to 14 ppm). The corresponding rush-hour (7:00 to 9:00 a.m., 4:00 to 6:00 p.m.) averages at fixed-site monitors were 3.1–3.5 mg/m^3 (2.7–3.1 ppm).

2.4   *Carbon monoxide exposures indoors*

People in developed countries spend a majority (~85%) of their time indoors (US EPA, 1989a,b); therefore, a comprehensive depiction of exposure to carbon monoxide must include this setting. The indoor sources, emissions and concentrations are sufficiently diverse, however, that only a few studies can be cited here as examples.

Targeted field studies that have monitored indoor carbon monoxide levels as a function of the presence or absence of combustion sources are described in more detail in section 5.6.

Early studies date back to before 1970, when it was found that indoor and outdoor carbon monoxide levels do not necessarily agree. For example, one study determined indoor–outdoor relationships for carbon monoxide over 2-week periods during summer, winter and fall in 1969 and 1970 in buildings in Hartford, Connecticut, USA (Yocom et al., 1971). With the exception of the private homes, which were essentially equal, there was a day-to-night effect in the fall and winter seasons; days were higher by about a factor of 2. These differences are consistent with higher traffic-related carbon monoxide levels outdoors in the daytime.

Indoor and outdoor carbon monoxide concentrations were measured in four homes, also in the Hartford, Connecticut, area, in 1973 and 1974 (Wade et al., 1975). All used gas-fired cooking stoves. Concentrations were measured in the kitchen, living room and bedroom. Stove use, as determined by activity diaries, correlated directly with carbon monoxide concentrations. Peak carbon monoxide concentrations in several of the kitchens exceeded 10 mg/m^3 (9 ppm), but average concentrations ranged from 2.3–3.4 mg/m^3 (2–3 ppm) to about 9 mg/m^3 (8 ppm). These results are in general agreement with results obtained in Boston, Massachusetts, USA (Moschandreas & Zabransky, 1982). In this study, the investigators found significant differences between carbon monoxide concentrations in rooms in homes where there were gas appliances.

Effects of portable kerosene-fired space heaters on indoor air quality were measured in an environmental chamber and a house (Traynor et al., 1982). Carbon monoxide emissions from white flame and blue flame heaters were compared. The white flame convective heater emitted less carbon monoxide than the blue flame radiant heater. Concentrations in the residence were <2.3 mg/m^3 (<2 ppm) and 2.3–8 mg/m^3 (2–7 ppm), respectively. The authors concluded that high levels might occur when kerosene heaters are used in small spaces or when air exchange rates are low.

A rapid method using an electrochemical PEM to survey carbon monoxide was applied in nine high-rise buildings in the San Francisco and Los Angeles, California, areas during 1980 and 1984 (Flachsbart & Ott, 1986). One building had exceptionally high carbon monoxide levels compared with the other buildings; average concentrations on various floors ranged from 6 to 41 mg/m³ (5 to 36 ppm). The highest levels were in the underground parking garage, which was found to be the source of elevated carbon monoxide within the building through transport via the elevator shaft.

The effect of residential wood combustion and specific heater type on indoor carbon monoxide levels has been investigated (Humphreys et al., 1986). Airtight and non-airtight heaters were compared in a research home in Tennessee, USA. Carbon monoxide emissions from the non-airtight heaters were generally higher than those from the airtight heaters. Peak indoor carbon monoxide concentration (ranging from 1.5 to 33.9 mg/m³ [1.3 to 29.6 ppm], depending on heater type) was related to fuel reloadings.

Two studies in the Netherlands have measured carbon monoxide levels in homes. Carbon monoxide levels in 254 Netherland homes with unvented gas-fired geysers (water heaters) were investigated during the winter of 1980 (Brunekreef et al., 1982). Concentrations at breathing height were grouped into the following categories: ≤11 mg/m³ (≤10 ppm; n = 154), 13–57 mg/m³ (11–50 ppm; n = 50), 58–110 mg/m³ (51–100 ppm; n = 25) and >110 mg/m³ (>100 ppm; n = 17). They found that a heater vent reduced indoor carbon monoxide concentrations and that the type of burner affected carbon monoxide levels. In another study, air pollution in Dutch homes was investigated by Lebret (1985). Carbon monoxide concentrations were measured in the kitchen (0–20.0 mg/m³ [0–17.5 ppm]), the living room (0–10.0 mg/m³ [0–8.7 ppm]) and the bedroom (0–4.0 mg/m³ [0–3.5 ppm]). Carbon monoxide levels were elevated in homes with gas cookers and unvented geysers. Kitchen carbon monoxide levels were higher than those in other locations as a result of peaks from the use of gas appliances. Carbon monoxide levels in living rooms were slightly higher in houses with smokers. The overall mean carbon monoxide level indoors was 0–3.1 mg/m³ (0–2.7 ppm) above outdoor levels.

In Zagreb, Yugoslavia, carbon monoxide was measured in eight urban institutions housing sensitive populations, including kindergartens, a children's hospital and homes for the elderly (Sisovic &

Fugas, 1985). Winter carbon monoxide concentrations ranged from 1.3 to 15.7 mg/m³ (1.1 to 13.7 ppm), and summer concentrations ranged from 0.7 to 7.9 mg/m³ (0.6 to 6.9 ppm). The authors attributed indoor carbon monoxide concentrations to nearby traffic density, general urban pollution, seasonal differences and day-to-day weather conditions. Indoor sources were not reported.

Toxic levels of carbon monoxide were also found in measurements at six ice skating rinks (Johnson et al., 1975b). This study was prompted by the reporting of symptoms of headache and nausea among 15 children who patronized one of the rinks. Carbon monoxide concentrations were found to be as high as 350 mg/m³ (304 ppm) during operation of a propane-powered ice-resurfacing machine. Depending on skating activity levels, the ice-resurfacing operation was performed for 10 min every 1–2 h. Because this machine was found to be the main source of carbon monoxide, using catalytic converters and properly tuning the engine greatly reduced emissions of carbon monoxide and, hence, reduced carbon monoxide concentrations. Similar findings have been reported by Spengler et al. (1978), Lévesque et al. (1990), Paulozzi et al. (1993) and Lee et al. (1994).

5.2.5   *Carbon monoxide exposures inside vehicles*

Studies of carbon monoxide concentrations inside automobiles have also been reported over the past decade.

Petersen & Sabersky (1975) measured pollutants inside an automobile under typical driving conditions. Carbon monoxide concentrations were generally less than 29 mg/m³ (25 ppm), with one 3-min peak of 52 mg/m³ (45 ppm). Average concentrations inside the vehicle were similar to those outside. No in-vehicle carbon monoxide sources were noted; however, a commuter's exposure is usually determined by other high-emitting vehicles, not by the driven vehicle itself (Chan et al., 1989; Shikiya et al., 1989).

Drowsiness, headache and nausea were reported by eight children who had ridden in school buses for about 2 h while travelling on a ski trip (Johnson et al., 1975a). The students reporting symptoms were seated in the rear of the bus, which had a rear-mounted engine and a leaky exhaust. The exhaust system was subsequently repaired. During a later ski outing for students, carbon monoxide concentrations were also monitored for a group of 66 school buses in the parking lot. The investigators found 5 buses with carbon monoxide concentrations of 6–29 mg/m³ (5–25 ppm) (mean 17 mg/m³ [15 ppm]), 24 buses showing concentrations in excess of 10 mg/m³ (9 ppm) for short periods and 2 buses showing up to 3 times the 10 mg/m³ (9 ppm) level for short periods. Drivers were advised to park so that exhausts from one bus would not be adjacent to the fresh air intake for another bus.

During a US cross-country trip in the spring of 1977, Chaney (1978) measured in-vehicle carbon monoxide concentrations. The carbon monoxide levels varied depending on traffic speed. On expressways in Chicago, Illinois, San Diego, California, and Los Angeles, California, when traffic speed was less than 16 km/h, carbon monoxide concentrations exceeded 17 mg/m³ (15 ppm). Levels increased to 52 mg/m³ (45 ppm) when traffic stopped. In addition, it was observed that heavily loaded vehicles (e.g., trucks) produced high carbon monoxide concentrations inside nearby vehicles, especially when the trucks were ascending a grade.

Colwill & Hickman (1980) measured carbon monoxide concentrations in 11 new cars as they were driven on a heavily trafficked route in and around London, United Kingdom. The inside mean carbon monoxide level for the 11 cars was 28.9 mg/m³ (25.2 ppm), whereas the outside mean level was 53.8 mg/m³ (47.0 ppm).

In a study mandated by the US Congress in the 1977 Clean Air Act Amendments, the EPA studied carbon monoxide intrusion into vehicles (Ziskind et al., 1981). The objective was to determine whether carbon monoxide was leaking into the passenger compartments of school buses, police cars and taxis and, if so, how prevalent the situation was. The study involved 1164 vehicles in Boston, Massachusetts, and Denver, Colorado. All vehicles were in use in a working fleet at the time of testing. The results indicated that all three types of vehicles often have multiple (an average of four to five) points of carbon monoxide intrusion — worn gaskets, accelerator pedals, rust spots in the trunk, etc. In 58% of the rides lasting longer than 8 h, carbon monoxide levels exceeded 10 mg/m³ (9 ppm). Thus, the study provided evidence that maintenance and possibly design of vehicles may be important factors in human exposure to carbon monoxide.

Petersen & Allen (1982) reported the results of carbon monoxide measurements taken inside vehicles under typical driving conditions in Los Angeles, California, over 5 days in October 1979. They found that the average ratio of interior to exterior carbon monoxide concentrations was 0.92. However, the hourly average interior carbon monoxide concentrations were 3.9 times higher than the nearest fixed-site measurements. In their analysis of the factors that influence interior carbon monoxide levels, they observed that traffic flow and traffic congestion (stop-and-go) are important, but "comfort state" (i.e., car windows open/closed, fan on/off, etc.) and meteorological parameters (i.e., wind speed, wind direction) have little influence on incremental exposures. Another study, carried out in Paris, France (Dor et al., 1995), also showed that the carbon monoxide concentration in cars can be 3 times that of the ambient air. A study in Hong Kong showed that the same may be true for concentrations in buses (Chan & Wu, 1993).

Flachsbart (1989) investigated the effectiveness of priority lanes on a Honolulu, Hawaii, arterial highway in reducing commuter travel time and exposure to carbon monoxide. The carbon monoxide concentrations and exposures of commuters in these lanes were substantially lower than in the non-priority lanes. Carbon monoxide exposure was reduced approximately 61% for express buses, 28% for high-occupancy vehicles and 18% for carpools when compared with that for regular automobiles. The higher speed associated with priority lanes helped reduce carbon monoxide exposure. These observations demonstrate that carbon monoxide concentrations have a high degree of spatial variability on roadways, associated with vehicle speed and traffic volume.

Ott et al. (1994) measured carbon monoxide exposures inside a car travelling on a major urban arterial highway in El Camino Real in the USA (traffic volume 30 500–45 000 vehicles per day) over a 13.5-month period. For 88 trips, the mean carbon monoxide concentration was 11.2 mg/m^3 (9.8 ppm), with a standard deviation of 6.6 mg/m^3 (5.8 ppm).

Fernadenz-Bremauntz & Ashmore (1995a,b) related exposure of commuters in vehicles to carbon monoxide to fixed-site concentrations at monitoring stations in Mexico City. The ambient levels were all more than 15 mg/m^3 (13 ppm). The highest median and 90th percentile in-vehicle concentrations were found in autos and minibuses (49–68 and 77–96 mg/m^3 [43–59 and 67–84 ppm]), with lower levels in buses (34 and 49 mg/m^3 [30 and 43 ppm]), trolleys (30 and 44 mg/m^3 [26 and 38 ppm]) and metro and light rail (24 and 33 mg/m^3 [21 and 29 ppm]). Average in-vehicle/ambient ratios for each mode of transport were as follows: automobile, 5.2; minivan, 5.2; minibus, 4.3; bus, 3.1; trolleybus, 3.0; and metro, 2.2.

Chan et al. (1991) investigated driver exposure to volatile organic chemicals, carbon monoxide, ozone and nitrogen dioxide under different urban, rural and interstate highway driving conditions using four different cars in Raleigh, North Carolina, USA. They found that the in-vehicle carbon monoxide concentrations did not vary significantly between the cars, and they were on average 4.5 times higher than the ambient carbon monoxide measurements. Car ventilation had little effect on the driver exposures.

Two Finnish studies of personal air pollution exposures of children showed that preschool children who commuted to a day care centre by bus or car were exposed to considerably higher peak carbon monoxide levels than children who went to the day care centre by walking or on a bike (Jantunen et al., 1995) and that the average carbon monoxide exposure of schoolchildren in Kuopio, Finland, in a car or bus was 4 times higher than in other microenvironments (Alm et al., 1995).

5.2.6 Carbon monoxide exposures outdoors

Carbon monoxide concentrations in outdoor settings (besides those measured at fixed monitoring stations) show considerable variability. Ott (1971) made 1128 carbon monoxide measurements at outdoor locations in San Jose, California, USA, at breathing height over a 6-month period and compared these results with the official fixed monitoring station data. This study simulated the measurements of the outdoor carbon monoxide exposures of pedestrians in downtown San Jose by having them carry personal monitoring pumps and sampling bags while walking standardized routes on congested sidewalks. If an outdoor measurement was made more than 100 m away from any major street, its carbon monoxide concentration was similar, suggesting the existence of a generalized urban background concentration in San Jose that was spatially uniform over the city (within a 33-km^2 grid) when one is sufficiently far away from mobile sources.

Because the San Jose monitoring station was located near a street with heavy traffic, it recorded concentrations approximately 100% higher than this background value. In contrast, outdoor carbon monoxide levels from personal monitoring studies of downtown pedestrians were 60% above the corresponding monitoring station values, and the correlation coefficient was low ($r = 0.20$). By collecting the pedestrian personal exposures over 8-h periods, it was possible to compare the levels with the air quality standards. On 2 of 7 days for which data were available, the pedestrian concentrations were particularly high (15 and 16.3 mg/m^3 [13 and 14.2 ppm]) and were 2–3 times the corresponding levels recorded at the same time (5.0 and 7.1 mg/m^3 [4.4 and 6.2 ppm]) at the air monitoring station (Ott & Eliassen, 1973; Ott & Mage, 1975). These results show that concentrations to which pedestrians were exposed on downtown streets could exceed recommended air quality standards, although the official air monitoring station record values were significantly less than that. It can be argued, however, that not many pedestrians spend a lot of time outdoors walking along downtown sidewalks, and that is one of the important reasons for including realistic human activity patterns in exposure assessments. It should also be noted that street-level carbon monoxide concentrations have significantly decreased over the last decade in countries where vehicular emission controls are in place.

Godin et al. (1972) conducted similar studies in downtown Toronto, Ontario, Canada, using 100-ml glass syringes in conjunction with NDIR spectrometry. They measured carbon monoxide concentrations along streets, inside passenger vehicles and at a variety of other locations. Like other investigators, they found that carbon monoxide concentrations were determined by very localized phenomena. In general, carbon monoxide concentrations in traffic and along streets were much higher than those observed at conventional fixed air monitoring stations. In a subsequent study in Toronto, Wright et al. (1975) used Ecolyzers to measure 4- to 6-min average carbon monoxide concentrations encountered by pedestrians and street workers and obtained similar results. Levels ranged from 11 to 57 mg/m^3 (10 to 50 ppm), varying with wind speed and direction, atmospheric stability, traffic density and height of buildings. They also measured carbon monoxide concentrations on the sidewalks of a street that subsequently was closed to traffic to become a pedestrian mall. Before the street was closed, the average concentrations at two intersections were 10.8 ± 4.6 mg/m^3 (9.4 ± 4.0 ppm) and 9.0 ± 2.2 mg/m^3 (7.9 ± 1.9 ppm) (mean, plus or minus standard deviation); after the street was closed, the averages dropped to 4.2 ± 0.6 mg/m^3 (3.7 ± 0.5 ppm) and 4.6 ± 1.1 mg/m^3 (4.0 ± 1.0 ppm), respectively, which were equivalent to the background level.

A large-scale field investigation was undertaken of carbon monoxide concentrations in indoor and outdoor locations in five California, USA, cities using personal monitors (Ott & Flachsbart, 1982). For outdoor commercial settings, the average carbon monoxide concentration was 5 mg/m^3 (4 ppm). This carbon monoxide level was statistically, but not substantially, greater than the average carbon monoxide concentration of 2.3 mg/m^3 (2 ppm) recorded simultaneously at nearby fixed monitoring stations.

The Organisation for Economic Co-operation and Development (OECD, 1997) recently published a study of trends and relative concentrations in Western Europe, the USA and Japan between 1988 and 1993, based on the annual maximum 8-h average concentration at urban traffic and urban residential sites. This showed that the average levels remained unchanged, but, at the most polluted locations in both residential and commercial (heavily trafficked) areas, levels of carbon monoxide have declined significantly. A short summary of results of different countries is given in Table 6, as published by the OECD (1997). Carbon monoxide measurements in atmospheric air were performed in many countries — for example, at approximately 300 sites in Germany using automatic devices.

Investigations in traffic routes of Amsterdam, Netherlands (van Wijnen et al., 1995), using personal air sampling resulted in much higher carbon monoxide concentrations in the samples of car drivers than in the personal air samples of cyclists. Similar results from Germany were cited.

5.3   Estimating population exposure to carbon monoxide

Accurate estimates of human exposure to carbon monoxide are a prerequisite for both a realistic appraisal of the risks posed by the pollutant and the design and implementation of effective control strategies. This section discusses the general concepts on which exposure assessment is based and approaches for estimating population exposure to carbon monoxide using exposure models. Because of

Table 6. Comparison of carbon monoxide concentrations, 1993

Site	Averaging period	Statistic	Carbon monoxide concentration					
			Western Europe		USA		Japan	
			mg/m^3	ppm	mg/m^3	ppm	mg/m^3	ppm
Urban traffic	8-h average	Average	8.3	7.2	6.8	5.9	5.5	4.8
	8-h average	95th percentile	13.1	11.4	11.6	10.1	9.1	7.9
Urban residential	8-h average	Average	5.7	5	6.4	5.6	4.3	3.8
	8-h average	95th percentile	9.5	8.3	10.1	8.8	7.1	6.2

problems in estimating population exposure solely from fixed-station data, several formal human exposure models have been developed. Some of these models include information on human activity patterns: the microenvironments people visit and the times they spend there. These models also contain submodels depicting the sources and concentrations likely to be found in each microenvironment, including indoor, outdoor and in-transit settings.

5.3.1 Components of exposure

Two aspects of exposure bear directly on the related health consequences. The first is the magnitude of the pollutant exposure. The second is the duration of the exposure. The magnitude is an important exposure parameter, because concentration typically is assumed to be directly proportional to dose and, ultimately, to the health outcome. But exposure implies a time component, and it is essential to specify the duration of an exposure. The health risks of exposure to a specific concentration for 5 min are likely to be different, all other factors being equal, from those of exposure to the same concentration for an hour.

The magnitude and duration of exposure can be determined by plotting an individual's air pollution exposure over time. The function $C_i(t)$ describes the air pollutant concentration to which an individual is exposed at any point in time t. Ott (1982) defined the quantity $C_i(t)$ as the *instantaneous exposure* of an individual. The shaded area under the graph represents the accumulation of instantaneous exposures over some period of time $(t_1 - t_0)$. This area is also equal to the integral of the air pollutant concentration function, $C_i(t)$, between t_0 and t_1. Ott (1982) defined the quantity represented by this area as the *integrated exposure*.

The *average exposure*, calculated by dividing the integrated exposure by the period of integration $(t_1 - t_0)$, represents the average air pollutant concentration to which an individual was exposed over the defined period of exposure. To facilitate comparison with established air quality standards, an averaging period is usually chosen to equal the averaging period of the standard. In this case, the average exposure is referred to as a standardized exposure.

As discussed above, exposure represents the joint occurrence of an individual being located at point (x,y,z) during time t with the simultaneous presence of an air pollutant at concentration $C_{xyz}(t)$. Consequently, an individual's exposure to an air pollutant is a function of location as well as time. If a volume at a location can be defined such that air pollutant concentrations within it are relatively homogeneous, yet potentially different from other locations, the volume may be considered a "microenvironment" (Duan, 1982). Microenvironments may be aggregated by location (e.g., indoor or outdoor) or activity performed at a location (e.g., residential or commercial) to form microenvironment types.

It is important to distinguish between individual exposures and population exposures. Sexton & Ryan (1988) defined the pollutant concentrations experienced by a specific individual during normal daily activities as "personal" or "individual" exposures. A personal exposure depends on the air pollutant concentrations that are present in the locations through which the person moves, as well as on the time spent at each location. Because time–activity patterns can vary substantially from person to person, individual exposures exhibit wide variability (Dockery & Spengler, 1981; Quackenboss et al., 1982; Sexton et al., 1984; Spengler et al., 1985; Stock et al., 1985; Wallace et al., 1985). Thus, although it is a relatively straightforward procedure to measure any one person's exposure, many such measurements may be needed to quantify the mean and variance of exposures for a defined group. The daily activities of a person in time and space define his or her activity pattern. Accurate estimates of air pollution exposure generally require that an exposure model account for the activity patterns of the population of interest. The activity patterns may be determined through "time budget" studies of the population. Studies of this type have been performed by Szalai (1972), Chapin (1974), Michelson & Reed (1975), Robinson (1977), Johnson (1987) and Schwab et al. (1990). The earlier studies may now be out of date because they were not designed to investigate human exposure questions and because lifestyles have changed over the past 25 years. Ongoing exposure studies have adopted the diary methods that were developed for sociological investigations and applied them to current exposure and time budget investigations. A few of these studies have been reported (e.g., Johnson, 1987; Schwab et al., 1990).

From a public health perspective, it is important to determine the "population exposure," which is the aggregate exposure for a specified group of people (e.g., a community or an identified occupational cohort). Because exposures are likely to vary substantially between individuals, specification of the distribution of personal exposures within a population, including the average value and the associated variance, is often the focus of exposure assessment studies. The upper tail of the distribution, which represents those individuals exposed to the highest concentrations, is frequently of special interest, because the determination of the number of individuals who experience elevated pollutant levels can be critical for health risk assessments. This is especially true for pollutants for which the relationship between dose and response is highly non-linear.

5.3.2   *Approaches to exposure modelling*

In recent years, the limitations of using fixed-site monitors alone to estimate public exposure to air pollutants have stimulated interest in using portable monitors to measure personal exposure. These PEMs were developed for carbon monoxide in the late 1970s by Energetics Science Incorporated and by General Electric. Wallace & Ott (1982) surveyed PEMs available then for carbon monoxide and other air pollutants.

The availability of these monitors has facilitated use of the direct and indirect approaches to assessing personal exposure (see section 5.2.2). Whether the direct or indirect approach is followed, the estimation of population exposure requires a "model" — that is, a mathematical or computerized approach of some kind. Sexton & Ryan (1988) suggested that most exposure models can be classified as one of three types: statistical, physical or physical-stochastic.

The statistical approach requires the collection of data on human exposures and the factors thought to be determinants of exposure. These data are combined in a statistical model, normally a regression equation or an analysis of variance, to investigate the relationship between air pollution exposure (dependent variable) and the factors contributing to the measured exposure (independent variables). An example of a statistical model is the regression model developed by Johnson et al. (1986) for estimating carbon monoxide exposures in Denver, Colorado, based on data obtained from the Denver Personal Monitoring Study. If the study group constitutes a representative sample, the derived statistical model may be extrapolated to the population defined by the sampling frame. It should also be noted that selection of factors thought to influence exposure has a substantial effect on the outcome of the analysis. Spurious conclusions can be drawn, for example, from statistical models that include parameters that are correlated with, but not causally related to, air pollution exposure.

In the physical modelling approach, the investigator makes an *a priori* assumption about the underlying physical processes that determine air pollution exposure and then attempts to approximate these processes through a mathematical formulation. Because the model is chosen by the investigator, it may produce biased results because of the inadvertent inclusion of inappropriate parameters or the improper exclusion of critical components. The National Ambient Air Quality Standards (NAAQS) Exposure Model as originally applied to carbon monoxide by Johnson & Paul (1983) is an example of a physical model.

The physical-stochastic approach combines elements of both the physical and statistical modelling approaches. The investigator begins by constructing a mathematical model that describes the physical basis for air pollution exposure. Then, a random or stochastic component that takes into account the imperfect knowledge of the physical parameters that determine exposure is introduced into the model. The physical-stochastic approach limits the effect of investigator-induced bias by the inclusion of the random component and allows for estimates of population distributions for air pollution exposure. Misleading results may still be produced, however, because of poor selection of model parameters. In addition, the required knowledge about distributional characteristics may be difficult to obtain. Examples of models based on this approach that have been applied to carbon monoxide include the Simulation of Human Activity and Pollutant Exposure model (Ott, 1984; Ott et al., 1988) and two models derived from the NAAQS Exposure Model, developed by Johnson et al. (1990).

5.4 Exposure measurements in populations and subpopulations

5.4.1 *Carboxyhaemoglobin measurements in populations*

Numerous studies have used the above-described methodologies to characterize the levels of carboxyhaemoglobin in the general population. These studies have been designed to determine frequency distributions of carboxyhaemoglobin levels in the populations being studied. In general, the higher the frequency of carboxyhaemoglobin levels above baseline in non-smoking subjects, the greater the incidence of significant carbon monoxide exposure.

Carboxyhaemoglobin levels in blood donors have been studied for various urban populations in the USA. Included have been studies of blood donors and sources of carbon monoxide in the metropolitan St. Louis, Missouri, population (Kahn et al., 1974); evaluation of smoking and carboxyhaemoglobin in the St. Louis metropolitan population (Wallace et al., 1974); carboxyhaemoglobin analyses of 16 649 blood samples provided by the Red Cross Missouri–Illinois blood donor programme (Davis & Gantner, 1974); a survey of blood donors for percent carboxyhaemoglobin in Chicago, Illinois, Milwaukee, Wisconsin, New York, New York, and Los Angeles, California (Stewart et al., 1976); a national survey for carboxy-haemoglobin in American blood donors from urban, suburban and rural communities across the USA (Stewart et al., 1974); and the trend in percent carboxyhaemoglobin associated with vehicular traffic in Chicago blood donors (Stewart et al., 1976). These extensive studies of volunteer blood donor populations show three main sources of exposure to carbon monoxide in urban environments: smoking, general activities (usually associated with internal combustion engines) and occupational exposures. For comparisons of sources, the populations are divided into two main groups — smokers and non-smokers. The main groups are often divided further into subgroups consisting of industrial workers, drivers, pedestrians and others, for example. Among the two main groups, smokers show an average of 4% carboxyhaemoglobin, with a usual range of 3–8%; non-smokers average about 1% carboxyhaemoglobin (Radford & Drizd, 1982). Smoking behaviour generally occurs as an intermittent diurnal pattern, but carboxyhaemoglobin levels can rise to a maximum of about 15% in some individuals who chain-smoke. Similar results were obtained in a more recent study in Bahrain (Madany, 1992) and in a study in Beijing, People's Republic of China (Song et al., 1984).

Aside from tobacco smoke, the most significant sources of potential exposure to carbon monoxide in the population are community air pollution, occupational exposures and household exposures (Goldsmith, 1970). Community air pollution comes mainly from automobile exhaust and has a typical intermittent diurnal pattern. Occupational exposures occur for up to 8 h per day, 5 days per week, producing carboxyhaemoglobin levels generally less than 10%.

More recent studies characterizing carboxyhaemoglobin levels in the population have appeared in the literature. Turner et al. (1986) used an IL 182 CO-Oximeter to determine percent carboxyhaemoglobin in venous blood of a study group consisting of both smoking and non-smoking hospital staff, inpatients and outpatients. Blood samples were collected for 3487 subjects (1255 non-smokers) during morning hours over a 5-year period. A detailed smoking history was obtained at the time of blood collection. Using 1.7% carboxyhaemoglobin as a normal cut-off value, the distribution for the population studied showed above-normal results for 94.7% of cigarette smokers, 10.3% of primary cigar smokers, 97.4% of those exposed to environmental smoke from cigars and 94.7% of those exposed to environmental smoke from pipes.

Zwart & van Kampen (1985) tested a blood supply using a routine spectrophotometric method for total haemoglobin and for carboxyhaemoglobin in 3022 samples of blood for transfusion in hospital patients in the Netherlands. For surgery patients over a 1-year period, the distribution of percent carboxyhaemoglobin in samples collected as part of the surgical protocol showed 65% below 1.5% carboxyhaemoglobin, 26.5% between 1.5% and 5% carboxyhaemoglobin, 6.7% between 5% and 10% carboxyhaemoglobin and 0.3% in excess of 10% carboxyhaemoglobin. This distribution of percent carboxyhaemoglobin was homogeneous across the entire blood supply, resulting in 1 in 12 patients having blood transfusions at 75% available haemoglobin capacity.

Radford & Drizd (1982) analysed blood carboxyhaemoglobin in approximately 8400 samples obtained from respondents in the 65 geographic areas of the second US National Health and Nutrition Examination Survey during the period 1976–1980. When the frequency distributions of blood carboxyhaemoglobin levels are plotted on a logarithmic probability scale to facilitate comparison of the results for different age groups and smoking habits, it is evident that adult smokers in the USA have carboxyhaemoglobin levels considerably higher than those of non-smokers, with 79% of the smokers' blood samples above 2% carboxyhaemoglobin and 27% of the

observations above 5% carboxyhaemoglobin. The nationwide distributions of persons aged 12–74 who have never smoked and who are ex-
smokers were similar, with 5.8% of the ex-smokers and 6.4% of the
never-smokers above 2% carboxyhaemoglobin. It is evident that a
significant proportion of the non-smoking US population had blood
levels above 2% carboxyhaemoglobin. For these two non-smoking
groups, blood levels above 5% were found in 0.7% of the never-
smokers and 1.5% of the ex-smokers. It is possible that these high
blood levels could be due, in part, to misclassification of some
smokers as either ex- or non-smokers. Children aged 2–11 had lower
carboxyhaemoglobin levels than the other groups, with only 2.3% of
the children's samples above 2% carboxyhaemoglobin and 0.2%
above 5% carboxyhaemoglobin.

Wallace & Ziegenfus (1985) utilized available data from the
second National Health and Nutrition Examination Survey to analyse
the relationship between measured carboxyhaemoglobin levels and the
associated 8-h carbon monoxide concentrations at nearby fixed monitors. Carboxyhaemoglobin data were available for a total of 1658 non-
smokers in 20 cities. The authors concluded that fixed outdoor carbon
monoxide monitors alone are, in general, not providing useful estimates of carbon monoxide exposure of urban residents.

5.4.2 Breath measurements in populations

In a study by Wallace (1983) in which breath measurements of
carbon monoxide were used to detect an indoor air problem,
65 workers in an office had been complaining for some months of
late-afternoon sleepiness and other symptoms, which they attributed
to the new carpet. About 40 of the workers had their breath tested for
carbon monoxide on a Friday afternoon and again on a Monday
morning. The average breath carbon monoxide levels decreased from
26 mg/m^3 (23 ppm) on Friday to 8 mg/m^3 (7 ppm) on Monday
morning, indicating a work-related condition. Non-working fans in the
parking garage and broken fire doors were identified as the cause of
the problem.

Wald et al. (1981) obtained measurements of percent carboxyhaemoglobin for 11 749 men, aged 35–64, who attended a medical
centre in London, United Kingdom, for comprehensive health screening examinations between 11:00 a.m. and 5:00 p.m. The time of
smoking for each cigarette, cigar or pipe smoked since waking was

recorded at the time of collection of a venous blood sample. Percent carboxyhaemoglobin was determined using an IL 181 CO-Oximeter. Using 2% carboxyhaemoglobin as a normal cut-off value, 81% of cigarette smokers, 35% of cigar and pipe smokers and 1% of non-smokers were found to be above normal. An investigation of carboxyhaemoglobin and alveolar carbon monoxide was conducted on a subgroup of 187 men (162 smokers and 25 non-smokers). Three samples of alveolar air were collected at 2-min intervals within 5 min of collecting venous blood for carboxyhaemoglobin estimation. Alveolar air was collected by having the subject hold his breath for 20 s and then exhale through a 1-m glass tube with an internal diameter of 17 mm and fitted with a 3-litre anaesthetic bag at the distal end. Air at the proximal end of the tube was considered to be alveolar air, and a sample was removed by a small side tube located 5 mm from the mouthpiece. The carbon monoxide content was measured using an Ecolyzer. The instrumental measurement is based on detection of the oxidation of carbon monoxide to carbon dioxide by a catalytically active electrode in an aqueous electrolyte. The mean of the last two readings to the nearest 0.29 mg/m^3 (0.25 ppm) was recorded as the alveolar carbon monoxide concentration. Subjects reporting recent alcohol consumption were excluded, because ethanol in the breath affects the response of the Ecolyzer. A linear regression equation of percent carboxyhaemoglobin on alveolar carbon monoxide level had a correlation coefficient of 0.97, indicating that a carboxyhaemoglobin level could be estimated reliably from an alveolar carbon monoxide level.

Honigman et al. (1982) determined alveolar carbon monoxide concentrations by end-expired breath analysis for athletes (joggers). The group included 36 non-smoking males and 7 non-smoking females, all conditioned joggers, covering at least 34 km per week for the previous 6 months in the Denver, Colorado, area. The participants exercised for a 40-min period each day over one of three defined courses in the Denver urban environment (elevation 1610 m). Samples of expired air were collected and analysed before start of exercise, after 20 min and again at the end of the 40-min exercise period. Heart rate measurements at 20 min and 40 min were 84 and 82% of mean age-predicted maxima, respectively, indicating exercise in the aerobic range. Relative changes in carbon monoxide concentrations in expired air were plotted and compared with carbon monoxide concentrations in ambient air measured at the time of collecting breath samples. Air and breath samples were analysed using an MSA Model 70. Relative changes in expired end-air carbon monoxide based on the concentration of carbon monoxide in breath before the start of exercise were plotted in terms of the ambient air concentrations measured during the exercise period, at both 20 and 40 min of exercise. For ambient concentrations of carbon monoxide below 7 mg/m^3 (6 ppm), the aerobic exercise served to decrease the relative amount of expired end-air carbon monoxide compared with the concentration measured before the start of exercise. For ambient concentrations in the range of 7–8 mg/m^3 (6–7 ppm), there was no net change in the carbon monoxide concentrations in the expired air. For ambient air concentrations in excess of 8 mg/m^3 (7 ppm), the aerobic exercise resulted in relative increases in expired carbon monoxide, with the increases after 40 min being greater than similar increases observed at the 20-min measurements. Sedentary controls at the measurement stations showed no relative changes. Thus, aerobic exercise, as predicted by the physiological models of uptake and elimination, is shown to enhance transport of carbon monoxide, thereby decreasing the time to reach equilibrium conditions.

Verhoeff et al. (1983) surveyed 15 identical residences that used natural gas for cooking and geyser units for water heating. Carbon monoxide concentrations in the flue gases were measured using an Ecolyzer (2000 series). The flue gases were diluted to the dynamic range of the instrument for carbon monoxide (determined by Draeger tube analyses for carbon dioxide dilution to 2.0–2.5%). Breath samples were collected from 29 inhabitants by having each participant hold a deep breath for 20 s and exhale completely through a glass sampling tube (225-ml volume). The sampling tube was stoppered and taken to a laboratory for analysis of carbon monoxide content using a gas–liquid chromatograph (Hewlett-Packard 5880A). The overall coefficient of variation for sampling and analysis was 7%, based on results of previous measurements. No significant differences were observed for non-smokers as a result of their cooking or dishwashing activities using the natural gas fixtures. There was a slight increase in carbon monoxide in expired air for smokers, but this may be due to the possibility of increased smoking during the dinner hour.

Wallace et al. (1984) reported data on measurements of expired end-air carbon monoxide and comparisons with predicted values based on personal carbon monoxide measurements for populations in

Denver, Colorado, and Washington, DC. Correlations between breath carbon monoxide and preceding 8-h average carbon monoxide exposures were high (0.6–0.7) in both cities. Correlation coefficients were calculated for 1-h to 10-h average personal carbon monoxide exposures in 1-h increments; the highest correlations occurred at 7–9 h. However, breath carbon monoxide levels showed no relationship with ambient carbon monoxide measurements at the nearest fixed-station monitor.

A major large-scale study employing breath measurements of carbon monoxide was carried out by the US EPA in Washington, DC, and Denver, Colorado, in the winter of 1982–83 (Hartwell et al., 1984; Johnson, 1984; Wallace et al., 1984, 1988; Akland et al., 1985). In Washington, DC, 870 breath samples were collected from 812 participants; 895 breath samples were collected from 454 Denver participants (two breath samples on 2 consecutive days in Denver). All participants also carried personal monitors to measure their exposures over a 24-h period in Washington, DC, or a 48-h period in Denver. The subjects in each city formed a probability sample representing 1.2 million adult non-smokers in Washington, DC, and 500 000 adult non-smokers in Denver.

The distributions of breath levels in the two cities appeared to be roughly lognormal, with geometric means of 6.0 mg carbon monoxide/m^3 (5.2 ppm) for Denver and 5.0 mg carbon monoxide/m^3 (4.4 ppm) for Washington, DC. Geometric standard deviations were about 1.8 mg/m^3 (1.6 ppm) for each city. Arithmetic means were 8.1 mg/m^3 (7.1 ppm) for Denver and 6.0 mg/m^3 (5.2 ppm) for Washington, DC.

Of greater regulatory significance is the number of people whose carboxyhaemoglobin levels exceeded the value of 2.1%, because the US EPA has determined that the current 10 mg/m^3 (9 ppm), 8-h average standard would keep more than 99.9% of the most sensitive non-smoking adult population below this level of protection (Federal Register, 1985). An alveolar carbon monoxide concentration of about 11 mg/m^3 (10 ppm) would correspond to a carboxyhaemoglobin level of 2%. The percentage of people with measured breath values exceeding this level was about 6% in Washington, DC. This percentage was increased to 10% when the correction for the effect of room air was applied. Of course, because the breath samples were taken on days and at times when carbon monoxide levels were not necessarily at their highest level during the year, these percentages are *lower limits* of the estimated number of people who may have incurred carboxy-haemoglobin levels above 2%. Yet the two central stations in Washington, DC, recorded a total of one exceedance of the 10 mg/m^3 (9 ppm) standard during the winter of 1982–83. Models based on fixed-station readings would have predicted that an exceedingly tiny proportion of the Washington, DC, population received exposures exceeding the standard. Therefore, the results from the breath measurements indicated that a much larger portion of both Denver and Washington, DC, residents were exceeding 2% carboxyhaemoglobin than was predicted by models based on fixed-station measurements.

It should also be noted that the number of people with measured maximum 8-h exposures exceeding the EPA outdoor standard of 10 mg/m^3 (9 ppm) was only about 3.5% of the Washington, DC, subjects. This value appears to disagree with the value of 10% obtained from the corrected breath samples. However, the personal monitors used in the study were shown to experience several different problems, including a loss of response associated with battery discharge towards the end of the 24-h monitoring period, which caused them to read low just at the time the breath samples were being collected. Therefore, Wallace et al. (1988) concluded that the breath measurements were correct and the personal air measurements were biased low. The importance of including breath measurements in future exposure and epidemiology studies is indicated by this study.

Hwang et al. (1984) described the use of expired air analysis for carbon monoxide in an emergency clinical setting to diagnose the presence and extent of carbon monoxide intoxication. The subjects were 47 Korean patients brought in for emergency treatment who showed various levels of consciousness: alertness (11), drowsiness (21), stupor (7), semicoma (5), coma (1) and unknown (2). The study group included 16 males, aged 16–57, and 31 females, aged 11–62. Exposure durations ranged from 2 to 10 h, with all exposures occurring in the evening and nighttime hours. The source of carbon monoxide was mainly charcoal fires used for cooking and heating. In order to estimate carbon monoxide concentrations in expired air, a detector tube (Gastec 1La containing potassium palladosulfite as both a reactant and colour-change indicator for the presence of carbon monoxide on silica gel) was fitted to a Gastec manual sampling pump.

One stroke of the sampling plunger represents 100 ml of air. A 100-ml sample of expired air was collected by inserting a detector tube at a nostril and slowly pulling back the plunger for one full stroke for expired air. A 10-ml sample of venous blood was also collected at this time for determining percent carboxyhaemoglobin using a CO-Oximeter. The subjects showed signs of acute intoxication, and significant relationships were found between carbon monoxide levels in expired air and percent carboxyhaemoglobin.

Cox & Whichelow (1985) analysed end-exhaled air (collected over approximately the last half of the exhalation cycle) for carbon monoxide concentrations for a random population of 168 adults — 69 smokers and 99 non-smokers. The results were used to evaluate the influence of home heating systems on exposures to and absorption of carbon monoxide. Ambient indoor concentrations of carbon monoxide were measured in the homes of study subjects. The subjects included 86 men and 82 women, ranging in age from 18 to 74. Interviews were usually conducted in the living room of the subject's home. The type of heating system in use was noted, and the indoor air concentration of carbon monoxide was measured using an Ecolyzer. After the ambient indoor carbon monoxide level was determined, a breath sample was collected from the subject. The subject was asked to hold a deep breath for 20 s and then to exhale completely into a trilaminate plastic bag. The bag was fitted to the port of the Ecolyzer, and the carbon monoxide content of the exhaled air was measured. For smokers, the time since smoking their last cigarette and the number of cigarettes per day were noted. For non-smokers, there was a strong correlation between carbon monoxide levels in ambient air and carbon monoxide levels in expired air. With smokers, the correlation was strongest with the number of cigarettes smoked per day. The data also supported the supposition that smokers are a further source of ambient carbon monoxide in the indoor environment.

Lambert et al. (1988) compared carbon monoxide levels in breath with carboxyhaemoglobin levels in blood in 28 subjects (including 2 smokers). Breath carbon monoxide was collected using the standard technique developed by Jones et al. (1958): maximal inspiration was followed by a 20-s breath-hold, and the first portion of the expired breath was discarded. Excellent precision ($\pm$ 0.23 mg/m^3 [$\pm$ 0.2 ppm]) was obtained in 35 duplicate samples. Blood samples were collected within 15 min of the breath samples using a gas-tight plastic syringe rinsed with sodium heparin. Carboxyhaemoglobin was measured using an IL 282 CO-Oximeter. Some samples were also measured using a gas chromatograph.

By using least squares regression, the null hypothesis of no difference in the slope and intercept estimates for non-smokers and smokers was not rejected (i.e., there was no association between blood carboxyhaemoglobin and breath carbon monoxide in either non-smokers or smokers).

5.4.3   Subject age

The relationship between age and carboxyhaemoglobin level is not well established. Kahn et al. (1974) reported that non-smoking subjects under the age of 19 years had a significantly lower percent carboxyhaemoglobin than older subjects, but there was no difference in carboxyhaemoglobin between the ages of 20 and 59 years. Kahn et al. (1974) also reported that there was a slight decrease in the carboxyhaemoglobin levels in non-smoking subjects over the age of 60 years. Radford & Drizd (1982) reported that younger subjects, 3–11 years old, had lower levels of carboxyhaemoglobin than did the older age group of 12–74 years. Goldsmith (1970) reported that expired carbon monoxide levels were unchanged with age in non-smokers; however, there was a steady decline in the expired carbon monoxide levels with age in smokers. The decrease in expired carbon monoxide is disproportionately large for the decrease in carboxy-haemoglobin levels measured by Kahn et al. (1974) in older subjects. Therefore, by comparison of the data from these two studies, it would appear that older subjects have higher levels of carboxyhaemoglobin than predicted from the expired carbon monoxide levels. It is not known how much of this effect is due to aging of the pulmonary system, resulting in a condition similar to that of subjects with obstructive pulmonary disease (see below).

5.4.4   Pulmonary disease

A major potential influence on the relationship between blood and alveolar partial pressures of carbon monoxide is the presence of significant lung disease. Hackney et al. (1962) demonstrated that the slow increase in exhaled carbon monoxide concentration in a rebreathing system peaked after 1.5 min in healthy subjects but required 4 min in a subject with lung disease. These findings have been substantiated by Guyatt et al. (1988), who reported that patients with pulmonary disease did not have the same relationship between percent carboxyhaemoglobin and breath-hold carbon monoxide concentrations. The group with pulmonary disease had a forced expiratory volume in 1 s (FEV_1)/forced vital capacity (FVC) percentage of <71.5%, whereas the healthy subjects had an FEV_1/FVC percentage of >86%. The linear regression for the healthy group was $COHb = 0.629 + 0.158(ppm \ CO)$; for the pulmonary disease group, the linear regression was $COHb = 0.369 + 0.185(ppm \ CO)$. This means that at low carbon monoxide levels, individuals with obstructive pulmonary disease would have a lower "alveolar" carbon monoxide level for any given percent carboxyhaemoglobin level than would the healthy subjects.

5.4.5 *Effects of smoking*

Studies evaluating the effect of cigarette smoking on end-expired carbon monoxide have found a phasic response that depends on smoking behaviour (Henningfield et al., 1980; Woodman et al., 1987). There is an initial rapid increase in the carbon monoxide concentration of expired air as a result of smoking. This is followed by a rapid (5-min) decrease after cessation of smoking and a slow decrease over the 5- to 60-min period after smoking. A comparison of the results from one study (Tsukamoto & Matsuda, 1985) showed that the carbon monoxide concentration in expired air increases by approximately 6 mg/m³ (5 ppm) after smoking one cigarette. This corresponds to an increase of 0.67% carboxyhaemoglobin based on blood–breath relationships developed by the authors. Use of cigarettes with different tar and nicotine yields or the use of filter-tip cigarettes showed no apparent effect on end-expired carbon monoxide concentrations (Castelli et al., 1982).

The relationship between breath-hold carbon monoxide and blood carbon monoxide is apparently altered as a result of smoking, making the detection of small changes difficult. Guyatt et al. (1988) showed that smoking one cigarette results in a variable response in the relationship between breath-hold alveolar carbon monoxide fraction [$F_ACO(Bh)$] and carboxyhaemoglobin levels. The range of $F_ACO(Bh)$ values for an increase of 1% carboxyhaemoglobin was from −6 to +6 mg/m³ (−5 to +5 ppm). The correlation between the change in $F_ACO(Bh)$ and the change in carboxyhaemoglobin in 500 subjects was only 0.705. This r value indicates that only 50% of the change in $F_ACO(Bh)$ was due to changes in carboxyhaemoglobin. It is not known how much of this residual error is due to subject compliance or to error in the method. Therefore, the results obtained with breath-holding in smoking subjects should be viewed with caution unless large differences in $F_ACO(Bh)$ are reported (i.e., considerable cigarette consumption is being evaluated).

5.5 Megacities and other major urban areas

Throughout the world, there are large urban areas that have serious air pollution problems and encompass large land areas and over 10 million people (total population of the 20 megacities in 1990 was estimated to be 275 million). The cities that have been designated as megacities include 2 in North America (Los Angeles, California, USA; New York, New York, USA), 4 in Central and South America (Buenos Aires, Argentina; Mexico City, Mexico; Rio de Janeiro, Brazil; São Paulo, Brazil), 1 in Africa (Cairo, Egypt), 11 in Asia (Bangkok, Thailand; Beijing, People's Republic of China; Bombay, India; Calcutta, India; Delhi, India; Jakarta, Indonesia; Karachi, Pakistan; Manila, Philippines; Seoul, South Korea; Shanghai, People's Republic of China; Tokyo, Japan) and 2 in Europe (London, United Kingdom; Moscow, Russia). However, many other cities are heading for megacity status.

Of the 20 megacities studied (WHO/UNEP, 1992; Mage et al., 1996), monitoring capabilities for carbon monoxide have been desig-nated as none or unknown in 9, rudimentary in 1, adequate in 4 and good in 6 (see Table 7). The air quality for carbon monoxide has been designated as no data available or insufficient data for assessment in six; serious problems, WHO guidelines exceeded by more than a factor of 2 in one; moderate to heavy pollution, WHO guidelines exceeded by up to a factor of 2 (short-term guidelines exceeded on a regular basis at certain locations) in seven; and low pollution, WHO guidelines are normally met (short-term guidelines may be exceeded occasionally) in six (see Table 7).

The problem of air pollution in Mexico City is confounded by geography. The metropolitan area of Mexico City is located at a mean altitude of 2240 m and is situated in the Mexican Basin. The reduced oxygen in the air at high altitude causes carbon monoxide emissions to increase because of incomplete combustion, and it exacerbates the health effects attributed to carbon monoxide, especially among highly

Table 7. Overview of carbon monoxide air quality and monitoring capabilities in 20 megacities[e]

Megacity	Air quality for carbon monoxide[b]	Status of monitoring capabilities[c]
Bangkok, Thailand	a	B
Beijing, People's Republic of China	d	B
Bombay, India	a	D
Buenos Aires, Argentina	d	D
Cairo, Egypt	b	D
Calcutta, India	d	D
Delhi, India	a	D
Jakarta, Indonesia	b	D
Karachi, Pakistan	d	D
London, United Kingdom	b	B
Los Angeles, USA	b	A
Manila, Philippines	d	D
Mexico City, Mexico	c	A
Moscow, Russia	b	C
New York, USA	b	A
Rio de Janeiro, Brazil	a	B
São Paulo, Brazil	b	A
Seoul, South Korea	a	A
Shanghai, People's Republic of China	d	D
Tokyo, Japan	a	A

[a] Adapted from WHO/UNEP (1992); Mage et al. (1996).
[b] a = Low pollution, WHO guidelines normally met (short-term guidelines may be exceeded occasionally).
  b = Moderate to heavy pollution, WHO guidelines exceeded by up to a factor of 2 (short-term guidelines exceeded on a regular basis at certain locations).
  c = Serious problem, WHO guidelines exceeded by more than a factor of 2.
  d = No data available or insufficient data for assessment.
[c] A = Good.
  B = Adequate.
  C = Rudimentary.
  D = None or not known.

susceptible population groups, including children and pregnant women.

Data for air quality trends in about 40 countries were collected in the WHO/United Nations Environment Programme (UNEP) Global Environmental Monitoring System (GEMS/Air) (UNEP/WHO, 1993). The GEMS/Air programme was terminated in 1995. WHO has set up a successive programme, the Air Management Information System (WHO, 1997a). Summary air quality data for carbon monoxide are currently available for many countries, such as Austria, Germany, Greece, Japan, New Zealand and Switzerland. For example, between 1986 and 1995, the annual mean carbon monoxide concentrations ranged from 5.1 to 8.4 mg/m^3 (4.5 to 7.3 ppm) in Greece, from 0.8 to 2.7 mg/m^3 (0.7 to 2.4 ppm) in Japan and from 0.9 to 1.9 mg/m^3 (0.8 to 1.7 ppm) in New Zealand. In Australia, the annual mean carbon monoxide concentrations from 1988 to 1995 ranged from 0.66 to 1.36 mg/m^3 (0.58 to 1.19 ppm). Air quality trends for carbon monoxide over the same period are shown in Fig. 3 for Athens, Greece, Chongqing, People's Republic of China, Frankfurt, Germany, Johannesburg, South Africa, London, United Kingdom, and Los Angeles, USA.

In a recent report (Eerens et al., 1995), carbon monoxide data from 105 cities in 35 European states are given. They include annual average 8-h maximum concentrations and number of days exceeding WHO Air Quality Guidelines and twice the WHO Air Quality Guidelines.

5.6 Indoor concentrations and exposures

Indoor concentrations of carbon monoxide are a function of outdoor concentrations, indoor sources (source type, source condition, source use, etc.), infiltration/ventilation and air mixing between and within rooms. In residences without sources, average carbon monoxide concentrations are approximately equal to average outdoor concentrations at the corresponding elevation, generally decreasing with height above the ground. Proximity to outdoor sources (i.e., structures near heavily travelled roadways or with attached garages or parking garages) can have a major impact on indoor carbon monoxide concentrations.

Johannesburg, South Africa

Concentration [µg m³]

3500
3000
2500
2000
1500
1000
500
0

☒ City centre

1986 1987 1988 1989 1990 1991 1992 1993 1994 1995
Year

Frankfurt, Germany

Concentration [µg m³]

5000
4000
3000
2000
1000
0

Carbon Monoxide

■ City centre /commercial
☒ Residential

1986 1987 1988 1989 1990 1991 1992 1993 1994 1995
Year

London, United Kingdom

Carbon monoxide annual

☒ Urban centre
■ Kerbside
☒ Urban background

Concentration [µg m³]

5000
4000
3000
2000
1000
0

1986 1987 1988 1989 1990 1991 1992 1993 1994 1995
Year

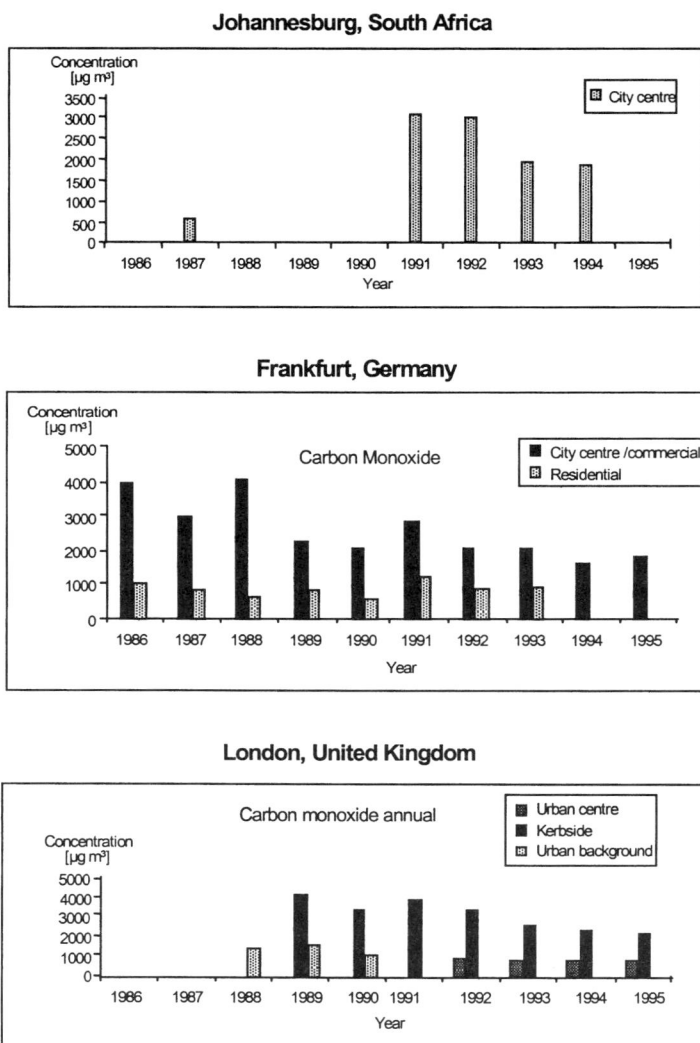

Fig. 3. Trends in carbon monoxide air quality between 1986 and 1995 in Johannesburg (South Africa), Frankfurt (Germany), London (United Kingdom), Athens (Greece), Los Angeles (USA) and Chongqing (China) (from WHO, 1997a).

Athens, Greece

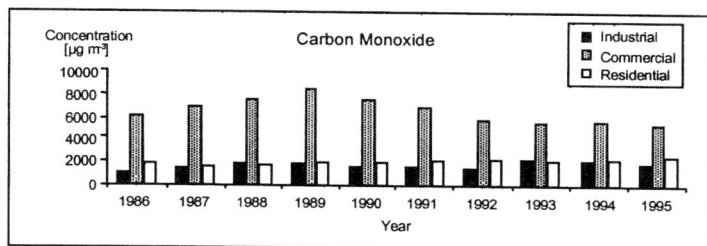

Concentration [μg m⁻³] — Carbon Monoxide

Legend: ■ Industrial, ▨ Commercial, □ Residential

Year

Los Angeles, USA

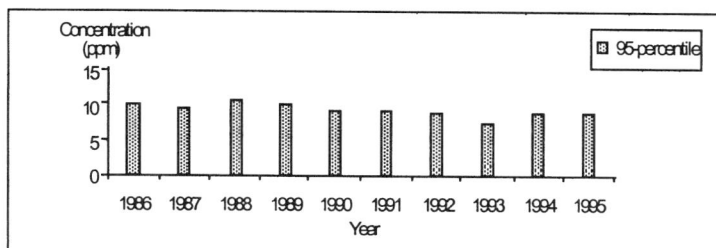

Concentration (ppm)

Legend: ▨ 95-percentile

Year

Chongqing, China

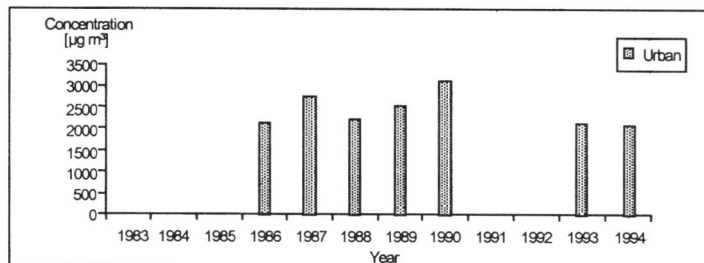

Concentration [μg m⁻³]

Legend: ▨ Urban

Year

Fig. 3. (contd).

The development of small lightweight and portable electrochemical carbon monoxide monitors over the past decade has permitted the measurement of personal carbon monoxide exposures and carbon monoxide concentrations in a number of indoor environments. The available data on indoor carbon monoxide concentrations have been obtained from total personal exposure studies or studies in which various indoor environments have been targeted for measurements.

The extensive total personal carbon monoxide exposure studies conducted by the US EPA in Washington, DC, and Denver, Colorado, have shown that the highest carbon monoxide concentrations occur in indoor microenvironments associated with transportation sources (parking garages, cars, buses, etc.). Concentrations in these environments were found to frequently exceed 10 mg/m^3 (9 ppm). Studies targeted towards specific indoor microenvironments have also identified the in-vehicle commuting microenvironment as an environment in which carbon monoxide concentrations frequently exceed 10 mg/m^3 (9 ppm) and occasionally exceed 40 mg/m^3 (35 ppm). Special environments or occurrences (indoor ice skating rinks, offices where emissions from parking garages migrate indoors, etc.) have been reported where indoor carbon monoxide levels can exceed the recommended air quality guidelines.

A majority of the targeted field studies (see section 5.6.2) monitored indoor carbon monoxide levels as a function of the presence or absence of combustion sources (gas ranges, unvented gas and kerosene space heaters, wood-burning stoves and fireplaces and tobacco combustion). The results of these studies indicate that the presence and use of an unvented combustion source result in indoor carbon monoxide levels above those found outdoors. The associated increase in carbon monoxide concentrations can vary considerably as a function of the source, source use, condition of the source and averaging time of the measurement. Intermittent sources such as gas cooking ranges can result in high peak carbon monoxide concentrations (in excess of 10 mg/m^3 [9 ppm]), whereas long-term average increases in concentrations (i.e., 24-h) associated with gas ranges are considerably lower (on the order of 1.1 mg/m^3 [1 ppm]). The contribution of tobacco combustion to indoor carbon monoxide levels is variable. Under conditions of high smoking and low ventilation, the contribution can be on the order of a few milligrams per cubic metre (parts per million). One study suggested that the contribution to residential carbon monoxide concentrations of tobacco combustion is on the order of 1.1 mg/m³ (1 ppm), whereas another study showed no significant increase in residential carbon monoxide levels.

Unvented or poorly vented combustion sources that are used for substantial periods of time (e.g., unvented gas and kerosene space heaters) appear to be the major contributors to residential carbon monoxide concentrations. One extensive study of unvented gas space heaters indicated that 12% of the homes had 15-h average carbon monoxide concentrations greater than 9 mg/m³ (8 ppm), with the highest concentration at 41.9 mg/m³ (36.6 ppm). Only very limited data are available on the contribution of kerosene heaters to the average carbon monoxide concentrations in residences, and these data indicate a much lower contribution than that of gas heaters. Peak carbon monoxide concentrations associated with both unvented gas and kerosene space heaters can exceed the current ambient 1- and 8-h standards (40 and 10 mg/m³ [35 and 9 ppm], respectively) in residences, and, because of the nature of the source (continuous), those peaks tend to be sustained for several hours.

Very limited data on carbon monoxide levels in residences with wood-burning stoves or fireplaces are available. Non-airtight stoves can contribute substantially to residential carbon monoxide concentrations, whereas airtight stoves can result in small increases. The available data indicate that fireplaces do not contribute measurably to average indoor concentrations. No information is available for samples of residences with leaky flues. In addition, there is no information available on short-term indoor carbon monoxide levels associated with these sources, nor are there studies that examine the impact of attached garages on residential carbon monoxide concentrations.

The available data on short-term (1-h) and long-term (8-h) indoor carbon monoxide concentrations as a function of microenvironments and sources in those microenvironments are not adequate to assess exposures in those environments. In addition, little is known about the spatial variability of carbon monoxide indoors. These indoor microenvironments represent the most important carbon monoxide exposures for individuals and as such need to be better characterized.

Concentrations of carbon monoxide in an enclosed environment are affected by a number of factors in addition to the source factors.

These factors include outdoor concentrations, proximity to outdoor sources (i.e., parking garages or traffic), volume of the space and mixing within and between indoor spaces.

Carbon monoxide measurements in enclosed spaces have been made either in support of total personal exposure studies or in targeted indoor studies. In the personal exposure studies, individuals wear the monitors in the course of their daily activities, taking them through a number of different microenvironments. In targeted studies, carbon monoxide measurements are taken in indoor spaces independent of the activities of occupants of those spaces.

5.6.1 *Indoor concentrations recorded in personal exposure studies*

Three studies have reported carbon monoxide concentrations in various microenvironments as part of an effort to measure total human exposure to carbon monoxide and to assess the accuracy of exposure estimates calculated from fixed-site monitoring data. In each study, subjects wore personal carbon monoxide exposure monitors for one or more 24-h periods. Carbon monoxide concentrations were recorded on data loggers at varying time intervals as a function of time spent in various microenvironments. Participants kept an activity diary in which they were asked to record time, activity (e.g., cooking), location (microenvironment type), presence and use of sources (e.g., smokers or gas stoves) and other pertinent information. Carbon monoxide concentrations by microenvironment were extracted from the measured concentrations by use of the activity diaries.

Two of the studies, conducted in Denver, Colorado, and Washington, DC, by the US EPA (Hartwell et al., 1984; Johnson, 1984; Whitmore et al., 1984; Akland et al., 1985), measured the frequency distribution of carbon monoxide exposure in a representative sample of the urban population. The study populations were selected using a multistage sampling strategy. The third study, also conducted in Washington, DC (Nagda & Koontz, 1985), utilized a convenience sample.

The first-mentioned Washington, DC, study obtained a total of 814 person-day samples for 1161 participants, whereas the Denver study obtained 899 person-day samples for 485 participants. The Denver study obtained consecutive 24-h samples for each participant, whereas the Washington, DC, study obtained one 24-h sample for each participant. Both studies were conducted during the winter of 1982–83.

A comparison of carbon monoxide concentrations measured in the Washington, DC, and Denver studies is shown in Table 8 (from Akland et al., 1985). Concentrations measured in all microenvironments for the Denver study were higher than those for the Washington, DC, study. This is consistent with the finding that daily maximum 1-h and 8-h carbon monoxide concentrations at outdoor fixed monitoring sites were about a factor of 2 higher in the Denver area than in the Washington, DC, area during the course of the studies (Akland et al., 1985). The highest concentrations in both studies were associated with indoor parking garages and commuting, whereas the lowest levels were measured in indoor environments without sources of carbon monoxide. Concentrations associated with commuting are no doubt higher owing to the proximity to and density of outside carbon monoxide sources (cars, buses and trucks), particularly during commuting hours when traffic is heaviest. Indoor levels, especially residential levels in the absence of indoor sources, are lower primarily because of the time of day of sampling (non-commuting hours with lower outdoor levels). A more detailed breakdown of carbon monoxide concentrations by microenvironments for the Denver study is shown in Table 9 (Johnson, 1984). Microenvironments associated with motor vehicles result in the highest concentrations, with concentrations reaching or exceeding the NAAQS 10 mg/m^3 (9 ppm) reference level.

No statistical difference ($P > 0.05$) in carbon monoxide concentrations was found between residences with and without gas ranges in the Washington, DC, study. The results of a similar analysis on the Denver data, according to the presence or absence of selected indoor sources, are shown in Table 10 (Johnson, 1984); in contrast to the Washington, DC, study, the presence of an operating gas stove in the Denver study resulted in a statistically significant increase of 2.97 mg/m^3 (2.59 ppm). Attached garages, use of gas ranges and presence of smokers were all shown to result in higher indoor carbon monoxide concentrations. Concentrations were well below the NAAQS 10 mg/m^3 (9 ppm) reference level but were substantially above concentrations in residences without the sources.

In the second Washington, DC, study (Nagda & Koontz, 1985), 197 person-days of samples were collected from 58 subjects,

Table 8. Summary of carbon monoxide exposure levels and time spent per day in selected microenvironments[a]

Microenvironment	Denver, Colorado				Washington, DC			
	n	CO concentration (mean ± SE)[b]		Median time (min)	n	CO concentration (mean ± SE)[b]		Median time (min)
		mg/m³	ppm			mg/m³	ppm	
Indoors, parking garage	31	21.5 ± 5.68	18.8 ± 4.96	14	59	11.9 ± 5.07	10.4 ± 4.43	11
In transit, car	643	9.2 ± 0.37	8.0 ± 0.32	71	592	5.7 ± 0.16	5.0 ± 0.14	79
In transit, other (bus, truck, etc.)	107	9.0 ± 0.70	7.9 ± 0.61	66	130	4.1 ± 0.34	3.6 ± 0.30	49
Outdoors, near roadway	188	4.5 ± 0.41	3.9 ± 0.36	33	164	3.0 ± 0.23	2.6 ± 0.20	20
In transit, walking	171	4.8 ± 0.52	4.2 ± 0.45	28	226	2.7 ± 0.33	2.4 ± 0.29	32
Indoors, restaurant	205	4.8 ± 0.33	4.2 ± 0.29	58	170	2.4 ± 0.37	2.1 ± 0.32	45
Indoors, office	283	3.4 ± 0.23	3.0 ± 0.20	478	349	2.2 ± 0.31	1.9 ± 0.27	428
Indoors, store/shopping mall	243	3.4 ± 0.25	3.0 ± 0.22	50	225	2.9 ± 0.56	2.5 ± 0.49	36
Indoors, residence	776	1.9 ± 0.11	1.7 ± 0.10	975	705	1.4 ± 0.11	1.2 ± 0.10	1048
Indoors, total	776	2.4 ± 0.10	2.1 ± 0.09	1243	705	1.6 ± 0.09	1.4 ± 0.08	1332

[a] From Akland et al. (1985).
[b] n = number of person-days with non-zero durations; SE = standard error.

Table 9. Indoor microenvironments in Denver, Colorado, listed in descending order of weighted mean carbon monoxide concentration[a]

Category	Number of observations	CO concentration (mean ± SD)[b]	
		mg/m³	ppm
Public garage	116	15.41 ± 20.77	13.46 ± 18.14
Service station or motor vehicle repair facility	125	10.50 ± 10.68	9.17 ± 9.33
Other location	427	8.47 ± 20.58	7.40 ± 17.97
Other repair shop	55	6.46 ± 8.78	5.64 ± 7.67
Shopping mall	58	5.61 ± 7.44	4.90 ± 6.50
Residential garage	66	4.98 ± 8.08	4.35 ± 7.06
Restaurant	524	4.25 ± 4.98	3.71 ± 4.35
Office	2 287	4.11 ± 4.79	3.59 ± 4.18
Auditorium, sports arena, concert hall, etc.	100	3.86 ± 5.45	3.37 ± 4.76
Store	734	3.70 ± 6.37	3.23 ± 5.56
Health care facility	351	2.54 ± 4.89	2.22 ± 4.25
Other public buildings	115	2.46 ± 3.73	2.15 ± 3.26
Manufacturing facility	42	2.34 ± 2.92	2.04 ± 2.55
Residence	21 543	2.34 ± 4.65	2.04 ± 4.06
School	426	1.88 ± 3.16	1.64 ± 2.76
Church	179	1.79 ± 3.84	1.56 ± 3.35

[a] Adapted from Johnson et al. (1984).
[b] SD = standard deviation.

representing three population subgroups: housewives, office workers and construction workers. A comparison of residential carbon monoxide concentrations from that study as a function of combustion sources and whether smoking was reported is shown in Table 11. Use of gas ranges and kerosene space heaters was found to result in higher indoor carbon monoxide concentrations. The statistical significance of the differences was not given. Concentrations were highest in micro-environments associated with commuting.

.6.2   Targeted microenvironmental studies

As demonstrated from the personal exposure studies discussed above, individuals, in the course of their daily activities, can encounter a wide range of carbon monoxide concentrations as a function of the microenvironments in which they spend time. A number of studies have been conducted over the last decade to investigate concentrations

Table 10. Weighted means of residential exposure grouped according to the presence or absence of selected indoor carbon monoxide sources in Denver, Colorado[a]

CO source	CO concentration (mean ± SD)[b]				Difference in means		Significance level of t-test[c]
	Source present		Source absent		mg/m³	ppm	
	mg/m³	ppm	mg/m³	ppm			
Attached garage	2.62 ± 6.11	2.29 ± 5.34	2.15 ± 3.44	1.88 ± 3.00	0.47	0.41	P < 0.0005
Operating gas stove	5.18 ± 6.98	4.52 ± 6.10	2.21 ± 4.49	1.93 ± 3.92	2.97	2.59	P < 0.0005
Smokers	3.98 ± 7.53	3.48 ± 6.58	2.16 ± 4.23	1.89 ± 3.69	1.82	1.59	P < 0.0005

[a] Adapted from Johnson (1984).
[b] SD = standard deviation.
[c] Student t-test was performed on logarithms of personal exposure monitor values.

Table 11. Average Washington, DC, residential carbon monoxide exposures: impact of combustion appliance use and tobacco smoking[a]

Appliances	Reported tobacco smoking[b]								
	No			Yes			All cases		
	mg/m³	ppm		mg/m³	ppm		mg/m³	ppm	
None	1.4	1.2	(66)	1.7	1.5	(12)	1.4	1.2	(78)
Gas stove	2.5	2.2	(15)	1.5	1.3	(1)	2.5	2.2	(16)
Kerosene space heater	5.8	5.1	(3)		ND[c]		5.8	5.1	(3)
Wood burning	0.8	0.7	(2)		ND		0.8	0.7	(2)
Multiple appliances	1.1	1	(1)		ND		1.1	1	(1)
All cases	1.7	1.5	(87)	1.7	1.5	(13)	1.7	1.5	(100)

[a] Adapted from Nagda & Koontz (1985).
[b] Percentage of subjects' time in their own residences indicated in parentheses for each category of appliance use and tobacco smoking.
[c] ND = No data available.

of carbon monoxide in indoor microenvironments. These "targeted" studies have focused on indoor carbon monoxide concentrations as a function of either the microenvironment or sources in specific microenvironments.

.2.1 Indoor microenvironmental concentrations

A number of studies have investigated carbon monoxide levels in various indoor environments, independent of the existence of specific indoor sources. Major foci of these studies are microenvironments associated with commuting. A wide range of carbon monoxide concentrations were recorded in these studies, with the highest concentrations found in the indoor commuting microenvironments. These concentrations are frequently higher than concentrations recorded at fixed-site monitors but lower than concentrations measured immediately outside the vehicles. Concentrations are generally higher in automobiles than in public transportation microenvironments. A number of the studies noted that carbon monoxide concentrations in commuting vehicles can exceed recommended air quality guidelines. Flachsbart et al. (1987) noted that the most important factors influencing carbon monoxide concentrations inside automobiles included link-to-link variability (a proxy for traffic density, vehicle mix and roadway setting), day-to-day variability (a proxy for variations in meteorological factors and ambient carbon monoxide concentrations) and time of day. This study noted that with increased automobile speed, interior carbon monoxide concentrations decreased, because grams of carbon monoxide emitted per kilometre travelled decrease with increasing vehicle speed, and the turbulence of vehicle wake increases with increasing vehicle speed.

Service stations, car dealerships, parking garages and office spaces that have attached garages can exhibit high concentrations of carbon monoxide as a result of automobile exhaust. In one case (Wallace, 1983), corrective measures reduced office space carbon monoxide concentrations originating from an attached parking garage from 22 mg/m³ (19 ppm) to approximately 5 mg/m³ (4 ppm). In an investigation of seven ice skating rinks in the Boston, Massachusetts, area, one study (Spengler et al., 1978) reported exceptionally high average carbon monoxide concentrations (61.4 mg/m³ [53.6 ppm]), with a high reading of 220 mg/m³ (192 ppm). Ice-cleaning machines and poor ventilation were found to be responsible.

Residential and commercial buildings were generally found to have low concentrations of carbon monoxide, but information is seldom provided on the presence of indoor sources or outdoor levels.

5.6.2.2   *Concentrations associated with indoor sources*

The major indoor sources of carbon monoxide in residences are gas ranges and unvented kerosene and gas space heaters, with properly operating wood-burning stoves and fireplaces (non-leaky venting system) and tobacco combustion of secondary importance. Properly used gas ranges (ranges used for cooking and not space heating) are used intermittently and thus would contribute to short-term peak carbon monoxide levels indoors but likely would not result in substantial increases in longer-term average concentrations. Unvented kerosene and gas space heaters typically are used for several hours at a time and thus are likely to result in sustained higher levels of carbon monoxide. The improper operation of gas ranges or unvented gas or kerosene space heaters (e.g., low-wick setting for kerosene heaters or yellow-tipping operation of gas ranges) could result in substantial increases in indoor carbon monoxide levels. Carbon monoxide levels indoors associated with tobacco combustion are, based upon source emission data, expected to be low unless there is a very high smoking density and low ventilation. In the absence of a leaky flue or leaky fire box, indoor carbon monoxide levels from fireplaces or stoves should be low, with short peaks associated with charging the fire when some backdraft might occur.

The majority of studies investigating carbon monoxide concentrations in residences, as a function of the presence or absence of a known carbon monoxide source, typically have measured carbon monoxide concentrations associated with the source's use over short periods (on the order of a few minutes to a few hours). Only two studies (Koontz & Nagda, 1987; Research Triangle Institute, 1990) have reported long-term average carbon monoxide concentrations (over several hours) as a function of the presence of a carbon monoxide source for large residential sample sizes, whereas one study (McCarthy et al., 1987) reported longer-term average indoor carbon monoxide concentrations for a small sample.

1)   Average indoor-source-related concentrations

As part of a study to determine the impact of combustion sources on indoor air quality, a sample of 382 homes in New York State, USA (172 in Onondaga County and 174 in Suffolk County), was monitored for carbon monoxide concentrations during the winter of 1986 (Research Triangle Institute, 1990). In this study, four combustion sources were examined: gas cooking appliances, unvented kerosene space heaters, wood-burning stoves and fireplaces and tobacco products.

Gas ranges and kerosene heaters were found to result in small increases in average carbon monoxide levels. Use of a wood-burning stove or fireplace resulted in lower average carbon monoxide levels, presumably owing to increased air exchange rates associated with use. The study found no effect on average carbon monoxide levels with tobacco combustion and no difference by location in the residence.

Koontz & Nagda (1987), utilizing census data for sample selection, monitored 157 homes in 16 neighbourhoods in north-central Texas, USA, over a 9-week period between January and March 1985. Unvented gas space heaters were used as the primary means of heating in 82 residences (13 had one unvented gas space heater, 36 had two, and 33 had three or more) and as a secondary heat source in

29 residences (17 had one unvented gas space heater, and 12 had two or more).

Residences in which unvented gas space heaters are the primary heat source exhibited the highest carbon monoxide concentrations. Carbon monoxide concentrations were greater than or equal to 10 mg/m^3 (9 ppm) in 12% of the homes, with the highest concentration measured at 41.9 mg/m^3 (36.6 ppm). No values were measured above 10 mg/m^3 (9 ppm) for residences in which an unvented gas space heater was not used at all or was used as a secondary heat source. Five of the residences exceeded the 1-h, 40 mg/m^3 (35 ppm) level, whereas seven of the residences exceeded the 8-h, 10 mg/m^3 (9 ppm) level. Higher carbon monoxide levels were associated with maltuned unvented gas appliances and the use of multiple unvented gas appliances.

In a study of 14 homes with one or more unvented gas space heaters (primary source of heat) in the area of Atlanta, Georgia, USA, McCarthy et al. (1987) measured carbon monoxide levels by continuous NDIR monitors in two locations in the homes (room with the heater and a remote room in the house) and outdoors. One out of the 14 unvented gas space heater homes exceeded 10 mg/m^3 (9 ppm) during the sampling period. Mean indoor values ranged from 0.30 to 10.87 mg/m^3 (0.26 to 9.49 ppm) and varied as a function of the use pattern of the heater. Only one of the homes used more than one heater during the air sampling. Outdoor concentrations ranged from 0.34 to 1.8 mg/m^3 (0.3 to 1.6 ppm).

Investigations of indoor air pollution by different heating systems in 16 private houses in Germany (Moriske et al., 1996) showed carbon monoxide concentrations up to 16 mg/m^3 (14 ppm) (98th percentile) during the heating period and up to 4.6 mg/m^3 (4.0 ppm) during the non-heating period in 8 homes with coal burning and open fireplace. In 8 homes with central heating, carbon monoxide concentrations were up to 2.3 mg/m^3 (2.0 ppm) during both heating and non-heating periods. In a home on the ground floor of a block of flats with a central stove in the basement below, carbon monoxide concentrations were up to 64 mg/m^3 (56 ppm) during the heating season.

2)   Peak indoor-source-related concentrations

Short-term or peak indoor carbon monoxide concentrations associated with specific sources were obtained for a few field studies. In these studies, a wide range of peak carbon monoxide concentrations was observed in various residences with different indoor carbon monoxide sources. The highest concentrations measured (>700 mg/m^3 [>600 ppm]) were associated with emissions from geysers (water heaters), found in a large study conducted in the Netherlands (Brunekreef et al., 1982). Peak levels of carbon monoxide associated with gas ranges were from 1.1 mg/m^3 (1.0 ppm) to more than 110 mg/m^3 (100 ppm). This broad range is somewhat consistent with the results of other studies evaluating carbon monoxide emissions from gas ranges. The variability is in part due to the number of burners used, flame condition, condition of the burners, etc. As might be expected, radiant kerosene heaters produced higher carbon monoxide concentrations than did convective heaters. Unvented gas space heaters were generally associated with higher carbon monoxide peaks than were gas ranges or kerosene heaters. As noted above, the peaks associated with gas or kerosene heaters are likely to be sustained over longer periods of time because of the long source-use times.

Test houses have been used by investigators to evaluate the impact of specific sources, modifications to sources and variations in their use on residential peak carbon monoxide concentrations.

In one of the earliest investigations of indoor air quality, Wade et al. (1975) measured indoor and outdoor carbon monoxide levels in four houses that had gas stoves. Indoor concentrations were found to be 1.7–3.8 times higher than the outdoor levels. Carbon monoxide levels in one house exceeded 10 mg/m^3 (9 ppm), the NAAQS reference level. As part of a modelling study of emissions from a gas range, Davidson et al. (1987) measured carbon monoxide concentrations in three residences. Peak carbon monoxide levels in excess of 6 mg/m^3 (5 ppm) were measured in one townhouse.

Indoor carbon monoxide levels associated with wood-burning stoves were measured in two test house studies. In one study (Humphreys et al., 1986), indoor carbon monoxide levels associated with the use of both airtight (conventional and catalytic) and non-airtight wood heaters were evaluated in a 337-m^3 weatherized home. Indoor carbon monoxide concentrations were higher than outdoor levels for all tests. Conventional airtight stoves produced indoor carbon monoxide levels typically about 1.1–2.3 mg/m^3 (1–2 ppm) above background level, with a peak concentration of 10.4 mg/m^3 (9.1 ppm). Use of non-airtight stoves resulted in average indoor carbon monoxide concentrations 2.3–3.4 mg/m^3 (2–3 ppm) above outdoor concentrations, with peak concentrations as high as 33.9 mg/m^3 (29.6 ppm). In a 236-m^3 house (Traynor et al., 1984), four wood-burning stoves (three airtight and one non-airtight) were tested. The airtight stoves generally resulted in small contributions to both average and peak indoor carbon monoxide levels (0.11–1.1 mg/m^3 [0.1–1 ppm] for the average and 0.23–3.1 mg/m^3 [0.2–2.7 ppm] for the peak). The non-airtight stove contributed as much as 10.4 mg/m^3 (9.1 ppm) to the average indoor level and 49 mg/m^3 (43 ppm) to the peak.

3) Indoor concentrations related to environmental tobacco smoke

Carbon monoxide has been measured extensively in chamber studies as a surrogate for environmental tobacco smoke (e.g., Bridge & Corn, 1972; Hoegg, 1972; Weber et al., 1976, 1979a,b; Leaderer et al., 1984; Weber, 1984; Clausen et al., 1985). Under steady-state conditions in chamber studies, where outdoor carbon monoxide levels are monitored and the tobacco brands and smoking rates are controlled, carbon monoxide can be a reasonably good indicator of environmental tobacco smoke and is used as such. Under such chamber conditions, carbon monoxide concentrations typically range from less than 1.1 mg/m^3 (1 ppm) to greater than 11 mg/m^3 (10 ppm).

A number of field studies have monitored carbon monoxide in different indoor environments with and without smoking occupants. Although carbon monoxide concentrations were generally higher in indoor spaces when smoking occurred, the concentrations were highly variable. The variability of carbon monoxide production from tobacco combustion, the variability in the number of cigarettes smoked and differences in ventilation and variability of outdoor concentrations make it difficult to assess the contribution of tobacco combustion to indoor carbon monoxide concentrations. The chamber studies and field studies conducted do indicate that under typical smoking conditions encountered in residences or offices, carbon monoxide concentrations can be expected to be above background outdoor levels, but lower than the levels resulting from other unvented combustion sources. In indoor spaces where heavy smoking occurs and in small indoor spaces, carbon monoxide emissions from tobacco combustion will be an important contributor to carbon monoxide concentrations.

5.7 Occupational exposure

Carbon monoxide is a ubiquitous contaminant occurring in a variety of occupational settings. The number of persons occupationally exposed to carbon monoxide in the working environment is greater than for any other physical or chemical agent (Hosey, 1970), with estimates as high as 975 000 occupationally exposed at high levels in the USA (NIOSH, 1972).

Two main sources for background exposures in both occupational and non-occupational settings appear to be smoking and the internal combustion engine (NAS, 1969). Also, endogenous carbon monoxide may be derived from certain halomethanes that are biotransformed by mixed-function oxidases *in vivo*. Smoking is a personal habit that must be considered in evaluating exposure in general, as well as those occurring in workplaces. In addition, work environments are often located in densely populated areas, and such areas frequently have a higher background concentration of carbon monoxide compared with less densely populated residential areas. Thus, background exposures may be greater during work hours than during non-work hours.

There are several sources other than smoking and the internal combustion engine that contribute to exposure during work hours. These include contributions to background levels by combustion of organic materials in the geographic area of the workplace; work in specific industrial processes that produce carbon monoxide; and exposure to halomethanes that give rise to endogenous carbon monoxide during biotransformation. Therefore, a number of potential sources of carbon monoxide should be considered when evaluating the risk associated with carbon monoxide exposure, including personal habits, living conditions and co-exposure to other potential sources of carbon monoxide or to xenobiotics that are metabolized to carbon monoxide or that interfere with biotransformation processes. Finally, the particular vulnerability of specific groups at increased risk because of some physiological (pregnancy) or pathological (angina, anaemia, respiratory insufficiency) conditions should be taken into account in the health surveillance of occupationally exposed workers.

5.7.1 Occupational exposure limits

Elevated concentrations of carbon monoxide occur in numerous settings, including those at work, at home or in the street. Acute effects related to production of anoxia from exposures to carbon monoxide have historically been a basis for concern. In recent years, however, this concern has grown to include concerns for potential effects from chronic exposure as well (Sammons & Coleman, 1974; Rosenstock & Cullen, 1986a,b).

Three kinds of occupational exposure limits are currently utilized in a number of countries (Table 12). The most common are average permissible concentrations for a typical 8-h working day (time-weighted average, TWA), concentrations for short-term exposures, generally of 15-min duration (short-term exposure limit, STEL), and maximum permissible concentrations not to be exceeded (ceiling limit).

Whereas some countries do not apply legally binding occupational exposure limits and others refer to maximum allowable concentrations, threshold limit values (TLVs) are the most widely used and accepted standards representing "conditions under which it is believed that nearly all workers may be repeatedly exposed day after day without adverse health effects." In the introduction to the list of TLVs published by the American Conference of Governmental Industrial Hygienists (ACGIH, 1995), it is also stressed that these limits are not fine lines separating safe from dangerous situations, but rather guidelines to be used by professionals with a specific training in industrial hygiene. Moreover, it is recognized that a small percentage of workers may be "unusually responsive to some industrial chemicals because of genetic factors, age, personal habits (smoking, alcohol, or other drugs), medication, or previous exposure. Such workers may not be adequately protected from adverse health effects at concentrations at or below the threshold limit. An occupational physician should evaluate the extent to which such workers require additional protection" (ACGIH, 1995).

A TWA concentration of 29 mg/m^3 (25 ppm) has been recommended by ACGIH as the TLV for carbon monoxide since 1991. Such a limit is likely to be adopted by other countries traditionally relying on ACGIH-recommended TLVs and by countries relying on the recommended exposure limits (RELs) adopted by the US National

Table 12. Worldwide occupational exposure limits for carbon monoxide[a]

	TWA		STEL		Ceiling	
	mg/m^3	ppm	mg/m^3	ppm	mg/m^3	ppm
Austria	33	30	–	–	–	–
Belgium	55	50	–	–	–	–
Brazil	43	39	–	–	–	–
Bulgaria	20	–	–	–	–	–
Chile	44	40	–	–	–	–
China, P.R. of	30	–	–	–	–	–
China (Taiwan)	55	50	–	–	–	–
Czechoslovakia	30	–	–	–	150	–
Denmark	40	35	–	–	–	–
Egypt	–	100	–	–	–	–
Finland	55	50	85	75	–	–
France	–	50	–	–	–	–
FRG	33	30	–	–	–	–
Holland	55	50	–	–	–	–
Hungary	20	–	100	–	–	–
India	55	50	440	400	–	–
Indonesia	115	100	–	–	–	–
Italy	55	50	–	–	–	–
Japan	55	50	–	–	–	–
Mexico	55	50	–	–	–	–
Poland	20	–	–	–	–	–
Romania	30	–	50	–	–	–
Sweden	40	35	120	100	–	–
Switzerland	33	30	–	–	–	–
United Kingdom	55	50	440	400	–	–
USA	–	25	–	400	–	200
USSR	33	30	–	–	20	–
Venezuela	55	50	–	–	440	400
Yugoslavia	58	50	–	–	–	–

[a] From ACGIH (1987); CEC (1993).

Institute for Occupational Safety and Health (NIOSH), currently set at 40 mg/m³ (35 ppm). Although ceiling limits and STELs are not expressly indicated, general rules advise that exposure levels "may exceed the TLV-TWA three times for no more than 30 min during a work-day, and under no circumstances should they exceed five times the TLV-TWA." Therefore, a ceiling of 143 mg/m³ (125 ppm) and a STEL of 86 mg/m³ (75 ppm) may be derived from such a guidance. Biotransformation into carbon monoxide rather than other toxic properties is the scientific basis for setting the TLV for methylene chloride at 177 mg/m³ (50 ppm).

In addition to ambient monitoring, biomarkers can be used to assess exposure, susceptibility and early effects. Carboxyhaemoglobin is a widely accepted biomarker — or biological exposure index (BEI), according to ACGIH nomenclature — of exposure to carbon monoxide. The ACGIH has recently proposed a carboxyhaemoglobin level of 3.5% as a BEI being most likely reached by a non-smoker at the end of an 8-h exposure to 29 mg carbon monoxide/m³ (25 ppm) (ACGIH, 1991). The biological tolerance limit (BAT) recommended by the DFG (1996) is 5% carboxyhaemoglobin, corresponding to a carbon monoxide TWA level of 34 mg/m³ (30 ppm).

Neither the BEI nor the BAT is applicable to tobacco smokers. According to Gilli et al. (1979), a serum thiocyanate level above 3.8 mg/litre and a carboxyhaemoglobin level in the range of 2.5–6% are a certain indication of cigarette smoking, whereas a thiocyanate concentration below 3.8 mg/litre and a carboxyhaemoglobin level above 5% indicate occupational exposure to carbon monoxide. Tobacco smoking can also be assessed measuring cotinine excretion, but studies relating cotinine in urine to carboxyhaemoglobin are not available. An alternative BEI is carbon monoxide concentration in end-expired air collected at the end of the working shift. A concentration of 23 mg/m³ (20 ppm) would be reached after exposure to an 8-h TWA level of 29 mg/m³ (25 ppm).

It ought to be noted that both BEIs and BATs are usually established taking into account the relationship between biomarkers and exposure levels rather than the relationship between biomarkers and adverse effects. Therefore, higher carboxyhaemoglobin levels are generally accepted for smokers, taking into account the fact that their higher levels result from a voluntary habit giving rise to an additional carbon monoxide burden. However, if adverse effects are expected to occur as a consequence of carboxyhaemoglobin and carbon monoxide body burden, then smokers should be advised about their additional risk.

NIOSH (1972) observed that "the potential for exposure to carbon monoxide for employees in the work place is greater than for any other chemical or physical agent" and recommended that exposure to carbon monoxide be limited to a concentration no greater than 40 mg/m^3 (35 ppm), expressed as a TWA for a normal 8-h workday, 5 days per week. A ceiling concentration was also recommended at a limit of 230 mg/m^3 (200 ppm), not to exceed an exposure time greater than 30 min. Occupational exposures at the proposed concentrations and conditions underlying the basis of the standard were considered to maintain carboxyhaemoglobin in blood below 5%.

Although it was not stated, the basis of the recommended NIOSH standard (i.e., maintaining carboxyhaemoglobin below 5% in blood) assumes that (1) the sensitive population group protected by air quality standards would not be occupationally exposed and (2) contributions from other non-occupational sources would also be less than a TWA concentration of 40 mg/m^3 (35 ppm). It was recognized that such a standard may not provide the same degree of protection to smokers, for example. Other particularly vulnerable groups of workers deserve special consideration. Among these, pregnant women should require special protection because of the potentially deleterious effects of carbon monoxide exposure on the fetuses (NIOSH, 1972). The same requirement for safety of pregnant women has been established by the DFG (1996).

Although recognizing that biological changes might occur at the low level of exposure recommended in the proposed standard, NIOSH concluded that subtle aberrations in the nervous system with exposures producing carboxyhaemoglobin concentrations in blood at or below 5% did not demonstrate significant impairments that would cause concern for the health and safety of workers. In addition, NIOSH observed that individuals with impairments that interfere with normal oxygen delivery to tissues (e.g., emphysema, anaemia, coronary heart disease) may not have the same degree of protection as have less impaired individuals. It was also recognized that work at higher altitudes (e.g., 1500–2400 m above sea level) would necessitate decreasing the exposure limit below 40 mg/m^3 (35 ppm), to compensate for a decrease in the oxygen partial pressure as a result of high-altitude environments and a corresponding decrease in oxygenation of the blood (NIOSH, 1972). High-altitude environments of concern include airline cabins at a pressure altitude of 1500 m or greater (NRC, 1986b) and high mountain tunnels (Miranda et al., 1967).

5.7.2 *Exposure sources*

The contribution of occupational exposures to carbon monoxide can be separated from other sources of carbon monoxide exposure, but there are at least two conditions to consider:

(1) When carbon monoxide concentrations at work are higher than the carbon monoxide equilibrium concentration associated with the percent carboxyhaemoglobin at the start of the work shift, there will be a net absorption of carbon monoxide and an increase in percent carboxyhaemoglobin. Non-smokers will show an increase that is greater than that for smokers because they start from a lower baseline carboxyhaemoglobin level. In some cases, non-smokers may show an increase and smokers a decrease in percent carboxyhaemoglobin.

(2) When carbon monoxide concentrations at work are lower than the equilibrium concentration necessary to produce the worker's current level of carboxyhaemoglobin, then the percent carboxyhaemoglobin will show a decrease. There will be a net loss of carbon monoxide at work.

As mentioned above, occupational exposures can stem from three sources: (1) through background concentrations of carbon monoxide, (2) through work in industrial processes that produce carbon monoxide as a product or by-product and (3) through exposure to some halomethanes that are metabolized to carbon monoxide *in vivo*. In addition, work environments that tend to accumulate carbon monoxide concentrations may result in occupational exposures. Rosenman (1984) lists a number of occupations in which the workers may be exposed to high carbon monoxide concentrations. This list includes acetylene workers, blast furnace workers, coke oven workers, diesel engine operators, garage mechanics, steel workers, metal oxide reducers, miners, nickel refining workers, organic chemical synthesizers, petroleum refinery workers and pulp and paper workers. In addition, because methylene chloride is metabolized to carbon monoxide in the body, aerosol packagers, anaesthetic makers, bitumen makers, degreasers, fat extractors, flavouring makers, leather finish workers, oil processors, paint remover makers, resin makers and workers exposed to dichloromethane-containing solvent mixtures and stain removers can also have high carboxyhaemoglobin levels.

5.7.3 *Combined exposure to xenobiotics metabolized to carbon monoxide*

Certain halomethanes, particularly dichloromethane, a frequently used organic solvent also known as methylene chloride (reviewed in EHC 164), are metabolized to carbon monoxide, carbon dioxide and chlorine (or iodine or bromine) in a reaction catalysed by cytochrome P-450 2E1.

Combined exposures frequently occur at the workplace, and their temporal sequence may result in opposite effects. Concurrent exposure to other substrates of cytochrome P-450 2E1 — including ethanol and a number of common organic solvents, such as benzene and its alkyl derivatives, trichloroethylene and acetone — may cause a competitive inhibition of the oxidation of methylene chloride to carbon monoxide. As the same substances may act as inducers, provided they are no longer present to compete with methylene chloride, increased rates of carbon monoxide formation are expected to occur in people chronically exposed to such solvents or to ethanol and then exposed to methylene chloride at the workplace. As a corollary of these observations, and taking into account the fact that the carbon monoxide produced from inhaled dichloromethane rather than the parent compound underlies non-cancer end-points and the fact that the carbon monoxide formation rate is modified in a different way by prior and concurrent exposure to other substances, it would seem logical to rely on the same BEIs proposed for carbon monoxide to monitor exposure to methylene chloride and other halomethanes rather than on ambient monitoring alone.

5.7.4 *Typical studies at the workplace*

Aircraft accidents involving 113 aircraft, 184 crew members and 207 passengers were investigated to characterize accident toxicology and to aid in the search for causation of a crash (Blackmore, 1974). Determinations of percent carboxyhaemoglobin in blood samples obtained from victims enabled differentiation of a variety of accident sequences involving fires. For example, percent carboxyhaemoglobin determinations combined with passenger seating information and crew assignments can help differentiate between fire in flight and fire after the crash, between survivability of the crash and death due to smoke inhalation, and between specific malfunctions in equipment operated by a particular crew member and defects in space heating in the crew cabin or passenger compartment. One accident in the series was associated with a defective space heater in the crew compartment. Another accident was also suspicious with regard to a space heater.

Carbon monoxide concentrations were used to classify workers from 20 foundries into three groups: those with definite occupational exposure, those with slight exposure and controls (Hernberg et al., 1976). Angina pectoris, electrocardiogram findings and blood pressures of foundry workers were evaluated in terms of carbon monoxide exposure for the 1000 workers who had the longest occupational exposures for the 20 foundries. Angina showed a clear dose–response with exposure to carbon monoxide either from occupational sources or from smoking, but there was no such trend in electrocardiogram findings. The systolic and diastolic pressures of carbon monoxide-exposed workers were higher than those for other workers, when age and smoking habits were considered.

Carboxyhaemoglobin and smoking habits were studied for a population of steelworkers and compared with those for blast furnace workers as well as employees not exposed at work (Jones & Walters, 1962). Carbon monoxide is produced in coke ovens, blast furnaces and sintering operations. Exhaust gases from these operations are often used for heating and as fuels for other processes. Fifty-seven volunteers working in the blast furnace area were studied for smoking habits, symptoms of carbon monoxide exposure and estimations of carboxyhaemoglobin levels by an expired air technique. The main increase in carboxyhaemoglobin for blast furnace personnel was 2.0% for both smokers and non-smokers in the group. For smokers in the unexposed control group, there was a decrease in percent carboxy-haemoglobin. A follow-up study found similar results (Butt et al., 1974). Virtamo & Tossavainen (1976) reported a study of carbon monoxide measurements in air of 67 iron, steel or copper alloy foundries. Blood carboxyhaemoglobin of iron workers exceeded 6% in 26% of the non-smokers and 71% of the smokers studied.

Poulton (1987) found that a medical helicopter with its engine running in a narrowed or enclosed helipad was a source of potential exposure to carbon monoxide, JP-4 fuel and possibly other combustion products for flight crews, medical personnel, bystanders and patients being evacuated. Measurements were made by means of a portable infrared analyser. Carbon monoxide concentrations were found to be greatest near the heated exhaust. Concentrations ranged from 9 to 49 mg/m^3 (8 to 43 ppm).

Exhaust from seven of the most commonly used chain-saws (Nilsson et al., 1987) was analysed under laboratory conditions to characterize emissions. The investigators conducted field studies on exposures of loggers using chain-saws in felling operations and also in limbing and bucking into lengths. In response to an inquiry, 34% of the loggers responded that they often experienced discomfort from the exhaust fumes of chain-saws, and another 50% complained of occasional problems. Sampling for carbon monoxide exposures was carried out for 5 days during a 2-week work period in a sparse pine stand at an average wind speed of 0–3 m/s, a temperature range of 1–16 °C and a snow depth of 50–90 cm. Carbon monoxide concentrations ranged from 10 to 23 mg/m^3 (9 to 20 ppm), with a mean value of 20.0 mg/m^3 (17.5 ppm). Carbon monoxide concentrations measured under similar, but snow-free, conditions ranged from 24 to 44 mg/m^3 (21 to 38 ppm), with a mean value of 32 mg/m^3 (28 ppm). In another study, carbon monoxide exposures were monitored for non-smoking chain-saw operators; average exposures recorded were from 23 to 63 mg/m^3 (20 to 55 ppm), with carboxyhaemoglobin levels ranging from 1.5 to 3.0% (Van Netten et al., 1987).

Forklift operators, stevedores and winch operators were monitored for carbon monoxide in expired air, using a Mine Safety Appliances analyser, to calculate percent carboxyhaemoglobin (Breysse & Bovee, 1969). Periodic blood samples were collected to validate the calculations. Forklift operators and stevedores, but not winch operators, work in the holds of ships. The ships to be evaluated were selected on the basis of their use of gasoline-powered forklifts for operations. To evaluate seasonal variations in percent carboxyhaemoglobin, analyses were performed for one 5-day period per month for a full year. Efforts were made to select a variety of ships for evaluation. In total, 689 determinations of percent carboxyhaemoglobin were made from samples of blood to compare with values from samples of expired air. The samples were collected on 51 separate days involving 26 different ships. Smoking was found to be a major contributing factor to the percent carboxyhaemoglobin levels found. Carboxyhaemoglobin values for non-smokers indicated that the use of gasoline-powered lifts in the holds of the ships did not produce a carbon monoxide concentration in excess of 57 mg/m^3 (50 ppm) for up to 8 h as a TWA under the work rules and operating conditions in practice during the study. Smoking behaviour confounded exposure evaluations.

Carbon monoxide concentrations have been measured in a variety of workplaces where potential exists for accumulation from outside sources. Exposure conditions in workplaces, however, are substantially different. The methods to be applied, group characteristics, jobs being performed, smoking habits and physical characteristics of the facilities themselves introduce considerable variety in the approaches used. Typical studies are discussed below.

Wallace (1983) investigated carbon monoxide in air and breath of employees working at various times over a 1-month period in an office constructed in an underground parking garage. Carbon monoxide levels were determined by use of a device containing a proprietary solid polymer electrolyte to detect electrons emitted in the oxidation of carbon monoxide to carbon dioxide. Variation in carbon monoxide measurements in ambient air showed a strong correlation with traffic activity in the parking garage. Initially, the office carbon monoxide levels were found to be at a daily average of 21 mg/m^3 (18 ppm), with the average from 12:00 to 4:00 p.m. at 25 mg/m^3 (22 ppm) and the average from 4:00 to 5:00 p.m. at 41 mg/m^3 (36 ppm). Analyses of expired air collected from a group of 20 non-smokers working in the office showed a strong correlation with concentrations of carbon monoxide in ambient air and traffic activity. For example, the average carbon monoxide in expired air for one series of measurements was 26.8 mg/m^3 (23.4 ppm), compared with simultaneous measurements of carbon monoxide concentrations in air of 25–30 mg/m^3 (22–26 ppm). After a weekend, carbon monoxide concentrations in breath on Monday morning were substantially decreased (around 8 mg/m^3 [7 ppm]), but they rose again on Monday afternoon to equal the air levels of 14 mg/m³ (12 ppm). Closing fire doors and using existing garage fans decreased carbon monoxide concentrations in the garage offices to 2.3 mg/m³ (2 ppm) or less, concentrations similar to those for other offices in the complex that were located away from the garage area.

Carboxyhaemoglobin levels (Ramsey, 1967) were determined over a 3-month period during the winter for 38 parking garage attendants, and the values for carboxyhaemoglobin were compared with values from a group of 27 control subjects. Blood samples were collected by finger stick on Monday mornings at the start of the work week, at the end of the work shift on Mondays and at the end of the work week on Friday afternoons. Hourly analyses were carried out on three different weekdays using potassium paladosulfite indicator tubes for the concentrations of carbon monoxide at three of the six garages in the study. Hourly levels ranged from 8 to 270 mg/m³ (7 to 240 ppm), and the composite mean of the 18 daily averages was 67.4 ± 28.5 mg/m³ (58.9 ± 24.9 ppm). Although the Monday versus Friday afternoon values for carboxyhaemoglobin were not significantly different, there were significant differences between Monday morning and Monday afternoon values. Smokers showed higher starting baseline levels, but there was no apparent difference in net increase in carboxyhaemoglobin body burden between smokers and non-smokers. Carboxyhaemoglobin levels for non-smokers ranged from a mean of 1.5 ± 0.83% for the morning samples to 7.3 ± 3.46% for the afternoon samples. For smokers, these values were 2.9 ± 1.88% for the morning and 9.3 ± 3.16% for the afternoon. The authors observed a crude correlation between daily average carbon monoxide levels in air and carboxyhaemoglobin levels observed for a 2-day sampling period.

In a study of motor vehicle examiners conducted by NIOSH (Stern et al., 1981), carbon monoxide levels of 5–24 mg/m³ (4–21 ppm) TWA were recorded in six outdoor motor vehicle inspection stations. In contrast, the semi-open and enclosed stations had levels of 11–46 mg/m³ (10–40 ppm) TWA. The levels exceeded the recommended NIOSH standard of 40 mg/m³ (35 ppm) TWA on 10% of the days sampled. In addition, all stations experienced peak short-term levels above 230 mg/m³ (200 ppm).

Carboxyhaemoglobin levels were measured for 22 employees of an automobile dealership during the winter months when garage doors were closed and ceiling exhaust fans were turned off (Andrecs et al., 1979). Employees subjected to testing included garage mechanics, secretaries and sales personnel. These included 17 males aged 21–37 and five females aged 19–36. Blood samples were collected on a Monday morning before the start of work and on Friday at the end of the work week. Smokers working in the garage area showed a Monday morning mean carboxyhaemoglobin value of 4.87 ± 3.64% and a Friday afternoon mean value of 12.9 ± 0.83%. Non-smokers in the garage showed a corresponding increase in carboxyhaemoglobin, with a Monday morning mean value of 1.50 ± 1.37% and a Friday afternoon mean value of 8.71 ± 2.95%. Non-smokers working in areas other than the garage had a Friday afternoon mean value of 2.38 ± 2.32%, which was significantly lower than the mean values for smokers and non-smokers in the garage area. Environmental concentrations or breathing zone samples for carbon monoxide were not collected. The authors concluded that smokers have a higher baseline level of carboxyhaemoglobin than do non-smokers, but both groups show similar increases in carboxyhaemoglobin during the work week while working in the garage area. The authors observed that the concentrations of carboxyhaemoglobin found in garage workers were the same as those reported to produce neurological impairment. These results are consistent with those reported by Amendola & Hanes (1984), who reported some of the highest indoor levels collected at automobile service stations and dealerships. Concentrations ranged from 18.5–126.9 mg/m^3 (16.2–110.8 ppm) in cold weather to 2.5–24.7 mg/m^3 (2.2–21.6 ppm) in warm weather.

A group of 34 employees, 30 men and 4 women, working in multistorey garages was evaluated for exposures to exhaust fumes (Fristedt & Akesson, 1971). Thirteen were service employees working at street level, and 21 were shop employees working either one storey above or one storey below street level. Six facilities were included in the study. Blood samples were collected on a Friday at four facilities, on Thursday and Friday at another and on a Thursday only at a sixth facility. The blood samples were evaluated for red blood cell and white blood cell counts, carboxyhaemoglobin, lead and δ-amino-levulinic acid. Work histories, medical case histories and smoking habits were recorded. Among the employees evaluated, 11 of 24 smokers and 3 of 10 non-smokers complained of discomfort from exhaust fumes. Smokers complaining of discomfort averaged 6.6% carboxyhaemoglobin, and non-smokers complaining averaged 2.2% carboxyhaemoglobin. The corresponding values for non-complaining workers averaged 4.2 and 1.1%, respectively.

Air pollution by carbon monoxide in underground garages was investigated as part of a larger study of traffic pollutants in Paris, France (Chovin, 1967). Work conducted between the hours of 8:00 a.m. and 10:00 p.m. resulted in exposures in excess of 57 mg/m³ (50 ppm) and up to 86 mg/m³ (75 ppm), on a TWA basis.

As part of a larger study of carbon monoxide concentrations and traffic patterns in Paris (Chovin, 1967), samples were taken in road tunnels. There was good correlation between the traffic volumes combined with the lengths of the tunnels and the carbon monoxide concentrations found. None of the tunnels studied had mechanical ventilation. The average carbon monoxide concentrations in the tunnels were 31 mg/m³ (27 ppm) and 34 mg/m³ (30 ppm) for 1965 and 1966, respectively, compared with an average of 27 mg/m³ (24 ppm) in the streets for both years. The average risk for a person working or walking in a street or tunnel was considered by the authors to be 3–4 times less than the maximal risk indicated by values for carbon monoxide from instantaneous air sample measurements. In the USA, Evans et al. (1988) studied bridge and tunnel workers in metropolitan New York City, New York. The average carboxyhaemoglobin concentration over the 11 years of study averaged 1.73% for non-smoking bridge workers and 1.96% for tunnel workers.

In a discussion of factors to consider in carbon monoxide control of high-altitude highway tunnels, Miranda et al. (1967) reviewed the histories of several tunnels. Motor vehicles were estimated to emit about 0.03 kg carbon monoxide/km at sea level. At 3350 m and a grade of 1.64%, emissions were estimated at 0.1 kg/km (for vehicles moving upgrade). Tunnels with ventilation are generally designed to control carbon monoxide concentrations at or below 110 mg/m³ (100 ppm). The Holland Tunnel in New York was reported to average 74 mg/m³ (65 ppm), with a recorded maximum of 418 mg/m³ (365 ppm) due to a fire. For the Sumner Tunnel in Boston, Massachusetts, ventilation is started at carbon monoxide concentrations of 110 mg/m³ (100 ppm), and additional fans are turned on and an alarm is sounded at 290 mg/m³ (250 ppm). The average carbon monoxide concentration is 57 mg/m³ (50 ppm). The Mont Blanc Tunnel is 11.6 km long at an average elevation of 1274 m. This tunnel is designed to maintain carbon monoxide concentrations at or below 110 mg/m³ (100 ppm). The Grand Saint Bernard Tunnel is 5.6 km long at an average elevation of 1830 m. The tunnel is designed to maintain carbon monoxide concentrations at or below 230 mg/m³ (200 ppm). For the tunnel at 3350 m, the authors recommended maintaining carbon monoxide concentrations at or below 29 mg/m³ (25 ppm) for long-term exposures and at or below 57 mg/m³ (50 ppm) for peaks of 1-h exposure. The recommendations are based on considerations of a combination of hypoxia from lack of oxygen due to the altitude and stress of carbon monoxide exposures of workers and motorists.

Carbon monoxide exposures of tollbooth operators were studied along the New Jersey, USA, Turnpike. The results reported by Heinold et al. (1987) indicated that peak exposures for 1 h ranged from 14 to 27 mg/m³ (12 to 24 ppm), with peak 8-h exposures of 7–17 mg/m³ (6–15 ppm).

Carboxyhaemoglobin levels were determined for 15 non-smokers at the start, middle and end of a 40-day submarine patrol (Bondi et al., 1978). Values found were 2.1%, 1.7% and 1.7%, respectively. The average carbon monoxide concentration in ambient air was 8 mg/m³ (7 ppm). The authors observed that the levels of percent carboxy-haemoglobin found would not cause significant impairment of the submariners.

In contrast, Iglewicz et al. (1984) found in a 1981 study that carbon monoxide concentrations inside ambulances in New Jersey were often above the US EPA 8-h standard of 10 mg/m³ (9 ppm). For example, measurements made at the head of the stretcher exceeded 10 mg/m³ (9 ppm) on nearly 27% of the 690 vehicles tested, with 4.2% (29 vehicles) exceeding 40 mg/m³ (35 ppm).

Environmental tobacco smoke has been reviewed (NRC, 1986b) for contributions to air contaminants in airliner cabins and to potential exposures for passengers and flight crew members. Environmental tobacco smoke is described as a complex mixture containing many components. Analyses of carbon monoxide content and particulate matter in cabin air were used as surrogates for the vapour phases and solid components of environmental tobacco smoke, respectively. A mathematical model was developed and used to calculate the dilution of contaminants by outside make-up air. The amount of carbon monoxide in the cabin environment depends on the rate and number of cigarettes smoked and on the rate of dilution by outside make-up air. An additional factor to consider is the influence of pressure altitude on the absorption of carbon monoxide and other gases. The legal limit for pressure altitude is 2440 m. The partial pressure of oxygen is 16 kPa assuming 20% oxygen in the cabin air, compared with 20 kPa at sea level. It is possible that the absorption rate for carbon monoxide would be increased under hypobaric conditions.

A study of municipal bus drivers in the San Francisco Bay, California, USA, area by Quinlan et al. (1985) showed a TWA of 1.1–26 mg/m³ (1–23 ppm), with a mean TWA of 6.3 mg/m³ (5.5 ppm) and standard deviation of 5.6 mg/m³ (4.9 ppm). The peak exposures ranged from 8 to 54 mg/m³ (7 to 47 ppm), with a mean of 29.0 mg/m³ (25.3 ppm) and standard deviation of 14.3 mg/m³ (12.5 ppm).

Cooke (1986) reported finding no significant increases outside normal ranges, compared with the general population, for levels of blood lead and carboxyhaemoglobin in a group of 13 roadside workers. Samples were collected in the afternoon of a workday. Among the subjects, 7 of 13 were smokers and showed percent carboxyhaemoglobin in blood ranging from 3.0 to 8.8% (mean of 5.5%). For non-smokers, percent carboxyhaemoglobin ranged from 0.5 to 1.4% (mean of 1.2%). Each smoker had smoked at least one cigarette in the 4 h preceding collection of blood samples. No samples were collected before the start of work, and no measurements of carbon monoxide in air at the work sites were presented.

6. TOXICOKINETICS AND METABOLISM

6.1   Introduction

The binding of carbon monoxide to haemoglobin, producing carboxyhaemoglobin, decreases the oxygen carrying capacity of blood and interferes with oxygen release at the tissue level; these two main mechanisms of action underlie the potentially toxic effects of low-level carbon monoxide exposure. Impaired delivery of oxygen can interfere with cellular respiration and result in tissue hypoxia. Hypoxia of sensitive tissues, in turn, can affect the function of many organs, including the lungs. The effects would be expected to be more pronounced under conditions of stress, as with exercise, for example. Although the principal cause of carbon monoxide-induced toxicity at low exposure levels is thought to be increased carboxyhaemoglobin formation, the physiological response to carbon monoxide at the cellular level and its related biochemical effects are still not fully understood. Other mechanisms of carbon monoxide-induced toxicity have been hypothesized and assessed, such as hydroxyl radical production (Piantadosi et al., 1997) and lipid peroxidation (Thom, 1990, 1992, 1993) in the brain of carbon monoxide-poisoned rats, but none has been demonstrated to operate at relatively low (near-ambient) carbon monoxide exposure levels.

6.2   Endogenous carbon monoxide production

In addition to exogenous sources, humans are also exposed to small amounts of carbon monoxide produced endogenously. In the process of natural degradation of haemoglobin to bile pigments, in concert with the microsomal reduced nicotinamide adenine dinucleotide phosphate (NADPH) cytochrome P-450 reductase, two haem oxygenase isoenzymes, HO-1 and HO-2, catalyse the oxidative breakdown of the α-methene bridge of the tetrapyrrol ring of haem, leading to the formation of biliverdin and carbon monoxide. The major site of haem breakdown, and therefore the major organ for production of endogenous carbon monoxide, is the liver (Berk et al., 1976). The spleen and the erythropoietic system are other important catabolic generators of carbon monoxide. Because the amount of porphyrin breakdown is stoichiometrically related to the amount of endogenously formed carbon monoxide, the blood level of carboxyhaemoglobin and the concentration of carbon monoxide in the alveolar air have been used with mixed success as quantitative indices of the rate of haem catabolism (Landaw et al., 1970; Solanki et al., 1988). Not all endogenous carbon monoxide comes from haemoglobin degradation. Other haemoproteins, such as myoglobin, cytochromes, peroxidases and catalase, contribute approximately 20–25% to the total amount of carbon monoxide generated (Berk et al., 1976). Approximately 0.4 ml carbon monoxide/h is formed by haemoglobin catabolism, and about 0.1 ml/h originates from non-haemoglobin sources (Coburn et al., 1964). Metabolic processes other than haem catabolism contribute only a very small amount of carbon monoxide (Miyahara & Takahashi, 1971). In both males and females, week-to-week variations in carbon monoxide production are greater than day-to-day or within-day variations. Moreover, in females, carboxyhaemoglobin levels fluctuate with the menstrual cycle; the mean rate of carbon monoxide production in the premenstrual, progesterone phase almost doubles (Delivoria-Papadopoulos et al., 1970; Lynch & Moede, 1972). Neonates and pregnant women also showed a significant increase in endogenous carbon monoxide production related to increased breakdown of red blood cells.

Any disturbance leading to increased destruction of red blood cells and accelerated breakdown of other haemoproteins would lead to increased production of carbon monoxide. Haematomas, intravascular haemolysis of red blood cells, blood transfusion and ineffective erythropoiesis will all elevate the carbon monoxide concentration in the blood. Degradation of red blood cells under pathological conditions such as anaemias (haemolytic, sideroblastic, sickle cell), thalassaemia, Gilbert's syndrome with haemolysis and other haematological diseases will also accelerate carbon monoxide production (Berk et al., 1974; Solanki et al., 1988). In patients with haemolytic anaemia, the carbon monoxide production rate was 2–8 times higher and blood carboxyhaemoglobin concentration was 2–3 times higher than in normals (Coburn et al., 1966). Increased carbon monoxide production rates have been reported after administration of phenobarbital, progesterone (Delivoria-Papadopoulos et al., 1970) and diphenylhydantoin (Coburn, 1970).

The biological significance of the haem oxygenases appears to be not limited to the above-mentioned role. Evidence is accumulating to suggest that the formed carbon monoxide functions as a gaseous signalling molecule (Verma et al., 1993; Sue-Matsu et al., 1994; Vincent et al., 1994). The discrete neuronal localization of HO-2 is essentially the same as that for soluble guanylate cyclase. Carbon monoxide acts by stimulating this enzyme in target cells. The soluble guanylate cyclase contains one haem per heterodimer. The haem moiety does not bind oxygen, but does bind carbon monoxide (or nitric oxide). The soluble guanylate cyclase activation by carbon monoxide results in an increase in cyclic guanosine monophosphate and subsequent relaxation of smooth muscle cells, activation of nitric oxide synthase in endothelial cells and inhibition of platelet aggregation.

Other minor sources of carbon monoxide include the oxidation of phenols, flavonoids and halomethanes, as well as the lipid peroxidation of membrane lipids (Rodgers et al., 1994).

6.3 Uptake of carbon monoxide and formation of carboxyhaemoglobin

The rate of formation and elimination of carboxyhaemoglobin, its concentration in blood as well as its catabolism are controlled by numerous physical and physiological mechanisms (Figs. 4 and 5). The relative contribution of these mechanisms to the overall carboxy-haemoglobin kinetics will depend on the environmental conditions (ambient carbon monoxide concentration, altitude, etc.), physical activity of an individual and many other physiological processes, some of which are complex and still poorly understood.

The mass transport of carbon monoxide between the airway opening (mouth and nose) and red blood cell haemoglobin is predominantly controlled by physical processes. The carbon monoxide transfer to the haemoglobin binding sites is accomplished in two sequential steps: (1) transfer of carbon monoxide in a gas phase, between the airway opening and the alveoli; and (2) transfer in a "liquid" phase, across the air–blood interface, including the red blood cell. Although the mechanical action of the respiratory system and the molecular diffusion within the alveoli are the key mechanisms of transport in the gas phase, the diffusion of carbon monoxide across the alveoli–capillary barrier, plasma and red blood cell is the virtual mechanism in the liquid phase.

Diffusion of gases across the alveolar air–haemoglobin barrier is an entirely passive process. In order to reach the haemoglobin binding

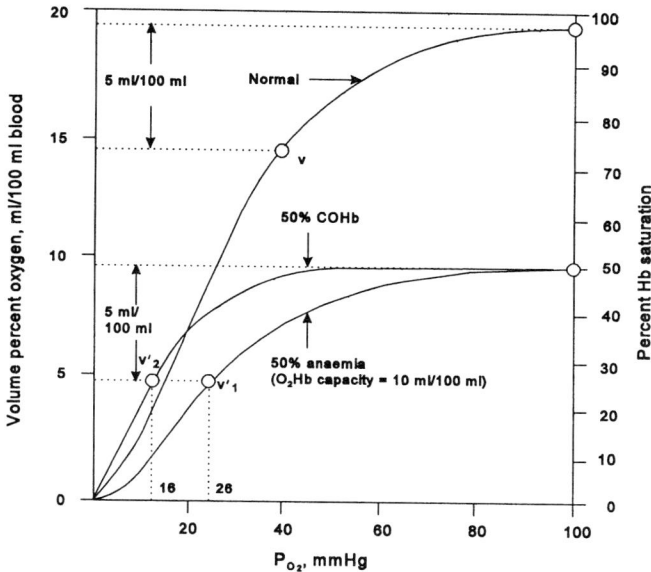

Fig. 4. Oxyhaemoglobin dissociation curves of normal human blood, of blood containing 50% carboxyhaemoglobin and of blood with a 50% normal haemoglobin concentration due to anaemia (adapted from Roughton & Darling, 1944; Rahn & Fenn, 1955; NRC, 1977).

sites, carbon monoxide and other gas molecules have to pass across the alveoli–capillary membrane, diffuse through the plasma, pass across the red blood cell membrane and finally enter the red blood cell stroma before reaction between carbon monoxide and haemoglobin can take place. The molecular transfer across the membrane and the blood phase is governed by general physicochemical laws, particularly Fick's first law of diffusion. The exchange and equilibration of gases between the two compartments (air and blood) are very rapid. The dominant driving force is a partial pressure differential of carbon monoxide across this membrane. For example, inhalation of a bolus of air containing high levels of carbon monoxide will rapidly increase blood carboxyhaemoglobin; by immediate and tight binding of carbon monoxide to haemoglobin, the partial pressure of carbon monoxide within the red blood cell is kept low, thus maintaining a high pressure differential between air and blood and consequent diffusion of carbon

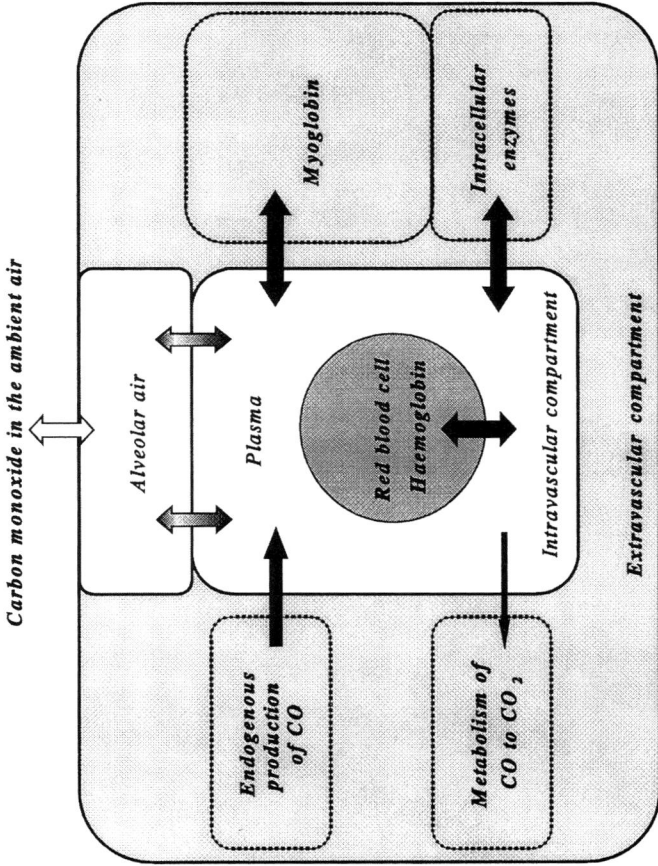

Fig. 5. Transport, distribution and metabolism of carbon monoxide in body compartments.

monoxide into blood. Subsequent inhalation of carbon monoxide-free air progressively decreases the gradient to the point of its reversal (higher carbon monoxide pressure on the blood side than in alveolar air), and carbon monoxide will be released into alveolar air. The air–blood pressure gradient for carbon monoxide is usually much higher than the blood–air gradient; therefore, the carbon monoxide uptake will be a proportionally faster process than carbon monoxide elimination. The rate of carbon monoxide release will be further affected by the products of tissue metabolism. Under pathological conditions, where one or several components of the air–blood interface might be severely affected, as in emphysema, fibrosis or oedema, both the uptake and elimination of carbon monoxide will be affected.

The rate of diffusion of gases might be altered considerably by many physiological factors acting concomitantly. Diurnal variations in carbon monoxide diffusion related to variations in haemoglobin have been reported in normal healthy subjects (Frey et al., 1987). Others found the changes to be related to physiological factors as well, such as oxyhaemoglobin, carboxyhaemoglobin, partial pressure of alveolar carbon dioxide, ventilatory pattern, oxygen consumption, blood flow and functional residual capacity (Forster, 1987). It has been confirmed repeatedly that diffusion is body position and ventilation dependent. Carbon monoxide diffusion is significantly higher in a supine position at rest than in a sitting position at rest. In both positions, carbon monoxide diffusion is greater during exercise than at rest (McClean et al., 1981). Carbon monoxide diffusion will increase with exercise, and, at maximum work rates, the diffusion will be maximal regardless of position. This increase is attained by increases in both the membrane-diffusing component and the pulmonary capillary blood flow (Stokes et al., 1981). The above physiological processes will minimally affect carboxyhaemoglobin formation in healthy individuals exposed to low and relatively uniform levels of carbon monoxide. Under such ambient conditions, these factors will be the most influential during the initial period of carbon monoxide distribution and exchange. If sufficient time is allowed for equilibration, the sole determinant of carboxyhaemoglobin concentration in blood will be the ratio of the partial pressures of carbon monoxide (P_{CO}) and oxygen (P_{O_2}). However, the shorter the half-time for equilibration (e.g., due to hyperventilation, high concentration of carbon monoxide, increased cardiac output, etc.), the more involved these mechanisms will become in modulating the rate of carbon monoxide uptake (Pace et al., 1950; Coburn et al., 1965). At high transient carbon monoxide exposures of resting individuals, both the cardiac and the lung function mechanisms will control the rate of carbon monoxide uptake. Incomplete mixing of blood might result in a substantial difference between the arterial and venous carboxy-haemoglobin concentrations (Godin & Shephard, 1972). In chronic bronchitics, asthmatics and other subpopulations at risk (pregnant women, the elderly, etc.), the kinetics of carboxyhaemoglobin formation will be even more complex, because any abnormalities of ventilation and perfusion and gas diffusion will aggravate carbon monoxide exchange between blood and air.

Although the rate of carbon monoxide binding with haemoglobin is about one-fifth slower and the rate of dissociation from haemoglobin is an order of magnitude slower than the respective rates for oxygen, the carbon monoxide chemical affinity (represented by the Haldane coefficient, M) for haemoglobin is about 245 (240–250) times greater than that for oxygen (Roughton, 1970). One part of carbon monoxide and 245 parts of oxygen would form equal parts of oxy-haemoglobin and carboxyhaemoglobin (50% of each), which would be achieved by breathing air containing 21% oxygen and 650 mg carbon monoxide/m^3 (570 ppm). Moreover, under steady-state conditions (gas exchange between blood and atmosphere remains constant), the ratio of carboxyhaemoglobin to oxyhaemoglobin is proportional to the ratio of their respective partial pressures. The relationship between the affinity constant M and P_{O_2} and P_{CO}, first expressed by Haldane (1898), has the following form:

$$COHb/O_2Hb = M\,(P_{CO}/P_{O_2}) \qquad (6\text{-}1)$$

At equilibrium, when haemoglobin is maximally saturated by oxygen and carbon monoxide at their respective gas tensions, the M value for all practical purposes is independent of pH and 2,3-diphospho-glycerate over a wide range of P_{CO}/P_{O_2} ratios. The M, however, is temperature dependent (Wyman et al., 1982).

Under dynamic conditions, competitive binding of oxygen and carbon monoxide to haemoglobin is complex; simply said, the greater the number of haems bound to carbon monoxide, the greater the affinity of free haems for oxygen. Any decrease in the amount of available haemoglobin for oxygen transport (carbon monoxide poisoning, bleeding, anaemia, blood diseases, etc.) will reduce the quantity of oxygen carried by blood to the tissue. However, carbon monoxide not only occupies oxygen binding sites, molecule for molecule, thus reducing the amount of available oxygen, but also alters the characteristic relationship between oxyhaemoglobin and P_{O_2}, which in normal blood is S-shaped. With increasing concentration of carboxy-haemoglobin in blood, the dissociation curve is shifted gradually to the left, and its shape is transformed into that of a rectangular hyperbola (Fig. 4). Because the shift occurs over a critical saturation range for release of oxygen to tissues, a reduction in oxyhaemoglobin by carbon monoxide poisoning will have more severe effects on the release of oxygen than the equivalent reduction in haemoglobin due to anaemia. Thus, in an anaemic patient (50%) at a tissue P_{O_2} of 3.5 kPa (26 torr) (v'_1), 5 vol % of oxygen (50% desaturation) might be extracted from blood, the amount sufficient to sustain tissue metabolism. In contrast, in a person poisoned with carbon monoxide (50% carboxyhaemo-globin), the tissue P_{O_2} will have to drop to 2.1 kPa (16 torr) (v'_2; severe hypoxia) to release the same 5 vol % oxygen (Fig. 4). Any higher demand on oxygen under these conditions (e.g., by exercise) might result in coma of the carbon monoxide-poisoned individual.

There are over 350 variants to normal human haemoglobin (Zink-ham et al., 1980). In the haemoglobin S variant, sickling takes place when deoxyhaemoglobin S in the red blood cell reaches a critical level and causes intracellular polymerization. Oxygenation of the haemo-globin S molecules in the polymer, therefore, should lead to a change in molecular shape, breakup of the polymer and unsickling of the cell. Carbon monoxide was considered at one time to be potentially beneficial, because it ultimately would reduce the concentration of deoxyhaemoglobin S by converting part of the haemoglobin to car-boxyhaemoglobin. Exposure to carbon monoxide, however, was not considered to be an effective clinical treatment, because high carboxy-haemoglobin levels (>20%) were required.

Other haematological disorders can cause elevated concentrations of carboxyhaemoglobin in the blood. Ko & Eisenberg (1987) studied a patient with Waldenström's macroglobulinaemia. Not only was the carboxyhaemoglobin saturation elevated, but the half-life of carboxy-haemoglobin was about 3 times longer than in a normal individual. Presumably, exogenous exposure to carbon monoxide, in conjunction with higher endogenous carbon monoxide levels, could result in critical levels of carboxyhaemoglobin. However, because carbon monoxide can also modify the characteristics of unstable haemo-globin, as demonstrated in patients with haemoglobin S, it is not known how ambient or near-ambient levels of carbon monoxide would affect individuals with these disorders.

Although the lung in its function as a transport system for gases is exposed continuously to carbon monoxide, very little carbon monoxide actually diffuses and is stored in the lung tissue itself, except for the alveolar region. The epithelium of the conductive zone (nasopharynx and large airways) presents a significant barrier to diffusion of carbon monoxide (Guyatt et al., 1981). Therefore, diffusion and gas uptake by the tissue, even at high carbon monoxide concentrations, will be exceedingly slow; most of this small amount of carbon monoxide will be dissolved in the mucosa of the airways. Diffusion into the submucosal layers and interstitium will depend on the concentration of carbon monoxide and duration of exposure. Experimental exposures of the orinasal cavity of monkeys to excep-tionally high concentrations of carbon monoxide for a short period of time increased their blood carboxyhaemoglobin level to only 1.5%. Comparative exposures of the whole lung, however, elevated carboxy-haemoglobin to almost 60% (Schoenfisch et al., 1980). Thus, diffusion of carbon monoxide across the airway mucosa will contribute extremely little, if at all, to overall carboxyhaemoglobin concentration. In the transitional zone ($\leq$20th generation), where both conductive and diffusive transport take place, diffusion of carbon monoxide into lung interstitium will be much easier, and at times more complete. In the respiratory zone (alveoli), which is the most effective interface for carbon monoxide transfer, diffusion into the lung interstitium will be complete. Because the total lung tissue mass is rather small compared with that of other carbon monoxide compartments, a relatively small amount of carbon monoxide (primarily dissolved carbon monoxide) will be distributed within the lung structures.

The role of myoglobin in oxygen transport is not yet fully understood. Myoglobin as a respiratory haemoprotein of muscular tissue will undergo a reversible reaction with carbon monoxide in a manner similar to that of oxygen. The greater affinity of oxygen for myoglobin than for haemoglobin (hyperbolic versus S-shaped dissoci-ation curve) is in this instance physiologically beneficial, because a small drop in tissue P_{O_2} will release a large amount of oxygen from oxymyoglobin. It is believed that the main function of myoglobin is to serve as a temporary store of oxygen and act as a diffusion facilitator between haemoglobin and the tissues.

Myoglobin has an affinity constant approximately 8 times lower than that of haemoglobin (M = 20–40 versus 245, respectively). As with haemoglobin, the combination velocity constant between carbon monoxide and myoglobin is only slightly lower than for oxygen, but the dissociation velocity constant is much lower than for oxygen. The combination of greater affinity (myoglobin is 90% saturated at P_{O_2} of 2.7 kPa [20 mmHg]) and lower dissociation velocity constant for carbon monoxide favours retention of carbon monoxide in the muscular tissue. Thus, a considerable amount of carbon monoxide can potentially be stored in the skeletal muscle. The ratio of carboxymyoglobin to carboxyhaemoglobin saturation for skeletal muscle of a resting dog and cat has been determined to be 0.4–0.9; for cardiac muscle, the ratio is slightly higher (0.8–1.2) (Coburn et al., 1973; Sokal et al., 1986). Prolonged exposures did not change this ratio in either muscle (Sokal et al., 1984). During exercise, the relative rate of carbon monoxide binding increases more for myoglobin than for haemoglobin, and carbon monoxide will diffuse from blood to skeletal muscle (Werner & Lindahl, 1980); consequently, the carboxymyoglobin/carboxyhaemoglobin will increase for both skeletal and cardiac muscles (Sokal et al., 1986). A similar shift in carbon monoxide has been observed under hypoxic conditions, because a fall in intracellular P_{O_2} below a critical level will increase the relative affinity of myoglobin for carbon monoxide (Coburn et al., 1971). Consequent reduction in the oxygen carrying capacity of myoglobin might have a profound effect on the supply of oxygen to the tissue.

In 1965, Coburn, Forster and Kane developed a differential equation to describe the major physiological variables that determine the concentration of carboxyhaemoglobin in blood ([$COHb$]) for the examination of the endogenous production of carbon monoxide. The equation, referred to as the Coburn-Forster-Kane, or CFK, model, is still much in use today for the prediction of blood carboxyhaemoglobin concentration consequent to inhalation of carbon monoxide (Fig. 6), for two reasons. First, the model is quite robust to challenges to the original assumptions. Second, the model can be relatively easily adapted to more specialized applications.

Eq. 6-2 represents the CFK model:

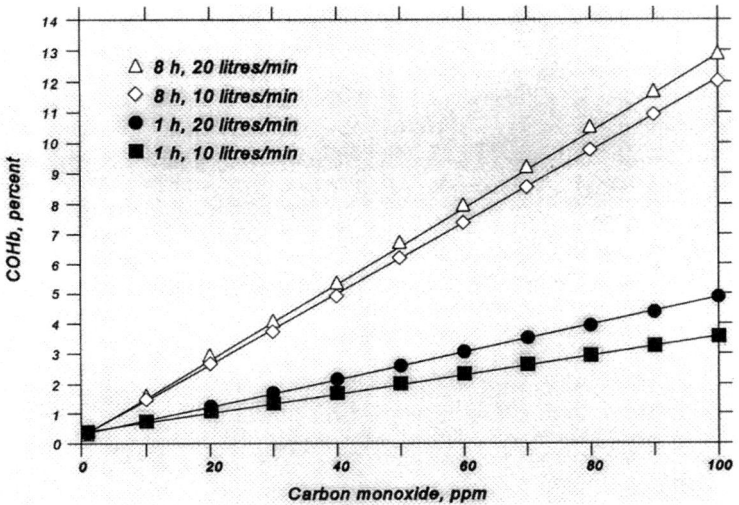

Fig. 6. Relationship between carbon monoxide concentration and carboxyhaemoglobin levels in blood at four different conditions of exposure. Predicted carboxyhaemoglobin levels resulting from 1- and 8-h exposures to carbon monoxide at rest (alveolar ventilation rate of 10 litres/min) and with light exercise (20 litres/min) are based on the CFK equation (Coburn et al., 1965) using the following assumed parameters for non-smoking adults: altitude = 0 ft (0 m), initial carboxyhaemoglobin level = 0.5%, Haldane coefficient = 218, blood volume = 5.5 litres, haemoglobin level = 15 g/100 ml, lung diffusivity = 30 ml/torr (0.23 ml/Pa) per minute, endogenous rate = 0.007 ml/min.

$$V_B d[COHb]/dt = \dot{V}_{CO} - [COHb]\bar{P}_c O_2/MB[O_2Hb] + P_I CO/B \quad (6\text{-}2)$$

where:

$$B = 1/D_L CO + P_L/\dot{V}_A$$

and V_B is the blood volume in millilitres (5500 ml), $[COHb]$ represents millilitres of carbon monoxide per millilitre of blood, $\dot{V}_{CO}$ is the endogenous carbon monoxide production in millilitres per minute (0.007 ml/min), $\bar{P}_c O_2$ is the average partial pressure of oxygen in the lung capillaries in millimetres of mercury (100 mmHg), M is the Haldane affinity ratio (218), $[O_2Hb]$ represents millilitres of oxygen per millilitre of blood (the maximum oxygen capacity of blood is 0.2), $P_I CO$ is the partial pressure of carbon monoxide in inhaled air in millimetres of mercury, $D_L CO$ is the pulmonary diffusing capacity for carbon monoxide in millilitres per minute per millimetre of mercury (30 ml/min per mmHg), P_L is the pressure of dry gases in the lungs in millimetres of mercury (713 mmHg) and $\dot{V}_A$ is the alveolar ventilation rate in millilitres per minute (6000 ml/min).

Under the assumption that $[O_2Hb]$ is constant, Eq. 6-2 is linear. In this case, the equation is restricted to relatively low carboxyhaemoglobin levels. For higher levels, the reduction in oxyhaemoglobin with increasing carboxyhaemoglobin must be taken into account, thus making Eq. 6-2 non-linear. The values in parentheses indicated for the variables of Eq. 6-2 are the values given in Peterson & Stewart (1970), although it is not clear whether a consistent set of conditions (i.e., body temperature and pressure, saturated with water vapour, or standard temperature and pressure, dry) was used. In addition, Peterson & Stewart (1970) assumed a constant value of $[O_2Hb]$, thus making Eq. 6-2 linear. Restricting the conditions to low carbon monoxide exposures allows the mathematical assumption of instant equilibration of (1) the gases in the lungs, (2) carboxyhaemoglobin concentrations between venous and arterial blood and (3) carboxyhaemoglobin concentrations between the blood and carbon monoxide stores in non-vascular tissues.

Not all physiological variables influence equally the rate of formation and the equilibrium value of blood carboxyhaemoglobin. Moreover, the measurement of estimate errors in these variables might produce errors in calculated blood carboxyhaemoglobin levels. By performing sensitivity analysis for both the linearized and the non-linear forms of the CFK equation, assuming several different work-loads, McCartney (1990) determined the effects that the errors in the variable values might have on the calculated value of $[COHb]$. Fig. 7 graphs a temporal behaviour of fractional sensitivities (F) of the key determinants of $[COHb]$. In this figure, the linearized form of the CFK equation was used to develop the curves for a series of variables, assuming a fixed minute ventilation (34.5 litres/min) and constant ambient concentration of carbon monoxide (110 mg/m^3 [100 ppm]). The fractional sensitivity of ± 1 means that a 1% error in the value of a variable will induce a 1% error in the calculated value of $[COHb]$ at equilibrium. As seen from the figure, the values of V_B, $\dot{V}_A$ and D_LCO (not graphed) have no influence on the initial or the equilibrium carboxyhaemoglobin concentration but affect the rate at which the

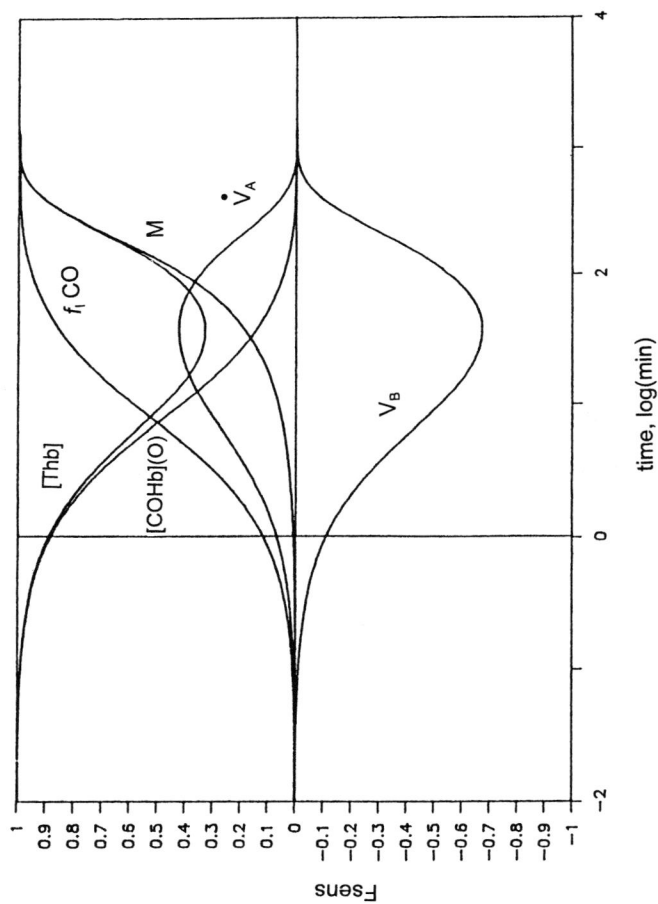

Fig. 7. Fractional sensitivity of carboxyhaemoglobin to the key carboxyhaemoglobin determinants as a function of time (see text for details) (from McCartney, 1990).

equilibrium is achieved. The error in these variables does not affect the calculated equilibrium blood carboxyhaemoglobin concentration. All the other variables, however, affect both the carboxyhaemoglobin equilibrium and the rate at which it is approached. Incorrect determination of the initial value of $[COHb]$ and $[THb]$ (total blood concentration of haemoglobin) will induce carboxyhaemoglobin concentration errors during buildup. While an error in the initial value of $[COHb]$ progressively loses its influence on the calculated $[COHb]$, for $[THb]$ the impact of this error on $[COHb]$ will be diminished only transiently, regaining its full effect at equilibrium. In contrast to the initial value of $[COHb]$, an error in M, $\bar{P}_cO_2$ or F_lCO will attain its maximum influence at equilibrium.

Tikuisis et al. (1992) studied the rate of formation of carboxyhaemoglobin in healthy young males at a low (approximately 45 W) and moderate (approximately 90 W) exercise load. Individuals were exposed to 3400 mg carbon monoxide/m^3 (3000 ppm) for 3 min at rest followed by three intermittent exposures, ranging from 3400 mg carbon monoxide/m^3 (3000 ppm) for 1 min at low exercise to 764 mg carbon monoxide/m^3 (667 ppm) at moderate exercise. The CFK equation underpredicted the increase in carboxyhaemoglobin for the exposures at rest and the first exposure at exercise, whereas it overpredicted the carboxyhaemoglobin increase for the latter two exposures at exercise. The carboxyhaemoglobin concentration after all exposures reached approximately 10%. The measured and predicted carboxyhaemoglobin values differed by <1 percentage point. The slight shift of the measured carboxyhaemoglobin dissociation curve from the predicted curve was attributed to a delay in the delivery of carboxyhaemoglobin to the blood sampling point, a dorsal hand vein.

Benignus et al. (1994) extended the observations of Tikuisis et al. (1992). They exposed 15 men to 7652 mg C^{18}O/m^3 (6683 ppm) for 3.1–6.7 min at rest. Both arterial and venous blood carboxyhaemoglobin levels were determined frequently during and for 10 min following the exposures. Except for the Haldane constant (M), which was assumed to be 245, all other physiological parameters of the CFK equation were measured for each individual from the very beginning. Arterial carboxyhaemoglobin was considerably higher than the venous carboxyhaemoglobin. The rate of increase in blood carboxyhaemoglobin and the arterial–venous carboxyhaemoglobin differences varied widely among individuals. The peak arterial carboxyhaemoglobin concentration at the end of exposure ranged from 13.9 to 20.9%. The peak venous carboxyhaemoglobin concentration reached during the recovery period ranged from 12.4 to 18.1%. The arterial–venous carboxyhaemoglobin difference ranged from 2.3 to 12.1% carboxy-haemoglobin. These increases in venous and arterial carboxyhaemo-globin were not predicted accurately by the CFK equation. Venous blood carboxyhaemoglobin levels were overestimated, whereas arterial blood carboxyhaemoglobin levels were significantly and consistently underestimated by the CFK equation. Thus, exposure of such organs as brain or heart to carboxyhaemoglobin may be sub-stantially higher than expected during transient carbon monoxide exposure.

Recently, Singh et al. (1991) developed a mathematical model for carbon monoxide uptake. The model was extended by Selvakumar et al. (1993) to include the elimination phase as well. A fixed-point interactive approach was used to solve the non-linear functions. The authors found good agreement between experimental data and predicted values. They also claimed that their model gives better approximation of real carboxyhaemoglobin values than the CFK model.

To more accurately predict differences between arterial and venous carboxyhaemoglobin during transient exposures to carbon monoxide, Smith et al. (1994) modified the CFK equation to take into account differences in regional blood flow, particularly the arm. The predicted carboxyhaemoglobin values for both venous and arterial blood based on the Smith model were very close to the measured values obtained from experimental subjects exposed to carbon monoxide (up to 22% carboxyhaemoglobin).

Hill et al. (1977) provided a figure of predicted values of blood carboxyhaemoglobin (%) in a mother and fetus during prolonged exposures of the mother to carbon monoxide (34–340 mg/m^3 [30–300 ppm]) and subsequent washout with no carbon monoxide in the air (Fig. 8).

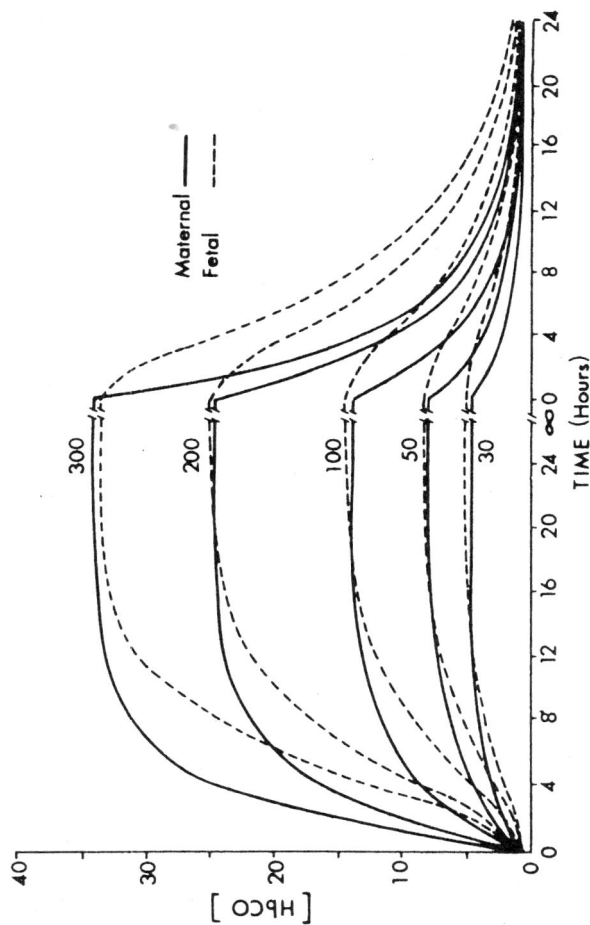

Fig. 8. Predicted (CFK model) buildup of blood carboxyhaemoglobin (%) in a mother and fetus during prolonged exposures of mother to carbon monoxide (34–340 mg/m³ [30–300 ppm]) and subsequent washout with no carbon monoxide in the air (from Hill et al., 1977).

6.4 Distribution of carbon monoxide

6.4.1 *Intracellular effects of carbon monoxide*

The principal cause of carbon monoxide toxicity at low exposure levels is thought to be tissue hypoxia due to carbon monoxide binding to haemoglobin; however, certain physiological aspects of carbon monoxide exposure are not well explained by decreases in intracellular P_{O_2} related to the presence of carboxyhaemoglobin. For many years, it has been known that carbon monoxide is distributed to extravascular sites such as skeletal muscle (Coburn et al., 1971, 1973) and that 10–50% of the total body store of carbon monoxide may be extravascular (Luomanmaki & Coburn, 1969). Furthermore, extravascular carbon monoxide is metabolized slowly to carbon dioxide *in vivo* (Fenn, 1970). Consequently, secondary mechanisms of carbon monoxide toxicity related to intracellular uptake of carbon monoxide have been the focus of a great deal of research interest. Carbon monoxide binding to many intracellular compounds has been well documented both *in vitro* and *in vivo*; however, it is still uncertain whether or not intracellular uptake of carbon monoxide in the presence of haemoglobin is sufficient to cause either acute organ system dysfunction or long-term health effects. The virtual absence of sensitive techniques capable of assessing intracellular carbon monoxide binding under physiological conditions has resulted in a variety of indirect approaches to the problem, as well as many negative studies.

Carbon monoxide is known to react with a variety of metal-containing proteins found in nature. Carbon monoxide-binding metalloproteins present in mammalian tissues include oxygen carrier proteins such as haemoglobin (Douglas et al., 1912) and myoglobin (Antonini & Brunori, 1971) as well as metalloenzymes (oxidoreductases) such as cytochrome *c* oxidase (Keilin & Hartree, 1939), cytochromes of the P-450 type (Omura & Sato, 1964), tryptophan oxygenase (Tanaka & Knox, 1959) and dopamine hydroxylase (Kaufman, 1966). These metalloproteins contain iron and/or copper centres at their active sites that form metal–ligand complexes with carbon monoxide in competition with molecular oxygen. Carbon monoxide and oxygen form complexes with metalloenzymes only when the iron and copper are in their reduced forms (Fe(II), Cu(I)). Caughey (1970) reviewed the similarities and differences in the physicochemical characteristics of carbon monoxide and oxygen binding to these transition metal ions. The competitive relationship between carbon monoxide and oxygen for the active site of intracellular haemoproteins is usually described by the Warburg partition coefficient (R), which is the carbon monoxide/oxygen ratio that produces 50% inhibition of the oxygen uptake of the enzyme, or, in the case of myoglobin, a 50% decrease in the number of available oxygen binding sites.

The measured Warburg coefficients of various mammalian carbon monoxide-binding proteins have been tabulated recently by Coburn & Forman (1987) (see Table 13). These R values range from approximately 0.025 for myoglobin to 0.1–12 for cytochrome P-450. Warburg coefficient values of 2–28 have been reported for cytochrome c oxidase (Keilin & Hartree, 1939; Wohlrab & Ogunmola, 1971; Wharton & Gibson, 1976). By comparison, the R value for human haemoglobin of 0.0045 is some 3 orders of magnitude less than that of cytochrome c oxidase. This means, for example, that carbon monoxide would bind to cytochrome oxidase *in vivo* only if oxygen gradients from red blood cells in the capillary to the mitochondria were quite steep. Application of R values for intracellular haemoproteins in this way, however, merits caution, because most measurements of carbon monoxide binding have not been made at physiological temperatures or at relevant rates of electron transport.

Apart from questions about the relevance of extrapolating *in vitro* partition coefficients to physiological conditions, experimental problems arise that are related to determining actual carbon monoxide/oxygen ratios in intact tissues. Reasonably good estimates of tissue P_{CO} may be obtained by calculating the value in mean capillary blood from the Haldane relationship (NRC, 1977), neglecting the low rate of carbon monoxide metabolism by the tissue. Experimental estimates of the P_{CO} in animal tissues have been found to be in close agreement with these calculations and average slightly less than alveolar P_{CO} (Goethert et al., 1970; Goethert, 1972). In general, steady-state estimates for tissue P_{CO} range from 3 to 70 Pa (0.02 to 0.5 torr) at carboxyhaemoglobin concentrations of 5–50%. Therefore, at 50% carboxyhaemoglobin, a carbon monoxide/oxygen ratio of 5 may be achieved at sites of intracellular oxygen uptake only if tissue P_{O_2} in the vicinity of the carbon monoxide-binding proteins is approximately 10 Pa (0.1 torr).

Table 13. *In vitro* inhibition ratios for metalloproteins that bind carbon monoxide[a]

Haemoprotein	Source	R[b]	M[c]	Temperature (°C)
Haemoglobin	Human red blood cell	0.0045	218	37
Myoglobin	Sperm whale	0.025–0.040	25–40	25
Cytochrome c oxidase	Bovine heart	5–15	0.1–0.2	25
Cytochrome P-450	Rat liver	0.1–12	10–0.1	30–37
Dopamine β-hydroxylase	Bovine adrenal	2	0.5	–
Tryptophan oxygenase	*Pseudomonas*	0.55	1.8	25

[a] Adapted from Coburn & Forman (1987).
[b] $R = CO/O_2$ at 50% inhibition.
[c] $M = 1/R$.

Whether such low intracellular P_{O_2} values exist in target tissues such as brain and heart during carbon monoxide exposure is difficult to determine from the existing scientific literature. Experimental measurements of tissue P_{O_2} using polarographic microelectrodes indicate a significant range of P_{O_2} values in different tissues and regional differences in P_{O_2} within a given tissue. This normal variability in tissue P_{O_2} is related to differences in capillary perfusion, red blood cell spacing, velocity and path length, and local requirements for oxygen. Normal P_{O_2} values obtained from such recordings are generally in the range of 0–4 kPa (0–30 torr) (Leniger-Follert et al., 1975). These P_{O_2} values usually represent average interstitial values, although it is often difficult to determine the exact location of the electrode and the effect of oxygen consumption by the electrode on the P_{O_2} measurement. Furthermore, the gradient between the capillary and the intracellular sites of oxygen utilization is thought to be quite steep (Sies, 1977). A major component of the gradient arises between the red blood cell and interstitium (Hellums, 1977), but the P_{O_2} gradient between the cell membrane and respiring mitochondria and other oxygen-requiring organelles remains undetermined in intact normal tissues. Even less is known about intracellular P_{O_2} in the presence of carboxyhaemoglobin. It has been determined, however, that both P_{O_2} in brain tissue (Zorn, 1972) and cerebrovenous P_{O_2} (Koehler et al., 1984) decrease linearly as a function of carboxyhaemoglobin concentration. Presumably, then, intracellular P_{O_2} declines with increasing carboxyhaemoglobin concentration, and, at certain locations, carbon monoxide forms ligands with the oxygen-dependent, intracellular haemoproteins. As the intracellular P_{O_2} decreases, the carbon monoxide/oxygen ratio in the tissue increases at constant P_{CO}, and an increasing fraction of the available intracellular oxygen binding sites become occupied by carbon monoxide.

The intracellular uptake of carbon monoxide behaves generally according to the preceding principles; most of the experimental evidence for this line of reasoning was derived from *in vivo* studies of carboxyhaemoglobin formation by Coburn and colleagues (1965). For all intracellular haemoproteins, however, two crucial quantitative unknowns remain: (1) the fraction of intracellular binding sites in discrete tissues inhibited by carbon monoxide at any level of carboxyhaemoglobin saturation, and (2) the critical fraction of inhibited sites necessary to amplify or initiate a deleterious physiological effect or trigger biochemical responses with long-term health effects. In general, then, the activities of certain intracellular haemoproteins may be altered at physiologically tolerable levels of carboxyhaemoglobin. The problem is in determining what level of intracellular reserve is available during carbon monoxide hypoxia.

6.4.2 Carbon monoxide binding to myoglobin

The red protein myoglobin is involved in the transport of oxygen from capillaries to mitochondria in red muscles. The binding of carbon monoxide to myoglobin in heart and skeletal muscle *in vivo* has been demonstrated at levels of carboxyhaemoglobin below 2% in heart and 1% in skeletal muscle (Coburn & Mayers, 1971; Coburn et al., 1973). The carboxymyoglobin/carboxyhaemoglobin saturation ratio has been found to be approximately 1 in cardiac muscle and less than 1 in skeletal muscle. These ratios did not increase with increases in carboxyhaemoglobin up to 50% saturation. In the presence of hypoxaemia and hypoperfusion, the amount of carbon monoxide uptake by myoglobin has been measured and was shown to increase (Coburn et al., 1971, 1973). A similar conclusion has been reached during maximal exercise in humans, where carbon monoxide shifts from haemoglobin to the intracellular compartment (i.e., myoglobin, at carboxyhaemoglobin levels of 2–2.5%) (Clark & Coburn, 1975). The significance of carbon monoxide uptake by myoglobin is uncertain, because our understanding of the functional role of myoglobin in working muscle is incomplete. Myoglobin undoubtedly enhances the uptake of oxygen by muscle cells so that the continuous oxygen demand of working muscle is satisfied (Wittenberg et al., 1975). Myoglobin may contribute to muscle function by serving as an oxygen store, by enhancing intracellular diffusion of oxygen or by acting as an oxygen buffer to maintain a constant mitochondrial P_{O_2} during changes in oxygen supply. Functional myoglobin has been found to be necessary for maintenance of maximum oxygen uptake and mechanical tension in exercising skeletal muscle (Cole, 1982). The binding of carbon monoxide to myoglobin would therefore be expected to limit oxygen availability to mitochondria in working muscle. This possibility has been verified theoretically by computer simulations of Hoofd & Kreuzer (1978) and Agostoni et al. (1980). The three-compartment (arterial and venous capillary blood and myoglobin) computer model of Agostoni et al. (1980) predicted that carboxymyoglobin formation in low P_{O_2} regions of the heart (e.g., subendocardium) could be sufficient to impair intracellular oxygen transport to mitochondria at carboxyhaemoglobin saturations of 5–10%. The concentration of carboxymyoglobin was also predicted to increase during conditions of hypoxia, ischaemia and increased oxygen demand.

Wittenberg & Wittenberg (1993) investigated the effect of increased carboxymyoglobin in cardiac muscle cells and subsequent decreases in myoglobin-mediated oxidative phosphorylation. Isolated rat cardiac myocytes were maintained near the intracellular P_{O_2} of the working heart (0.3–0.7 kPa [2–5 torr]) and near the end-venous P_{O_2} (2.7 kPa [20 torr]) and exposed to low-pressure carbon monoxide (0.009–9 kPa [0.07–70 torr]; 100–100 000 mg/m^3 [90–90 000 ppm]). The fraction of intracellular myoglobin, determined spectrophotometrically, was in agreement with the fraction predicted from the ratio of carbon monoxide to oxygen partial pressures. When carboxymyoglobin was at least 40% of the total intracellular myoglobin, the rate of oxidative phosphorylation was significantly lower than that in control cells from the same preparation. The authors estimated that this result would be achieved when arterial carboxyhaemoglobin levels reached approximately 20–40%. In previous studies by the authors (Wittenberg & Wittenberg, 1985, 1987), spectrophotometry showed no evidence of carbon monoxide binding to cytochrome oxidase, the only mitochondrial component known to bind with carbon monoxide, until very high levels of carboxymyoglobin were reached (almost

100%). Therefore, the effect of carbon monoxide on cellular respiration in these experiments was apparently not due to inhibition of cytochrome oxidase.

Increases in cardiac carboxymyoglobin levels have been measured after heavy workloads in carbon monoxide-exposed rats, independent of changes in carboxyhaemoglobin concentration (Sokal et al., 1986). These investigators reported that exercise significantly increased cardiac carboxymyoglobin at carboxyhaemoglobin saturations of approximately 10, 20 and 50%, although metabolic acidosis worsened only at 50% carboxyhaemoglobin. It remains unknown, however, whether or not low carboxymyoglobin could be responsible for decreases in maximal oxygen uptake during exercise reported at carboxyhaemoglobin levels of 4–5%.

6.4.3 *Carbon monoxide uptake by cytochrome P-450*

Mixed-function oxidases (cytochrome P-450) are involved in the detoxification of a number of drugs and steroids by "oxidation." These enzymes are distributed widely throughout mammalian tissues; the highest concentrations are found in the microsomes of the liver, adrenal gland and lungs of some species (Estabrook et al., 1970). These oxidases are also present in low concentrations in kidney and brain tissues. Mixed-function oxidases catalyse a variety of reactions (e.g., hydroxylation) involving the uptake of a pair of electrons from NADPH with reduction of one atom of oxygen to water and incorporation of the other into substrates (White & Coon, 1980). These enzymes bind carbon monoxide, and their binding constant values range from 0.1 to 12 *in vitro* (see Coburn & Forman, 1987). The sensitivity of cytochrome P-450 to carbon monoxide is increased under conditions of rapid electron transport (Estabrook et al., 1970); however, previous calculations have indicated that tissue P_{CO} is too low to inhibit the function of these haemoproteins *in vivo* at less than 15–20% carboxyhaemoglobin (Coburn & Forman, 1987). There have been few attempts to measure carbon monoxide binding coefficients for these enzymes in intact tissues. In isolated rabbit lung, the effects of carbon monoxide on mixed-function oxidases are consistent with a *K* of approximately 0.5 (Fisher et al., 1979). Carbon monoxide exposure decreases the rate of hepatic metabolism of hexobarbital and other drugs in experimental animals (Montgomery & Rubin, 1973; Roth & Rubin, 1976a,b). These effects of carbon monoxide on xenobiotic metabolism appear to be attributable entirely to carboxyhaemoglobin-related tissue hypoxia, because they are no greater than the effects of "equivalent" levels of hypoxic hypoxia. Three optical studies of rat liver perfused *in situ* with haemoglobin-free buffers have demonstrated uptake of carbon monoxide by cytochrome P-450 at carbon monoxide/oxygen ratios of 0.03–0.10 (Sies & Brauser, 1970; Iyanagi et al., 1981; Takano et al., 1985). In the study by Takano et al. (1985), significant inhibition of hexobarbital metabolism was found at a carbon monoxide/oxygen ratio of about 0.1. This carbon monoxide/oxygen ratio, if translated directly to carboxyhaemoglobin concentration, would produce a carboxyhaemoglobin level that is incompatible with survival (~95%). At present, there is no scientific evidence that carbon monoxide significantly inhibits the activity of mixed-function oxidases at carboxyhaemoglobin saturations below 15–20%. Although most studies do not indicate effects of carbon monoxide on cytochrome P-450 activity at physiologically relevant carbon monoxide concentrations, specific P-450 isoenzymes may have higher affinities for carbon monoxide. Also, the rate of substrate metabolism and substrate type may increase carbon monoxide binding by P-450 enzymes.

6.4.4 Carbon monoxide and cytochrome c oxidase

Cytochrome c oxidase, also known as cytochrome aa_3, is the terminal enzyme in the mitochondrial electron transport chain that catalyses the reduction of molecular oxygen to water.

Although the enzyme complex binds carbon monoxide, three reasons are often cited for why there are still considerable uncertainties as to this occurrence only under conditions of severe hypoxia. The reasons for this difficulty centre around differences in the redox behaviour of cytochrome oxidase *in vivo* relative to its *in vitro* behaviour. The enzyme has a high resting reduction level at normal P_{O_2} in brain (Jobsis et al., 1977) and other tissues, and its oxidation state varies directly with P_{O_2} *in vitro* (Kreisman et al., 1981). These findings may indicate that the oxidase operates near its effective K_M *in vivo* or that the availability of oxygen to each mitochondrion or respiratory chain is not continuous under most physiological circumstances. There may also be differences in or regulation of the K_M for oxygen of the enzyme according to regional metabolic conditions. For example, the apparent K_M for oxygen of cytochrome oxidase increases several times during rapid respiration (Oshino et al., 1974); in isolated cells, it varies as a function of the cytosolic phosphorylation potential (Erecinska & Wilson, 1982). Conditions of high respiration and/or high cytosolic phosphorylation potential *in vitro* increase the concentration of carbon monoxide–cytochrome oxidase at any carbon monoxide/oxygen value. This concept is particularly relevant for tissues like the heart and brain.

Enhanced sensitivity of cytochrome oxidase to carbon monoxide has been demonstrated in uncoupled mitochondria, where carbon monoxide/oxygen ratios as low as 0.2 delay the oxidation of reduced cytochrome oxidase in transit from anoxia to normoxia (Chance et al., 1970). Several studies of respiring tissues, however, have found carbon monoxide/oxygen ratios of 12–20 to be necessary for 50% inhibition of oxygen uptake (Coburn et al., 1979; Kidder, 1980; Fisher & Dodia, 1981). In this context, it is important to note that in a given tissue, the carbon monoxide/oxygen ratio necessary to inhibit one-half of the oxygen uptake does not necessarily correspond to carbon monoxide binding to one-half of the oxidase molecules. This is because unblocked cytochrome oxidase molecules may oxidize respiratory complexes of blocked chains, thus causing the oxygen consumption to fall more slowly than predicted for strictly linear systems. The capacity of tissues to compensate for electron transport inhibition by branching has not been investigated systematically as a function of P_{O_2}, carbon monoxide/oxygen ratio, cytosolic phosphorylation potential or rate of electron transport *in vivo*.

The contention that intracellular carbon monoxide uptake by cytochrome oxidase occurs is supported by a few experiments. It has been known for many years, primarily through the work of Fenn (Fenn & Cobb, 1932; Fenn, 1970), that carbon monoxide is slowly oxidized in the body to carbon dioxide. This oxidation occurs normally at a much lower rate than the endogenous rate of carbon monoxide production; however, the rate of oxidation of carbon monoxide increases in proportion to the carbon monoxide body store (Luomanmaki & Coburn, 1969). The oxidation of carbon monoxide to carbon dioxide was shown in 1965 by Tzagoloff & Wharton to be catalysed by reduced cytochrome oxidase. In addition, Young et al. (1979) demonstrated that oxidized cytochrome oxidase promotes carbon monoxide oxidation and, subsequently, that cytochrome oxidase in intact heart and brain mitochondria was capable of catalysing the reaction at a carbon monoxide/oxygen ratio of approximately 4 (Young &

Caughey, 1986). The physiological significance of this reaction is unknown.

Optical evidence suggesting that cytochrome oxidase is sensitive to carbon monoxide *in vivo* comes from studies of the effects of carbon monoxide on cerebrocortical cytochromes in fluorocarbon-perfused rats (Piantadosi et al., 1985, 1987). In these studies, carbon monoxide/oxygen ratios of 0.006–0.06 were associated with spectral evidence of carbon monoxide binding to reduced cytochrome oxidase. The spectral data also indicated that the intracellular uptake of carbon monoxide produced increases in the reduction level of *b*-type cytochromes in the brain cortex. At a carbon monoxide/oxygen ratio of 0.06, most (>80%) of the cytochrome *b* became reduced in the cerebral cortex. The cytochrome *b* response is not well understood; it is thought to represent an indirect (e.g., energy-dependent) response of mitochondrial *b*-cytochromes to carbon monoxide, because these cytochromes are not known to bind carbon monoxide *in situ*. The carbon monoxide/oxygen ratio used in the studies of Piantadosi et al. (1985, 1987) would produce carboxyhaemoglobin levels in the range of 50–90%. The venous P_{O_2} in those experiments, however, was about 13 kPa (100 torr); thus, at tissue P_{O_2}s that are significantly lower, this effect should occur at lower carboxyhaemoglobin saturations. It is unlikely, however, that cerebral uptake of carbon monoxide is significant at carboxyhaemoglobin below 5%, because tissue P_{CO} is so low in the presence of haemoglobin. The physiological significance of these effects of carbon monoxide has not yet been determined.

Direct effects of carbon monoxide on mitochondrial function have been suggested by several studies that indicate decreases in cytochrome oxidase activity by histochemistry in brain and heart after severe carbon monoxide intoxication in experimental animals (Savolainen et al., 1980; Somogyi et al., 1981; Pankow & Ponsold, 1984). The magnitude of the decrease in cytochrome oxidase activity may exceed that associated with severe hypoxia, although problems of determining "equivalent" levels of carbon monoxide hypoxia and hypoxic hypoxia have not been addressed adequately by these studies.

Dichloromethane belongs to the group of xenobiotics that are metabolized to carbon monoxide. Dichloromethane-derived carbon monoxide is produced by means of cytochrome P-450 2E1, bound to membranes of the endoplasmic reticulum of liver cells. The carbon monoxide may be distributed among various cell fractions, where it reacts with haemoproteins within the cells. Subsequently, it forms carboxyhaemoglobin. The cytochrome oxidase activities in different tissues of rats were studied by oral administration of 12.4 mmol dichloromethane/kg body weight (about 10% carboxyhaemoglobin 6 h after gavage) and by inhalation exposure to 883 000 mg dichloromethane/m^3 (250 000 ppm) for 20 s (3–4% carboxyhaemoglobin after 2 h). Six hours after dichloromethane ingestion, the cytochrome oxidase activity was reduced in the brain, lung and muscle by 28–42%. Twenty minutes after inhalation of dichloromethane, it was reduced by 42–51% in brain, liver, kidney and muscle. These effects were reversible. Pretreatment with the specific inhibitor of cytochrome P-450 2E1, diethyldithiocarbamate, prevented these effects. These findings indicate that the effect of dichloromethane is caused by the dichloromethane metabolite, carbon monoxide (Lehnebach et al., 1995).

The effects of passive cigarette smoking on oxidative phosphorylation in myocardial mitochondria have been studied in rabbits (Gvozdjakova et al., 1984). Mitochondrial respiratory rate (State 3 and State 4) and rates of oxidative phosphorylation were found to be decreased significantly by carboxyhaemoglobin concentrations of 6–7%. These data, however, are not definitive with respect to carbon monoxide, because they include effects of nicotine, which reached concentrations of 5.7 µg/litre in blood. A recent study by Snow et al. (1988) in dogs with prior experimental myocardial infarction indicated that a carboxyhaemoglobin concentration of 9.4% increased the resting reduction level of cytochrome oxidase in the heart. The carbon monoxide exposures were also accompanied by more rapid cytochrome oxidase reductions after coronary artery occlusion and less rapid reoxidation of the enzyme after release of the occlusion. The authors concluded that carbon monoxide trapped the oxidase in the reduced state during transient cardiac ischaemia. There is also evidence that formation of the carbon monoxide–cytochrome oxidase ligand occurs in the brain of the rat at carboxyhaemoglobin saturations of 40–50% (Brown & Piantadosi, 1990). This binding appears to be related to hypotension and probable cerebral hypoperfusion during carbon monoxide exposure. This effect is in concert with experimental evidence that carbon monoxide produces direct vasorelaxation of smooth muscle. This vasodilation occurs in rabbit aorta (Coburn et al., 1979), in the coronary circulation of isolated perfused rat heart

(McFaul & McGrath, 1987) and in the cerebral circulation of the fluorocarbon-perfused rat (Piantadosi et al., 1987). The mechanism of this vasodilator effect is unclear.

6.5 Elimination

The large number of data available on the rate of carbon monoxide uptake and the formation of carboxyhaemoglobin contrasts sharply with the limited information available on the dynamics of carbon monoxide washout from body stores and blood. Although the same factors that govern carbon monoxide uptake will affect carbon monoxide elimination, the relative importance of these factors might not be the same (Peterson & Stewart, 1970; Landaw, 1973). Both the formation and the decline of carboxyhaemoglobin fit a second-order function best, increasing during the uptake period and decreasing during the elimination period. Hence, an initial rapid decay will gradually slow down (Stewart et al., 1970; Landaw, 1973; Wagner et al., 1975). The elimination rate of carbon monoxide from an equilibrium state will follow a monotonically decreasing second-order (logarithmic or exponential) function (Pace et al., 1950). The rate, however, might not be constant following transient exposures to carbon monoxide where, at the end of exposure, the steady-state conditions were not reached yet. In this situation, particularly after very short and high carbon monoxide exposures, it is possible that carboxyhaemoglobin decline could be biphasic, and it can be approximated best by a double-exponential function: the initial rate of decline or "distribution" might be considerably faster than the later "elimination" phase (Wagner et al., 1975). The reported divergence of the carboxyhaemoglobin decline rate in blood and in exhaled air suggests that the carbon monoxide elimination rates from extravascular pools are slower than that reported for blood (Landaw, 1973). Although the absolute elimination rates are associated positively with the initial concentration of carboxyhaemoglobin, the relative elimination rates appear to be independent of the initial concentration of carboxyhaemoglobin (Wagner et al., 1975).

The half-time of carbon monoxide disappearance from blood under normal recovery conditions while breathing air showed considerable between-individual variance. For carboxyhaemoglobin concentrations of 2–10%, the half-time ranged from 3 to 5 h (Landaw, 1973); others reported the range to be 2–6.5 h for slightly higher initial concentrations of carboxyhaemoglobin (Peterson & Stewart, 1970).

The half-time of carboxyhaemoglobin elimination was prolonged in dogs with acute lung injury (Wu, 1992). The elimination of carbon monoxide following exposure of rabbits to 2–2.9% carbon monoxide for 40, 60 or 300 s (69–82% carboxyhaemoglobin at the end of the 300-s exposure) was not linear and showed a typical three-phase elimination pattern. The authors explained the third phase, about 90 min after exposure, by endogenous carbon monoxide production (Wazawa et al., 1996).

Increased inhaled concentrations of oxygen accelerated elimination of carbon monoxide; by breathing 100% oxygen, the half-time was shortened by almost 75% (Peterson & Stewart, 1970). The elevation of P_{O_2} to 300 kPa (3 atm) reduced the half-time to about 20 min, which is approximately a 14-fold decrease over that seen when breathing room air (Landaw, 1973; Britten & Myers, 1985). Although the washout of carbon monoxide can be somewhat accelerated by an admixture of 5% carbon dioxide in oxygen, hyperbaric oxygen treatment is more effective in facilitating displacement of carbon monoxide.

7. EFFECTS ON LABORATORY ANIMALS

7.1 Introduction

Concerns about the potential health effects of exposure to carbon monoxide have been addressed in extensive studies with various animal species. Under varied experimental protocols, considerable information has been obtained on the toxicity of carbon monoxide, its direct effects on the blood and other tissues, and the manifestations of these effects in the form of changes in organ function. Many of these studies, however, have been conducted at extremely high levels of carbon monoxide (i.e., levels not found in ambient air). Although severe effects from exposure to these high levels of carbon monoxide are not directly germane to the effects from exposure to current ambient levels of carbon monoxide, they can provide valuable information about potential effects of accidental exposure to carbon monoxide, particularly those exposures occurring indoors. These higher-level studies, therefore, are being considered in this chapter only if they extend dose–response information or if they provide clues to other potential health effects of carbon monoxide that have not been identified already. In this document, emphasis has been placed on studies conducted at ambient or near-ambient concentrations of carbon monoxide that have been published in the peer-reviewed literature.

The effects observed from non-human experimental studies have provided some insight into the role that carbon monoxide plays in cellular metabolism. Caution must be exercised, however, in extrapolating the results obtained from these data to humans. Not only are there questions related to species differences, but exposure conditions differ markedly in the studies conducted by different investigators. Information on the carboxyhaemoglobin levels achieved and the duration of exposure utilized in the studies will be provided in the text or tables if it was available in the original manuscript. Where this information is lacking, only the carbon monoxide levels (milligrams per cubic metre and parts per million) will be reported.

7.2 Cardiovascular system and blood

The cardiovascular system is sensitive to alterations in oxygen supply. Because inhaled carbon monoxide limits oxygen supply, it might be expected to adversely affect the cardiovascular system; the degree of hypoxia and the extent of tissue injury will be determined by the dose of carbon monoxide. The effect of carbon monoxide on the cardiovascular system has been the subject of several reviews (Turino, 1981; McGrath, 1982; Penney, 1988). This section will discuss studies in animals that have evaluated the effects of carbon monoxide on ventricular fibrillation, haemodynamics, cardiomegaly, haematology and atherosclerosis.

7.2.1 Disturbances in cardiac rhythm

Data obtained from animal studies suggest that carbon monoxide can disturb cardiac conduction and cause cardiac arrhythmias. In dogs exposed intermittently or continuously to carbon monoxide (57 and 110 mg/m^3 [50 and 100 ppm]; 2.6–12.0% carboxyhaemoglobin) for 6 weeks in environmental chambers, Preziosi et al. (1970) reported abnormal electrocardiograms; the changes appeared during the second week and continued throughout the exposure. The blood cytology, haemoglobin and haematocrit values were unchanged from control values. DeBias et al. (1973) studied the effects of breathing carbon monoxide (110–120 mg/m^3 [96–102 ppm]; 12.4% carboxyhaemoglobin) continuously (23 h per day for 24 weeks) on the electrocardiograms of healthy monkeys and monkeys with myocardial infarcts induced by injecting microspheres into the coronary circulation. The authors observed higher P-wave amplitudes in both the infarcted and non-infarcted monkeys and a higher incidence of T-wave inversion in the infarcted monkeys. The authors concluded that there was a greater degree of ischaemia in the infarcted animals breathing carbon monoxide. Although there was a greater incidence of T-wave inversion in the infarcted monkeys, the effects were transient and of such low magnitude that accurate measurements of amplitude were not possible.

In other long-term studies, however, several groups have reported no effects of carbon monoxide either on the electrocardiogram or on cardiac arrhythmias. Musselman et al. (1959) observed no changes in the electrocardiogram of dogs exposed continuously to carbon monoxide (57 mg/m^3 [50 ppm]; 7.3% carboxyhaemoglobin) for 3 months. These observations were confirmed by Malinow et al. (1976), who reported no effects on the electrocardiogram in cynomolgus monkeys (*Macaca fascicularis*) exposed to carbon monoxide (570 mg/m^3 [500 ppm], pulsed; 21.6% carboxyhaemoglobin) for 14 months.

Several research groups have investigated the effects of carbon monoxide on the vulnerability of the heart to induced ventricular fibrillation. DeBias et al. (1976) reported that carbon monoxide (110 mg/m^3 [100 ppm] inhaled for 16 h; 9.3% carboxyhaemoglobin) reduced the threshold for ventricular fibrillation induced by an electrical stimulus applied to the myocardium of monkeys during the final stage of ventricular repolarization. The voltage required to induce fibrillation was highest in normal animals breathing air and lowest in infarcted animals breathing carbon monoxide. Infarction alone and carbon monoxide alone each required significantly less voltage for fibrillation; when the two were combined, the effects on the myocardium were additive. These observations were confirmed in both anaesthetized, open-chested dogs with acute myocardial injury (Aronow et al., 1978) and normal dogs (Aronow et al., 1979) breathing carbon monoxide (110 mg/m^3 [100 ppm]; 6.3–6.5% carboxyhaemoglobin) for 2 h. However, Kaul et al. (1974) reported that anaesthetized dogs inhaling 570 mg carbon monoxide/m^3 (500 ppm; 20–35% carboxyhaemoglobin) for 90 min were resistant to direct electrocardiographic changes. At 20% carboxyhaemoglobin, there was evidence of enhanced sensitivity to digitalis-induced ventricular tachycardia, but there was no increase in vulnerability of the ventricles to hydrocarbon/ epinephrine or to digitalis-induced fibrillation following exposure to carbon monoxide resulting in 35% carboxyhaemoglobin.

Several workers have investigated the effect of breathing carbon monoxide shortly after cardiac injury on the electrical activity of the heart. Becker & Haak (1979) evaluated the effects of carbon monoxide (five sequential exposures to 5700 mg/m^3 [5000 ppm], producing 4.9–17.0% carboxyhaemoglobin) on the electrocardiograms of anaesthetized dogs 1 h after coronary artery ligation. Myocardial ischaemia, as judged by the amount of ST-segment elevation in epicardial electrocardiograms, increased significantly at the lowest carboxyhaemoglobin levels (4.9%) and increased further with increasing carbon monoxide exposure; there were no changes in heart rate, blood pressure, left atrial pressure, cardiac output or blood flow to the ischaemic myocardium. Similar results were noted by Sekiya et al. (1983), who investigated the influence of carbon monoxide (3400 mg/m^3 [3000 ppm] for 15 min followed by 150 mg/m^3 [130 ppm] for 1 h; 13–15% carboxyhaemoglobin) on the extent and severity of myocardial ischaemia in dogs. This dose of carbon monoxide inhaled prior to coronary artery ligation increased the severity and extent of ischaemic injury and the magnitude of ST-segment elevation more than did ligation alone. There were no changes in heart rate or arterial pressure. On the other hand, several groups have reported no effects of carbon monoxide on the electrocardiogram or on cardiac arrhythmias (Musselman et al., 1959; Malinow et al., 1976). Foster (1981) concluded that carbon monoxide (110 mg/m^3 [100 ppm] for 6–9 min; 10.4% carboxyhaemoglobin) is not arrhythmogenic in dogs during the early minutes of acute myocardial infarction following occlusion of the left anterior descending coronary artery. This level of carbon monoxide did not either effect slowing of conduction through the ischaemic myocardium or affect the incidence of spontaneous ventricular tachycardia. These results were confirmed by Hutcheon et al. (1983) in their investigation of the effects of carbon monoxide on the electrical threshold for ventricular arrhythmias and the effective refractory period of the heart. They concluded that carbon monoxide (230 mg/m^3 [200 ppm] for 60 and 90 min; 5.1–6.3% carboxyhaemo-globin) does not alter the effective refractory period or the electrical threshold for ventricular arrhythmias in dogs. These results are consistent with those of Mills et al. (1987), who studied the effects of 0–20% carboxyhaemoglobin on the electrical stability of the heart in chloralose-anaesthetized dogs during coronary occlusion. There were no major effects on heart rate, mean arterial blood pressure, effective refractory period, vulnerable period or ventricular fibrillation thresh-old.

The effects of acute carbon monoxide exposure on cardiac electrical stability were studied in several canine heart models (Vanoli et al., 1989; Verrier et al., 1990). These workers examined the direct effects of carbon monoxide on the normal and ischaemic heart in the anaesthetized dog as well as possible indirect effects mediated by changes in platelet aggregability or central nervous system activity in the conscious dog. In anaesthetized dogs, exposure to carbon monox-ide resulting in carboxyhaemoglobin levels of up to 20% (570 mg carbon monoxide/m^3 [500 ppm] for 90–120 min) had no effect on ventricular electrical stability in the normal or acutely ischaemic heart. In a second study using anaesthetized dogs, these workers evaluated the effects of carbon monoxide on platelet aggregability and its effect on coronary flow during partial coronary artery stenosis. Concentra-tions of carboxyhaemoglobin up to 20% (570 mg carbon monoxide/m^3 [500 ppm] for 60–120 min) did not alter platelet aggregability or its effect on coronary blood flow during stenosis. In a third model using conscious dogs, these workers studied the effects on the heart of carbon monoxide-elicited changes in central nervous system activity. They observed no adverse effects on cardiac excitability in response to carboxyhaemoglobin levels of up to 20% (230–570 mg carbon monoxide/m³ [200–500 ppm] for 90–120 min) or to 9.7 ± 1.6% carboxyhaemoglobin (29–57 mg carbon monoxide/m³ [25–50 ppm]) for 24 h.

Farber et al. (1990) studied the effects of acute exposure to carbon monoxide on ventricular arrhythmias in a dog model of sudden cardiac death. In this model, 60% of dogs with a healed anterior myocardial infarction will experience ventricular fibrillation during acute myocardial ischaemia with mild exercise. Dogs that develop ventricular fibrillation during acute myocardial ischaemia with exercise are considered at high risk for sudden death and are defined as "susceptible." Dogs that survive the test without a fatal arrhythmia are considered at low risk for sudden death and are defined as "resistant." Using this model, Farber et al. (1990) tested the effects of carboxyhaemoglobin levels ranging from 5% to 15% (1700 mg carbon monoxide/m³ [1500 ppm] for varying times) in resistant and susceptible dogs. Heart rates increased with increasing carboxyhaemoglobin levels, but the increase did not become significant until carboxyhaemoglobin levels reached 15%. This trend was observable at rest as well as during exercise in both resistant and susceptible dogs. In resistant animals, in which acute myocardial ischaemia is typically associated with bradycardia, this reflex response occurred earlier and was augmented by exposure to carbon monoxide. In both resistant and susceptible dogs, carbon monoxide induced a worsening of ventricular arrhythmias in a minority of cases. The ventricular arrhythmias were not reproducible in subsequent trials. The authors concluded that acute exposure to carbon monoxide is seldom arrhythmogenic in dogs that have survived myocardial infarction.

7.2.2 *Haemodynamic studies*

The effects of carbon monoxide on coronary flow, heart rate, blood pressure, cardiac output, myocardial oxygen consumption and blood flow to various organs have been investigated in laboratory animals. The results are somewhat contradictory (partly because exposure regimes differed); however, most workers agree that carbon monoxide in sufficiently high doses can affect many haemodynamic variables.

Adams et al. (1973) described increased coronary flow and heart rate and decreased myocardial oxygen consumption in anaesthetized dogs breathing 1700 mg carbon monoxide/m^3 (1500 ppm) for 30 min (23.1% carboxyhaemoglobin). The decreased oxygen consumption indicates that the coronary flow response was not great enough to compensate for the decreased oxygen availability. The authors noted that although there was a positive chronotropic response, there was no positive inotropic response. The authors speculated that (1) the carbon monoxide may have caused an increase in the endogenous rhythm or blocked the positive inotropic response or (2) the response to carbon monoxide was mediated through the cardiac afferent receptors to give a chronotropic response without the concomitant inotropic response. When they used β-adrenergic blocking agents, the heart rate response to carbon monoxide disappeared, suggesting possible reflex mediation by the sympathetic nervous system.

In a later study in chronically instrumented, awake dogs exposed to 1100 mg carbon monoxide/m^3 (1000 ppm), producing carboxyhaemoglobin levels of 30%, Young & Stone (1976) reported an increase in coronary flow with no change in myocardial oxygen consumption. The increased coronary flow occurred in animals with hearts paced at 150 beats/min, as well as in non-paced animals, and in animals with propranolol and atropine blockade. Because the changes in coronary flow with arterial oxygen saturation were similar whether the animals were paced or not, these workers concluded that the increase in coronary flow is independent of changes in heart rate. Furthermore, the authors reasoned that if the coronary vasodilation was caused entirely by the release of a metabolic vasodilator, associated with decreased arterial oxygen saturation, the change in coronary flow in animals with both β-adrenergic and parasympathetic blockade should be the same as in control dogs. Young & Stone (1976) concluded that coronary vasodilation observed with an arterial oxygen saturation reduced by carbon monoxide is mediated partially through an active neurogenic process.

Increased myocardial blood flow after carbon monoxide inhalation in dogs was confirmed by Einzig et al. (1980), who also demonstrated the regional nature of the blood flow response. Using labelled microspheres, these workers demonstrated that whereas both right and left ventricular beds were dilated maximally at carboxyhaemoglobin levels of 41% (17 000–23 000 mg carbon monoxide/m^3 [15 000–

20 000 ppm] for 10 min), subendocardial/subepicardial blood flow ratios were reduced. The authors concluded that in addition to the global hypoxia associated with carbon monoxide poisoning, there is also an underperfusion of the subendocardial layer, which is most pronounced in the left ventricle.

The results on the endocardium were confirmed by Kleinert et al. (1980), who reported the effects of lowering oxygen content by about 30% with low oxygen or carbon monoxide gas mixtures (11 000 mg carbon monoxide/m^3 [10 000 ppm] for 3 min; 21–28% carboxyhaemoglobin). Regional myocardial relative tissue partial pressure of oxygen (P_{O_2}), perfusion and small vessel blood content were evaluated in anaesthetized, thoracotomized rabbits. Both carbon monoxide and hypoxic hypoxia increased regional blood flow to the myocardium and also, but to a lesser extent, in the endocardium. Relative endocardial P_{O_2} fell more markedly than epicardial P_{O_2} in both conditions. Small vessel blood content increased more with carbon monoxide than with low P_{O_2}, whereas regional oxygen consumption increased under both conditions. The authors concluded that the response to lowered oxygen content (whether by inhaling low oxygen or carbon monoxide gas mixtures) is an increase in flow, metabolic rate and the number of open capillaries, and the effects of both types of hypoxia appear more severe in the endocardium.

A decrease in tissue P_{O_2} with carbon monoxide exposure has also been reported by Weiss & Cohen (1974). These workers exposed anaesthetized rats to 90 and 180 mg carbon monoxide/m^3 (80 and 160 ppm) for 20-min periods and measured tissue oxygen tension as well as heart rate. A statistically significant decrease in brain P_{O_2} occurred with inhalation of 180 mg carbon monoxide/m^3 (160 ppm), but there was no change in heart rate.

Horvath (1975) investigated the coronary flow response in dogs with carboxyhaemoglobin levels of 6.2–35.6% produced by continuous administration of precisely measured volumes of carbon monoxide. Coronary flow increased progressively as blood carboxyhaemoglobin increased and was maintained for the duration of the experiment. However, when animals with complete atrioventricular block were maintained by cardiac pacemakers and were exposed to carboxyhaemoglobin levels of 6–7%, there was no longer an increase in coronary blood flow. These results are provocative, because they suggest an increased danger from low carboxyhaemoglobin levels in cardiac-disabled individuals.

The effects of carbon monoxide hypoxia and hypoxic hypoxia on arterial blood pressure and other vascular parameters were also studied in carotid baroreceptor- and chemoreceptor-denervated dogs (Traystman & Fitzgerald, 1977). Arterial blood pressure was unchanged by carbon monoxide hypoxia but increased with hypoxic hypoxia. Similar results were seen in carotid baroreceptor-denervated animals with intact chemoreceptors. Following carotid chemodenervation, arterial blood pressure decreased equally with both types of hypoxia.

In a subsequent report from the same laboratory (Sylvester et al., 1979), the effects of carbon monoxide hypoxia (11 000 mg carbon monoxide/m^3 [10 000 ppm] followed by 1100 mg carbon monoxide/ m^3 [1000 ppm] for 15–20 min; 61–67% carboxyhaemoglobin) and hypoxic hypoxia were compared in anaesthetized, paralysed dogs. Cardiac output and stroke volume increased during both carbon monoxide and hypoxic hypoxia, whereas heart rate was variable. Mean arterial pressure decreased during carbon monoxide hypoxia but increased during hypoxic hypoxia. Total peripheral resistance fell during both hypoxias, but the decrease was greater during the carbon monoxide hypoxia. After resection of the carotid body, the circulatory effects of hypoxic and carbon monoxide hypoxia were the same and were characterized by decreases in mean arterial pressure and total peripheral resistance. In a second series of experiments with closed-chest dogs, hypoxic and carbon monoxide hypoxia caused equal catecholamine secretion before carotid body resection. After carotid body resection, the magnitude of the catecholamine response was doubled with both hypoxias. These workers concluded that the responses to hypoxic and carbon monoxide hypoxia differ and that the difference is dependent on intact chemo- and baroreflexes and on differences in arterial oxygen tension, but not on differences in catecholamine secretion or ventilatory response.

In cynomolgus monkeys exposed to 570 mg carbon monoxide/m^3 (500 ppm) intermittently for 12 h per day for 14 months (21.6% carboxyhaemoglobin), Malinow et al. (1976) reported no changes in arterial pressure, left ventricular pressure, time derivative of pressure (*dP/dt*) and ventricular contractility. On the other hand, Kanten et al. (1983) studied the effects of carbon monoxide (170 mg/m^3 [150 ppm];

carboxyhaemoglobin up to 16%) for 0.5–2 h on haemodynamic parameters in open-chest, anaesthetized rats and reported that heart rate, cardiac output, cardiac index, time derivative of maximal force (aortic) and stroke volume increased significantly, whereas mean arterial pressure, total peripheral resistance and left ventricular systolic pressure decreased. These effects were evident at carboxyhaemoglobin levels as low as 7.5% (0.5 h). There were no changes in stroke work, left ventricular dP/dt maximum and stroke power.

The effects of carbon monoxide on blood flow to various vascular beds has been investigated in several animal models, but most of the studies have been conducted at rather high carbon monoxide or carboxyhaemoglobin levels. In general, carbon monoxide increases cerebral blood flow.

In recent studies, Oremus et al. (1988) reported that in the anaesthetized rat breathing carbon monoxide (570 mg/m^3 [500 ppm]; 23% carboxyhaemoglobin) for 1 h, carbon monoxide reduces mean arterial pressure through peripheral vasodilation predominantly in the skeletal muscle vasculature. There were no differences in heart rate or mesenteric or renal resistance between the carbon monoxide-exposed and control groups. This was confirmed by Gannon et al. (1988), who reported that in the anaesthetized rat breathing carbon monoxide (570 mg/m^3 [500 ppm]; 24% carboxyhaemoglobin) for 1 h, carbon monoxide increased inside vessel diameter (36–40%) and flow rate (38–54%) and decreased mean arterial pressure to 79% of control in the cremaster muscle. There was no change in the response of 3A vessels to topical applications of phenylephrine as a result of carbon monoxide exposure.

King et al. (1984, 1985) compared whole-body and hindlimb blood flow responses in anaesthetized dogs exposed to carbon monoxide or anaemic hypoxia. Arterial oxygen content was reduced by moderate (50%) or severe (65%) carbon monoxide hypoxia (produced by dialysis with 100% carbon monoxide) or anaemic hypoxia (produced by haemodilution). These workers noted that cardiac output was elevated in all groups at 30 min and in the severe carbon monoxide group at 60 min. Hindlimb blood flow remained unchanged during carbon monoxide hypoxia in the animals with intact hindlimb innervation but was greater in animals with denervated hindlimbs. There was a decrease in mean arterial pressure in all groups associated with a fall in total peripheral resistance. Hindlimb resistance remained unchanged during moderate carbon monoxide hypoxia in the intact groups but was increased in the denervated group. The authors concluded that the increase in cardiac output during carbon monoxide hypoxia was directed to non-muscle areas of the body and that intact sympathetic innervation was required to achieve this redistribution. However, aortic chemoreceptor input was not necessary for the increase in cardiac output during severe carbon monoxide hypoxia or for the diversion of the increased flow to non-muscle tissues.

King et al. (1987) investigated the effects of high carbon monoxide (1100–11 000 mg/m³ [1000–10 000 ppm] to lower arterial oxygen content to 5–6 vol %) and hypoxic hypoxia on the contracting gastrocnemius muscle of anaesthetized dogs. Oxygen uptake decreased from the normoxic level in the carbon monoxide group but not in the hypoxic hypoxia group. Blood flow increased in both groups during hypoxia, but more so in the carbon monoxide group. Oxygen extraction increased further during contractions in the hypoxic group but fell in the carbon monoxide group. The authors observed that the oxygen uptake limitation occurring during carbon monoxide hypoxia and isometric contractions was associated with a reduced oxygen extraction and concluded that the leftward shift in the oxyhaemoglobin dissociation curve during carbon monoxide hypoxia may have impeded oxygen extraction.

Melinyshyn et al. (1988) investigated the role of β-adrenoreceptors in the circulatory responses of anaesthetized dogs to severe carbon monoxide (about a 63% decrease in arterial oxygen content obtained by dialysing with 100% carbon monoxide). One group was β-blocked with propranolol (β₁ and β₂ blockade), a second was β-blocked with ICI 118,551 (β₂ blockade) and a third was a time control. Cardiac output increased in all groups during carbon monoxide hypoxia, with the increase being greatest in the unblockaded group. Hindlimb blood flow rose during carbon monoxide hypoxia only in the unblockaded group. The authors concluded that 35% of the rise in cardiac output occurring during carbon monoxide hypoxia depended on peripheral vasodilation mediated through β₂-adrenoreceptors.

.2.3   Cardiomegaly

The early investigations of cardiac enlargement following prolonged exposure to carbon monoxide have been confirmed in different animal models and extended to characterize the development and regression of the cardiomegaly. Theodore et al. (1971) reported cardiac hypertrophy in rats breathing 460 mg carbon monoxide/m^3 (400 ppm) for the first 71 days, followed by 570 mg/m^3 (500 ppm) for the remaining 97 days, but not in dogs, baboons or monkeys receiving the same carbon monoxide exposure. Penney et al. (1974a) also noted cardiomegaly in rats breathing 570 mg carbon monoxide/m^3 (500 ppm); heart weights were one-third greater than predicted for controls within 14 days of exposure and were 140–153% of controls after 42 days of exposure. The cardiomegaly was accompanied by changes in cardiac lactate dehydrogenase isoenzyme composition that were similar to those reported in other conditions that cause cardiac hypertrophy (e.g., aortic and pulmonary artery constriction, coronary artery disease, altitude acclimation, severe anaemia).

To further characterize the hypertrophy and determine its threshold, Penney et al. (1974b) measured heart weights in rats exposed continuously to 110, 230 or 570 mg carbon monoxide/m^3 (100, 200 or 500 ppm; 9.26, 15.82 and 41.14% carboxyhaemoglobin) for various times (1–42 days); they noted significant increases in heart weights at 230 and 570 mg carbon monoxide/m^3 (200 and 500 ppm), with changes occurring in the left ventricle and septum, right ventricle and especially the atria. The authors concluded that whereas the threshold for an increased haemoglobin response is 110 mg carbon monoxide/m^3 (100 ppm; 9.26% carboxyhaemoglobin), the threshold for cardiac enlargement is near 230 mg carbon monoxide/m^3 (200 ppm; 12.03% carboxyhaemoglobin); unlike cardiac hypertrophy caused by altitude, which primarily involves the right ventricle, cardiac hypertrophy caused by carbon monoxide involves the whole heart.

The regression of cardiac hypertrophy in rats exposed continuously to moderate (460 mg/m^3 [400 ppm]; 35% carboxyhaemoglobin) or severe (570–1300 mg/m^3 [500–1100 ppm]; 58% carboxyhaemoglobin) carbon monoxide for 6 weeks was followed by Styka & Penney (1978). Heart weight to body weight ratio increased from 2.65 in controls to 3.52 and 4.01 with moderate and severe carbon monoxide exposure, respectively. Myocardial lactate dehydrogenase M subunits were elevated 5–6% by moderate and 12–14% by severe carbon monoxide exposure. Forty-one to 48 days after terminating the carbon monoxide exposure, haemoglobin concentrations among groups did not differ significantly; heart weight to body weight values were similar in the control and moderately exposed animals, but remained significantly elevated in the severely exposed animals.

In addition to cardiomegaly, Kjeldsen et al. (1972) reported ultrastructural changes in the myocardium of rabbits breathing 210 mg carbon monoxide/m^3 (180 ppm; 16.7% carboxyhaemoglobin) for 2 weeks. The changes included focal areas of necrosis of myofibrils and degenerative changes of the mitochondria. In addition, varying degrees of injury were noted in the blood vessels. These included oedema in the capillaries, stasis and perivascular haemorrhages on the venous side and endothelial swelling, subendothelial oedema and degenerative changes in myocytes on the arterial side.

The haemodynamic consequences of prolonged carbon monoxide exposure have been examined in rats breathing 570 mg carbon monoxide/m^3 (500 ppm; 38–42% carboxyhaemoglobin) for 1–42 days (Penney et al., 1979) and in goats breathing 180–250 mg carbon monoxide/m^3 (160–220 ppm; 20% carboxyhaemoglobin) for 2 weeks (James et al., 1979). In rats, cardiomegaly developed; stroke index, stroke power and cardiac index increased; and total systemic and pulmonary resistances decreased. Left and right ventricular systolic pressures, mean aortic pressure, maximum left ventricular *dP/dt* and heart rate did not change significantly. Penney et al. (1979) concluded that enhanced cardiac output, via an increased stroke volume, is a compensatory mechanism to provide tissue oxygenation during carbon monoxide intoxication and that increased cardiac work is the major factor responsible for the development of cardiomegaly. In chronically instrumented goats, James et al. (1979) noted that cardiac index, stroke volume, left ventricular contractility and heart rate were all unchanged during exposure to carbon monoxide but were depressed significantly during the first week following termination of the exposure. Discrepancies between the Penney et al. (1979) and James et al. (1979) studies may be the result of differences in the carbon monoxide concentrations or in the species used.

Penney et al. (1984) studied the compliance and measured the dimensions of hypertrophied hearts from rats breathing 570 mg carbon monoxide/m^3 (500 ppm; 38–40% carboxyhaemoglobin) for 38–47 days. Heart weight to body weight ratios increased from 2.69 to 3.34. Although compliance of the right and left ventricles was higher in the carbon monoxide group, the differences disappeared when the heart weight was normalized by body weight. Left ventricular apex-to-base length and left ventricular outside diameter increased 6.4 and 7.3%, respectively; there were no changes in left ventricle, right ventricle or septum thickness. The authors concluded that chronic carbon monoxide exposure produces eccentric cardiomegaly with no intrinsic change in wall stiffness.

The consequences of breathing carbon monoxide have also been investigated in perinatal animals. Postnatal exposures are described here; the effects of prenatal exposure to carbon monoxide are discussed in section 7.6.4.2.

Penney & Weeks (1979) examined the effects of inhaling 570 mg carbon monoxide/m^3 (500 ppm; 38–42% carboxyhaemoglobin) until 50 days of age on cardiac growth in young (5 days) and old (25 days) rats. They observed that the younger rats experienced the greater change in heart weight and DNA synthesis and concluded that the potential for cardiac DNA synthesis and muscle cell hyperplasia ends in rats during the 5th through 25th days of postnatal development.

Penney et al. (1980) compared the effects of prenatal carbon monoxide exposure in Sprague-Dawley rats at a dose of 230 mg/m^3 (200 ppm) with exposure both prenatally and neonatally until age 29 days. Neonatal carbon monoxide concentrations were elevated to 570 mg/m^3 (500 ppm). Cardiomegaly and depressed haemoglobin, haematocrit and red blood cell counts were found following carbon monoxide exposure. In subjects allowed to survive until young adulthood, the heart weight to body weight ratios of subjects receiving carbon monoxide both prenatally and neonatally were still elevated, whereas those in the prenatal carbon monoxide condition did not differ from control subjects in this measure.

Ventricular weights (wet and dry) and myocyte size and volume were measured in perinatal rats exposed to 230 mg carbon monoxide/m^3 (200 ppm) by Clubb et al. (1986). Pregnant rats were exposed to air or carbon monoxide, and, at birth, pups from these two groups were subdivided into four groups: (1) control group (air/air), which was maintained in air *in utero* and postpartum; (2) air/carbon monoxide group, which received carbon monoxide postpartum only; (3) carbon monoxide/carbon monoxide group, which received carbon monoxide *in utero* and postpartum; and (4) carbon monoxide/air group, which received carbon monoxide *in utero* but was maintained in air postpartum. Right ventricle weights were increased in animals exposed to carbon monoxide during the fetal period, but left ventricle weights were increased by carbon monoxide during the neonatal period. Although heart weight to body weight ratios increased to those of the carbon monoxide/carbon monoxide group by 12 days of age in animals exposed to carbon monoxide postnatally only (air/carbon monoxide), heart weight to body weight ratios decreased to those of controls (air/air) by 28 days of age in animals exposed to air post-natally following fetal carbon monoxide exposure (carbon monoxide/ air). There was no difference in myocyte volume between groups at birth. Left ventricle plus septum and right ventricle cell volumes of the carbon monoxide/carbon monoxide group were smaller than those of the controls at 28 days of age despite the heavier wet and dry weights of the carbon monoxide/carbon monoxide neonates. At birth, the carbon monoxide-exposed animals had more myocytes in the right ventricle than the air-exposed controls; carbon monoxide exposure after birth resulted in left ventricular hyperplasia.

Clubb et al. (1986) concluded that the increased haemodynamic load caused by carbon monoxide during the fetal period results in cardiomegaly, characterized by myocyte hyperplasia, and this cellular response is sustained throughout the early neonatal period in animals exposed to carbon monoxide postpartum.

Penney et al. (1993) investigated whether carbon monoxide expo-sure in the developing heart produces long-lasting alterations in coronary vasculature. One-day-old male rat pups exposed to 530 mg carbon monoxide/m^3 (500 ppm) continuously for 30 days were sacrificed at 61 and 110 days of age. Morphometric analysis was performed on the fixed hearts. There were no significant differences in the number of small (27–114 µm) or larger (>114 µm) vessels in any heart region. The septum in the carbon monoxide-exposed rats was an exception; there were more small veins at 61 days of age and more larger veins at 110 days of age. There was a significant increase in the number of small arteries at both ages in the carbon monoxide-exposed rats across all heart regions and in the smaller veins at 61 days of age. The large vessels in the septum at 61 days of age had a significantly greater diameter in the carbon monoxide-exposed compared with the air-exposed rats. This was also true for the large arteries in the septum and right ventricle of the 110-day-old rats. Taking all heart regions together, the large arteries in carbon monoxide-exposed rats were larger than those in air-exposed rats. Previous carbon monoxide exposure significantly increased large artery and total cross-sectional area in the septum and right ventricle at 61 days of age and in the right ventricle at 110 days of age. Total cross-sectional area of veins in the septum was also increased. Taking all heart regions together, carbon monoxide significantly increased small artery cross-sectional area at 61 days of age and small, large and total artery cross-sectional area at 110 days of age. With one exception (small veins, 110 days of age), there was no effect of carbon monoxide on vein cross-sectional area. These changes resulted in an increase in the percentage of total cross-sectional area contributed by the larger vessels. Pathological examination showed nothing abnormal. The results suggest profound and persistent changes in coronary vessel architecture following chronic neonatal carbon monoxide exposure.

Penney et al. (1992) determined whether the cardiomegaly produced by carbon monoxide is related to the development of pulmonary hypertension, as is known to occur at high altitude. Newborn rats were exposed to 570 mg carbon monoxide/m^3 (500 ppm) for 32 days at Detroit, Michigan, USA, or to 3500 m simulated altitude at Fort Collins, Colorado, USA. Right ventricle and left ventricle plus septum masses in the carbon monoxide-exposed rats were increased 38.0% and 37.4%, respectively, relative to air-maintained controls. Right ventricle and left ventricle plus septum masses in the altitude group were increased 55.7% and 9.3%, respectively, relative to controls. Cardiac hypertrophy declined in the carbon monoxide and altitude groups post-exposure but remained significantly above the controls for the right ventricle at 145 days of age. By use of an *in vitro* preparation, pulmonary vascular resistance and pulmonary arterial pressure were significantly increased immediately after altitude but not after carbon monoxide exposure and remained elevated in adulthood after altitude exposure. Pulmonary vascular resistance was correlated with haematocrit in altitude-exposed but not in carbon monoxide-exposed rats.

Bambach et al. (1991) performed *in situ* assessment of cardiomegaly in rats exposed to 570 mg carbon monoxide/m^3 (500 ppm) for 0–62 days using nuclear magnetic resonance imaging. Following final imaging, the rats were sacrificed and heart mass was determined. The mean outside diameter of the left ventricle plus septum showed a strong correlation with the duration of carbon monoxide exposure, whereas the correlation coefficients for the left ventricle plus septum lumen diameter and wall thickness were marginally significant. The mean pleural space diameter also increased significantly ($r = 0.64$, $P < 0.05$) with duration of carbon monoxide exposure. The results achieved with nuclear magnetic resonance imaging are consistent with past morphological studies of carbon monoxide-induced cardiomegaly, where heart dimensions were determined in relaxed frozen tissue, and corroborate the eccentric nature of the cardiomegaly.

The stimulus for carbon monoxide-induced cardiomegaly was investigated by Penney & Formolo (1993). Rats were exposed to 570 mg carbon monoxide/m^3 (500 ppm) for 30 days, some receiving the β_1-adrenergic blocker atenolol, others the α_1-adrenergic blocking agent prazosin. Systolic blood pressure was significanly lowered by carbon monoxide and by atenolol at a high dose. Carbon monoxide alone resulted in 30–43% and 18–25% weight increases in right ventricle and left ventricle plus septum, respectively. Neither low- nor high-dose prazosin significantly decreased right ventricle and left ventricle plus septum weights in the carbon monoxide-exposed rats. Low-dose atenolol failed to alter right ventricle and left ventricle plus septum weights in the carbon monoxide-exposed rats, whereas high-dose atenolol significantly increased right ventricle weight in the carbon monoxide-exposed rats. Thus, carbon monoxide-induced cardiomegaly develops in spite of lowered systolic blood pressure and the blockade of either α_1 or β_1 receptors. The results suggest the potentially powerful role of enhanced preload in driving cardiomegaly.

7.2.4 Haematology studies

Increase in haemoglobin concentration, as well as haematocrit ratio, is a well-documented response to hypoxia that serves to increase the oxygen carrying capacity of the blood. Guyton & Richardson (1961) and Smith & Crowell (1967), however, suggested that changes in haematocrit ratio not only affect the oxygen carrying capacity of the blood, but affect blood flow as well. Therefore, when haematocrit ratios increase much above normal, oxygen delivery to the tissues may be reduced, because the resultant decrease in blood flow can more than offset the increased oxygen carrying capacity of the blood. Smith & Crowell (1967) concluded that there is an optimum haematocrit ratio at sea level that shifts to a higher value with altitude acclimation.

Presumably a similar compensation would occur when oxygen transport is reduced by carbon monoxide; however, experimental evidence is needed to confirm these relationships.

Changes in haemoglobin concentration and haematocrit ratio have been reported in numerous animal studies. In dogs exposed to 57 mg carbon monoxide/m³ (50 ppm; 7.3% carboxyhaemoglobin) for 3 months, Musselman et al. (1959) reported a slight increase in haemoglobin concentration (12%), haematocrit ratio (10%) and red blood cells (10%). These observations were extended by Jones et al. (1971) to include several species of animals exposed to 58 mg carbon monoxide/m³ (51 ppm) or more (3.2–20.2% carboxyhaemoglobin), intermittently or continuously, for up to 90 days. There were no significant increases in the haemoglobin or haematocrit values observed in any of the species at 58 mg carbon monoxide/m³ (51 ppm; 3.2–6.2% carboxyhaemoglobin). At 110 mg carbon monoxide/m³ (96 ppm; 4.9–12.7% carboxyhaemoglobin), significant increases were noted in the haematocrit value for monkeys (from 43% to 47%) and in the haemoglobin (from 14.0g% to 16.49g%) and haematocrit values (from 46% to 52%) for rats. Haemoglobin and haematocrit values were elevated in rats (by 14% and 10%, respectively), guinea-pigs (8% and 10%, respectively) and monkeys (34% and 26%, respectively) exposed to 230 mg carbon monoxide/m³ (200 ppm; 9.4–12.0% carboxyhaemoglobin); they were also elevated in dogs, but there were too few animals to determine statistical significance. However, in dogs exposed to carbon monoxide (220 mg/m³ [195 ppm]; 30% carboxyhaemoglobin) for 72 h, Syvertsen & Harris (1973) reported that haematocrit and haemoglobin increased from 50.3% to 57.8% and from 15.0g% to 16.2g%, respectively. The differences in haematocrit and haemoglobin occurred after a 72-h exposure and were attributed to increased erythropoiesis. Because no measurements were made, however, the possibility of splenic contraction cannot be excluded. Penney et al. (1974b) observed significant increases in haemoglobin (from 15.6g% to 16.7g%) in rats exposed to 110 mg carbon monoxide/m³ (100 ppm) over several weeks and concluded that the threshold for an increased haemoglobin response is close to 110 mg/m³ (100 ppm; 9.26% carboxyhaemoglobin).

Several groups have reported no change in haemoglobin or haematocrit following carbon monoxide exposure. Thus, Preziosi et al. (1970) observed no significant change in haemoglobin concentration in dogs exposed to 57 or 110 mg carbon monoxide/m^3 (50 or 100 ppm; 2.6–12.0% carboxyhaemoglobin) for 6 weeks. In monkeys exposed to 23 or 74 mg carbon monoxide/m^3 (20 and 65 ppm; 1.9–10.2% carboxyhaemoglobin) for 2 years, Eckardt et al. (1972) noted no compensatory increases in haemoglobin concentration or haematocrit ratio. In mice exposed 5 days per week to 57 mg carbon monoxide/m^3 (50 ppm) for 1–3 months, Stupfel & Bouley (1970) observed no significant increase in haemoglobin.

Interestingly, in fetuses removed from pregnant rats after 21 days of exposure to carbon monoxide, Prigge & Hochrainer (1977) reported a significant increase in fetal haematocrit (from 33.3% to 34.5%) at 69 mg/m^3 (60 ppm) and a significant decrease in haemoglobin and haematocrit at 290 mg/m^3 (250 ppm) (from 9.1g% to 8.0g% and from 33.3% to 28.4%, respectively) and 570 mg/m^3 (500 ppm) (from 9.1g% to 6.5g% and from 33.3% to 21.9%, respectively). These results were confirmed by Penney et al. (1980), who reported significantly lower haemoglobin (12.6g% versus 15.8g%), haematocrit (46.2% versus 54.4%) and red blood cell counts (27.2% versus 29.1%) in newborns from pregnant rats exposed to 230 mg carbon monoxide/m^3 (200 ppm; 27.8% carboxyhaemoglobin) for the final 18 days of development than in controls. However, in a later study, Penney et al. (1983) reported that although red blood cell counts were depressed in neonates from pregnant rats exposed to 180, 190 or 230 mg carbon monoxide/m^3 (157, 166 or 200 ppm; 21.8–33.5% carboxyhaemoglobin) for the last 17 of 22 gestation days, mean corpuscular haemoglobin and volume were elevated.

The results from animal studies indicate that inhaled carbon monoxide can increase haemoglobin concentration and haematocrit ratio. Small increases in haemoglobin and haematocrit probably represent a compensation for the reduction in oxygen transport caused by carbon monoxide; excessive increases in haemoglobin and haematocrit may impose an additional workload on the heart and compromise blood flow to the tissue. The oxygen transport system of the fetus is especially sensitive to carbon monoxide inhaled by the mother.

.2.5   Atherosclerosis and thrombosis

There is little evidence to conclusively support a relationship between carbon monoxide exposure and atherosclerosis in animal models.

Astrup et al. (1967) described atheromatosis as well as increased cholesterol accumulation in aortas from rabbits fed cholesterol and exposed to carbon monoxide (190–400 mg/m³ [170–350 ppm]; 17–33% carboxyhaemoglobin) for 10 weeks. These observations were not supported, however, by Webster et al. (1970), who observed no changes in the aorta or carotid arteries or in serum cholesterol levels in squirrel monkeys fed cholesterol and exposed intermittently to carbon monoxide (110–340 mg/m³ [100–300 ppm]; 9–26% carboxyhaemoglobin) for 7 months; they did note enhanced atherosclerosis in the coronary arteries. Davies et al. (1976) confirmed that coronary artery atherosclerosis was significantly higher in rabbits fed cholesterol and exposed intermittently to carbon monoxide for 10 weeks (290 mg/m³ [250 ppm]; 20% carboxyhaemoglobin), but they also reported no significant differences between groups in aortic concentrations of triglycerides, cholesterol, phospholipids or plasma cholesterol.

In cynomolgus monkeys fed cholesterol and exposed intermittently to carbon monoxide for 14 months (57–570 mg/m³ [50–500 ppm]; 21.6% carboxyhaemoglobin), Malinow et al. (1976) observed no differences in plasma cholesterol levels or in coronary or aortic atherosclerosis. Armitage et al. (1976) confirmed that intermittent carbon monoxide (170 mg/m³ [150 ppm] for 52 and 84 weeks; 10% carboxyhaemoglobin) did not enhance the extent or severity of atherosclerosis in the normal white Carneau pigeon. Although carbon monoxide exposure did increase the severity of coronary artery atherosclerosis in birds fed cholesterol, the difference between groups, noted at 52 weeks, was not present after 84 weeks.

Stender et al. (1977) exposed rabbits that were fed high levels of cholesterol to carbon monoxide for 6 weeks continuously and intermittently (230 mg/m³ [200 ppm]; 17% carboxyhaemoglobin). In the cholesterol-fed group, carbon monoxide had no effect on free and esterified cholesterol concentrations in the inner layer of the aortic wall. In the normal group, carbon monoxide increased the concentration of cholesterol in the aortic arch, but there was no difference in the cholesterol content of the total aorta.

Hugod et al. (1978), using a blind technique and the same criteria to assess intimal damage as were used in earlier studies (Kjeldsen et al., 1972; Thomsen, 1974; Thomsen & Kjeldsen, 1975), noted no histological changes in the coronary arteries or aorta in rabbits exposed to carbon monoxide (230, 2300 or 4600 mg/m^3 [200, 2000 or 4000 ppm]) for 0.5–12 weeks. These workers suggested that the positive results obtained earlier were due to the non-blind evaluation techniques and the small number of animals used. Later, Hugod (1981) confirmed these negative results using electron microscopy.

Only a few studies published since 1979 have demonstrated a significant atherogenic effect of low-level carbon monoxide exposure. Turner et al. (1979) showed that carbon monoxide enhanced the development of coronary artery lesions in white Carneau pigeons that were fed a diet of 0.5% and 1%, but not 2%, cholesterol. The exposure was to 170 mg carbon monoxide/m^3 (150 ppm) for 6 h per day, 5 days per week, for 52 weeks (10–20% carboxyhaemoglobin). Plasma cholesterol levels may have been increased slightly by the carbon monoxide, but this was significant ($P < 0.5$) only at week 11. Marshall & Hess (1981) exposed mini-pigs to 180, 210 or 480 mg carbon monoxide/m^3 (160, 185 or 420 ppm) for 4 h per day for 1–16 days (5–30% carboxyhaemoglobin). The higher concentrations were associated with adhesion of platelets to arterial endothelium and to fossae of degenerated endothelial cells. Additional changes at the higher concentration included an increased haematocrit, an increase in blood viscosity and an increase in platelet aggregation.

Alcindor et al. (1984) studied rabbits with induced hypercholesterolaemia. Three sets of rabbits were studied. The first was a control group receiving a normal diet and breathing air. The second group was fed a 2% cholesterol diet. The third group was fed the same diet and was exposed to 170 mg carbon monoxide/m^3 (150 ppm). Carboxyhaemoglobin levels were not reported. Low-density lipoprotein particles in the carbon monoxide-exposed rabbits were richer in cholesterol and had a higher cholesterol-to-phospholipid molar ratio than did the particles from the non-exposed rabbits after 45 days ($P < 0.01$).

Other animal studies have given generally negative results. Bing et al. (1980) studied cynomolgus monkeys. Four animals were used as controls, and seven were exposed to carbon monoxide at a level of 460 mg/m^3 (400 ppm) for 10 alternate half-hours of each day during a 12-month period. Carboxyhaemoglobin levels showed a gradual increase to a peak at 5 h of 20%. The monkeys had no histological evidence of atherosclerosis, vessel wall damage or fat deposition in the arterial wall. There was no significant change in cholesterol or in lipoprotein levels. High-density to total cholesterol ratios did not differ between the carbon monoxide-exposed and air-exposed animals. These animals were on a normal diet with no augmentation of cholesterol or fat content. The study demonstrated that even high levels of carbon monoxide exposure are not invariably followed by arterial injury or abnormal lipid accumulation.

Similar negative results were reported by Sultzer et al. (1982), who studied swine. Pigs with and without von Willebrand's disease were divided into groups that were exposed to intermittent, low-level carbon monoxide or to air. Carbon monoxide was delivered at 115 mg/m^3 (100 ppm) for 8 h each weekday for 4 months. Carboxy-haemoglobin levels averaged 7% after 5 h of exposure. The degree of coronary and aortic atherosclerotic lesion development in response to a 2% cholesterol diet was similar in the two exposure groups. There was no effect of ambient carbon monoxide on the degree of hyper-cholesterolaemia induced by the diet. The findings showed no obvious effect of carbon monoxide on atherogenesis in hypercholesterolaemic pigs.

Penn et al. (1992) and Penn (1993) reported negative results in the cockerel model, which is sensitive to modulation by environmental agents. The plaque volume percentage in the aortic walls of experi-mental and control animals was determined after exposures to 57, 110, 170 or 230 mg carbon monoxide/m^3 (50, 100, 150 or 200 ppm), 2 h per day, 5 days per week, for 16 weeks. Chronic carbon monoxide exposures, either alone or in combination with a low-cholesterol diet, did not stimulate arteriosclerotic plaque development in cockerels. Exposure to carbon monoxide after large plaques had developed, as a result of either a high-cholesterol diet or injections with dimethyl-benz[a]anthracene, also did not stimulate any further plaque develop-ment. Although levels of carboxyhaemoglobin measured 10 min after the highest exposures were approximately 11–12%, it was apparent that birds eliminate carbon monoxide more rapidly than mammals. As a result, it was not clear if carboxyhaemoglobin levels were sufficient-ly elevated for a long enough period of time to draw any definitive conclusions from these experiments.

A number of studies have examined the contribution of carbon monoxide in cigarette smoke to the purported effects of smoking on atherogenesis and thrombosis. Rogers et al. (1980) fed a high-cholesterol diet to 36 baboons for up to 81 weeks. The animals were taught to puff either cigarette smoke or air from smoking machines in a human-like manner using operant conditioning with a water reward. Half of the baboons smoked 43 cigarettes each day. The baboons were given a cigarette or sham every 15 min during a 12-h day, except during times of blood drawing. Average carboxyhaemoglobin levels in smokers were about 1.9%. Only slight differences in the very-low-density and high-density lipoprotein levels were noted between the smokers and non-smokers. Additionally, platelet aggregation with adenosine 5'-phosphate and collagen was similar in the two groups.

Rogers et al. (1988) extended their previous study of male baboons for an additional 1.2 years of diet and smoking (total diet, 3.2 years; total smoking, 2.8 years). They also studied a separate group of 25 female baboons that received the diet for 2.6 years and were exposed to cigarette smoke for 1.6 years. Blood levels of carbon monoxide were determined by gas chromatography and were reported both as total concentration in millilitres per decilitre and as percent saturation of haemoglobin, as calculated by a validated linear regression equation. Levels of carboxyhaemoglobin in the male baboons averaged 0.64% at baseline, whereas carboxyhaemoglobin was on average 0.35% in female baboons at baseline. The weekly averages of carboxyhaemoglobin levels determined after 57 weeks were 2.01% and 1.13% in male and female baboons, respectively. The baseline cholesterol levels were 105 mg/dl and 88 mg/dl in the two groups of baboons. Levels at 16 weeks were 226 mg/dl in males and 291 mg/dl in females. There were no significant differences in total cholesterol, high-density lipoprotein cholesterol or low-density lipo-protein cholesterol between smokers and controls. There were slightly more fatty streaks and fibrous lesions in the carotid arteries of smokers than in controls. No differences in lesion prevalence, vascular content of lipids or prostaglandins were seen in aorta or coronary arteries.

The results reported by Rogers et al. (1980, 1988) suggest little, if any, effect of cigarette smoking on atherosclerotic lesion develop-ment in baboons. Whether these findings can be extrapolated to effects of smoking in humans is difficult to determine. The carboxyhaemo-globin levels attained in the experimental animals were barely 2%. Levels in human smokers are probably 4% or more during the waking hours of the day. On the other hand, the findings are consistent with most studies of the effects of low levels of carbon monoxide on atherogenesis.

A study of cockerels by Penn et al. (1983) has shown negative results as well. Three groups of cockerels, each including seven animals, were used to determine if the atherogenic effect of cigarette smoke could be separated from an effect due solely to carbon monoxide. Cockerels develop aortic fibromuscular atherosclerotic lesions spontaneously. Various agents, including some carcinogens, have been shown to accelerate the growth in thickness and extent of these lesions. The authors used this model by exposing one group of animals to the smoke from 40 cigarettes each day for 5 days per week. The cockerels were exposed from about 6 weeks of age until about 22 weeks of age. A similar group of animals was exposed to carbon monoxide calibrated to give a similar carboxyhaemoglobin level to that achieved in the animals exposed to cigarette smoke. The third group of animals was exposed to filtered air. Carboxyhaemoglobin levels following an exposure session were measured at 6, 9, 12 and 15 weeks. The average level in the air-exposed animals was 1.6%. Levels in the cigarette smoke and carbon monoxide groups were 6.7% and 7.2%, respectively. Atherosclerosis was quantified both by the extent of the aorta involved and by the cross-sectional area of the intimal thickening. The cigarette smoke-exposed group had more aortic lesions and lesions with greater cross-sectional area than did either the carbon monoxide-exposed group or the air-exposed group. This difference was significant at $P < 0.05$ in a one-tailed chi-square test. The data suggest that atherogenic effects of cigarette smoke are not solely attributable to carbon monoxide.

It has been postulated that a possible atherogenic effect of carbon monoxide may be mediated through an ability of carbon monoxide to enhance platelet aggregation or some other component of thrombosis. This possibility was raised in the study by Marshall & Hess (1981) noted above. Other studies, however, have demonstrated that the effect of carbon monoxide is to depress platelet aggregation. In one study (Mansouri & Perry, 1982), platelet aggregation in response to epinephrine and arachidonic acid was reduced in *in vitro* experiments in which carbon monoxide was bubbled through platelet-rich plasma. Similarly, platelets from smokers aggregated less well than platelets from non-smokers, although this inhibition of aggregation was not correlated with the level of carboxyhaemoglobin.

Madsen & Dyerberg (1984) extended these observations by studying the effects of carbon monoxide and nicotine on bleeding time in humans. Smoke from high-nicotine cigarettes caused a significant shortening of the bleeding time, whereas smoke from low-nicotine cigarettes caused no significant change in bleeding time. Carbon monoxide inhalation sufficient to raise carboxyhaemoglobin to 15% was followed by a shortening of the bleeding time (6.0–4.8 min), but for a short period of time (<1.5 h). After administration of acetyl-salicylic acid (Aspirin), neither nicotine nor carbon monoxide affected bleeding times or platelet aggregation. The findings suggest that the pro-aggregating effects of cigarette smoke are mediated through an inhibitory effect of nicotine on prostacyclin (PGI_2) production. Effects of carbon monoxide in the smoke seem to be minor and short-lived.

These findings were corroborated by Renaud et al. (1984), who studied the effects of smoking cigarettes of varying nicotine content on plasma clotting times and on aggregation of platelets with thrombin, adenosine diphosphate, collagen and epinephrine in 10 human subjects. Both the clotting functions and platelet aggregation were increased with increasing nicotine content in cigarettes. There was no correlation of these parameters, however, with carboxyhaemoglobin levels. The carboxyhaemoglobin levels were increased approximately 60% above baseline levels.

Effeney (1987) provided convincing evidence that these effects of nicotine and carbon monoxide on platelet function are mediated through opposing effects on PGI_2 production. Four rabbits were exposed to 460 mg carbon monoxide/m^3 (400 ppm) in an exposure chamber for 7–10 days. Carboxyhaemoglobin levels averaged about 20%. Ten rabbits received an infusion of nicotine for 7–10 days. Full-thickness samples of atrial and ventricular myocardium were incubated with arachidonic acid for determination of PGI_2 production by radioimmunoassay of 6-keto-$PGF_{1\alpha}$ (a stable metabolite of PGI_2) and by inhibition of platelet aggregation. Carbon monoxide exposure increased PGI_2 production, which was significant in the ventricular myocardium. Nicotine exposure reduced PGI_2 production in all tissues examined. The combination of nicotine and carbon monoxide caused a net increase in PGI_2 production. The effect of carbon monoxide may be to induce hypoxaemia, a known stimulant of PGI_2 production. This effect of carbon monoxide would serve to reduce aggregation.

Another explanation for the inhibition of platelet aggregation by carbon monoxide exposure has been provided by Bruene & Ullrich (1987). These investigators bubbled carbon monoxide through platelet-rich plasma and then challenged the platelets with various agonists. The carbon monoxide exposure was much greater than that encountered in physiological or even toxic states. The results, however, indicated that inhibition of aggregation was related to enhancement of guanylate cyclase action and associated increased cyclic guanosine monophosphate levels.

7.3 Respiratory system

7.3.1 Pulmonary morphology

Laboratory animal studies by Niden & Schulz (1965) and Fein et al. (1980) found that very high levels of carbon monoxide (5700–11 000 mg/m^3 [5000–10 000 ppm]) for 15–45 min were capable of producing capillary endothelial and alveolar epithelial oedema in rats and rabbits, respectively. These effects of carbon monoxide have not been reported, however, at lower levels of carbon monoxide exposure.

In a small number ($n = 5$) of New Zealand white rabbits, Fein et al. (1980) reported a significant increase in the permeability of ^{51}Cr-labelled ethylenediaminetetraacetic acid from alveoli to arterial blood within 15 min after the start of exposure to 0.8% (9200 mg/m^3 [8000 ppm]) carbon monoxide. Passage of this labelled marker persisted and increased throughout the remaining 30 min of the study. The mean carboxyhaemoglobin level after exposure was 63 ± 4% (mean ± standard error). Although morphometric examination was not performed, transmission electron microscopy showed evidence of capillary endothelial and alveolar epithelial swelling and oedema along with detachment of the endothelium from the basement membrane. Mitochondria were disintegrated, and alveolar Type 2 cells were depleted of lamellar bodies. None of these effects was found in four control animals exposed to air.

Despite an increase in gross lung weight, Penney et al. (1988) were unable to demonstrate any evidence of oedema in the lungs of male albino rats after 7.5 weeks of exposure to incrementally increasing concentrations of carbon monoxide ranging from 290 to 1500 mg/m^3 (250 to 1300 ppm). The authors also reported that this effect was not due to increased blood volume in the lung or fibrosis, as measured by lung hydroxyproline content. There was, therefore, no obvious explanation other than general growth for the lung hypertrophy reported in this study after chronic exposure to high concentrations of carbon monoxide.

Fisher et al. (1969) failed to find any histological changes in the lungs of mongrel dogs exposed to carbon monoxide concentrations of 9200–16 000 mg/m³ (8000–14 000 ppm) for 14–20 min (up to 18% carboxyhaemoglobin). Similarly, no morphological changes were found by Hugod (1980) in the lungs of adult rabbits continuously exposed to 230 mg carbon monoxide/m³ (200 ppm) for up to 6 weeks (11.9–19% carboxyhaemoglobin) or to 2200 mg carbon monoxide/m³ (1900 ppm) for 5 h (31–39% carboxyhaemoglobin).

Niden (1971) speculated about possible effects of low levels of carbon monoxide on cellular oxidative pathways when he reported that exposure of mice to concentrations of carbon monoxide ranging from 57 to 100 mg/m³ (50 to 90 ppm) for 1–5 days, resulting in carboxyhaemoglobin levels of <10%, produced increased cristae in the mitochondria and dilation of the smooth endoplasmic reticulum in the non-ciliated bronchiolar (Clara) cell. Minimal changes, consisting of fragmentation of lamellar bodies, were found in the Type 2 epithelial cell. The morphological appearance of the remaining cells of the terminal airways was normal. The results of this study were not presented in detail, however, and have not been confirmed at low concentrations of carbon monoxide. Thus, the significance, if any, of changes in the structure of cells lining the terminal airways is unknown.

Weissbecker et al. (1969) found no significant changes in the viability of alveolar macrophages exposed *in vitro* to high concentrations of carbon monoxide (up to 220 000 mg/m³ [190 000 ppm]). These results were later confirmed in more extensive *in vivo* exposure studies by Chen et al. (1982), who obtained alveolar macrophages by bronchoalveolar lavage from rats exposed to 570 mg carbon monoxide/m³ (500 ppm; 41–42% carboxyhaemoglobin) from birth through 33 days of age. Morphological and functional changes in the exposed cells were minimal. There were no statistically significant differences in cell number, viability, maximal diameter, surface area or acid phosphatase activity. The phagocytic ability of alveolar macrophages was enhanced by carbon monoxide exposure, as determined by a statistically significant ($P < 0.05$) increase in the percentage of spread forms and cells containing increased numbers of retained latex particles. The biological significance of this finding is questionable, however, because very few ($n = 5$) animals were evaluated and no follow-up studies have been performed.

7.3.2 Lung function

Laboratory animal studies of lung function changes associated with carbon monoxide exposure parallel the morphology studies described above, because high concentrations (1700–11 000 mg/m³ [1500–10 000 ppm]) of carbon monoxide were utilized.

Fisher et al. (1969) ventilated the left lung of seven dogs with 8–14% (92 000–160 000 mg/m³ [80 000–140 000 ppm]) carbon monoxide for 14–20 min. Femoral artery blood carboxyhaemoglobin levels ranged from 8% to 18% at the end of carbon monoxide ventilation. No changes in the diffusing capacity or pressure–volume characteristics of the lung were found.

Fein et al. (1980) measured lung function in nine New Zealand white rabbits exposed for 45 min to either 0.8% (9200 mg/m³ [8000 ppm]) carbon monoxide or air. After carbon monoxide exposure, carboxyhaemoglobin levels reached 63%. The dynamic lung compliance significantly decreased and the airway resistance significantly increased at 15 min and 30 min after the start of carbon monoxide exposure, respectively. The mean blood pressure fell to 62% of the baseline value by the end of exposure; the heart rate was not changed. The arterial pH decreased progressively throughout exposure, although there were no changes in the alveolar–arterial P_{O_2} difference.

Robinson et al. (1985) examined ventilation ($\dot{V}_A$) and perfusion ($\dot{Q}$) distribution in mongrel dogs during and following carbon monoxide exposure. A small number ($n = 5$) of mongrel dogs were exposed to 1% carbon monoxide (11 000 mg/m³ [10 000 ppm]) for 10 min, resulting in peak carboxyhaemoglobin levels of 59 ± 5.4%. Inert gas distributions were measured at peak exposure and 2, 4 and 24 h after exposure. No changes in $\dot{V}_A/\dot{Q}$ were found. Previous studies were unable to show accumulation of lung water in the same model (Halebian et al., 1984a,b). The authors concluded that other constituents of smoke, besides carbon monoxide, were responsible for the pulmonary oedema and $\dot{V}_A/\dot{Q}$ mismatching found in victims exposed to smoke in closed-space fires.

Little is known about the effects of carbon monoxide on ventilation in laboratory animals, and the few studies available are contradictory. No effects of carbon monoxide on ventilation were found in unanaesthetized rabbits (Korner, 1965) or cats (Neubauer et al., 1981), whereas large increases were reported in conscious goats (Santiago & Edelman, 1976; Doblar et al., 1977; Chapman et al., 1980). In anaesthetized cats, high concentrations of carbon monoxide (11 000 mg/m^3 [10 000 ppm]) increased ventilation (Lahiri & Delaney, 1976). Gautier & Bonora (1983) used cats to compare the central effects of hypoxia on control of ventilation under conscious and anaesthetized conditions. The cats were exposed for 60 min to either low inspired oxygen fraction (where the fraction of inspired oxygen $[F_IO_2]$ = 0.115) or carbon monoxide diluted in air. In conscious cats, 1700 mg carbon monoxide/m^3 (1500 ppm) caused decreased ventilation, whereas higher concentrations (2300 mg/m^3 [2000 ppm]) induced first a small decrease, followed by tachypnoea, which is typical of hypoxic hypoxia in carotid-denervated conscious animals. In anaesthetized cats, however, carbon monoxide caused only mild changes in ventilation.

Other respiratory effects of carbon monoxide hypoxia, such as the increased total pulmonary resistance estimated by tracheal pressure, have been reported in anaesthetized laboratory rats and guinea-pigs (Mordelet-Dambrine et al., 1978; Mordelet-Dambrine & Stupfel, 1979). The significance of this effect is unknown, however, particularly under the extremely high carbon monoxide exposure conditions utilized in these studies (4-min inhalation of 2.84% carbon monoxide [33 000 mg/m^3; 28 400 ppm]), which produced carboxyhaemoglobin concentrations above 60% (Stupfel et al., 1981). Similar increases in tracheal pressure were also seen with hypoxic hypoxia (F_IO_2 = 0.89), suggesting a possible general mechanism associated with severe tissue hypoxia.

7.4 Central nervous system and behavioural effects

7.4.1 Effects on cerebral blood flow and brain metabolism

7.4.1.1 Effects on cerebral blood flow

Kety & Schmidt (1948) demonstrated that cerebral oxygen consumption is about 3.5 ml of oxygen per 100 g of brain (cerebral hemispheres) per minute in normal adult humans. This consumption of oxygen is virtually unchanged under a variety of conditions. The cerebral vasculature responds to decreases in oxygen availability by increasing cerebral blood flow in order to maintain cerebral oxygen delivery and/or by increasing oxygen extraction in order to maintain cerebral oxygen utilization when cerebral oxygen delivery is limited.

An important point to emphasize when comparing carbon monoxide hypoxia with hypoxic hypoxia is that although arterial oxygen content is reduced with both hypoxic and carbon monoxide hypoxia, there is no reduction in arterial oxygen tension with carbon monoxide hypoxia. Comparisons of the equivalent effects of both carbon monoxide and hypoxic hypoxia on cerebral blood flow and cerebral oxygen consumption in the same animal preparations have been made by Traystman and co-workers in several studies (Traystman & Fitzgerald, 1977, 1981; Traystman et al., 1978; Fitzgerald & Traystman, 1980; Koehler et al., 1982, 1983, 1984, 1985). The concept of equivalent effects of both types of hypoxia (hypoxic and carbon monoxide) has been described by Permutt & Farhi (1969) and involves the comparison of physiological effects of elevated carboxyhaemoglobin and low oxygen at equal reductions in haemoglobin, arterial oxygen content, arterial or venous oxygen tension or blood flow.

Traystman et al. (1978) previously reported that the increase in cerebral blood flow in dogs during a reduction in arterial oxygen content, produced by exposing the animal to a low oxygen gas mixture (hypoxic hypoxia), was not different from the increase in cerebral blood flow when oxygen content was decreased by adding carbon monoxide to the breathing gas mixture (carbon monoxide hypoxia) (Fig. 9). This was true both before and after carotid sinus chemo-denervation. This study also demonstrated that mean arterial blood pressure increased with hypoxic hypoxia, whereas it decreased with carbon monoxide hypoxia. Because the cerebral blood flow increase

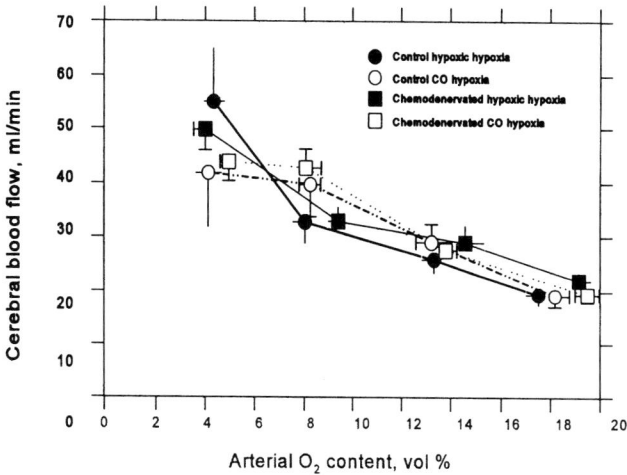

Fig. 9. Effect of hypoxic hypoxia and carbon monoxide hypoxia on cerebral blood flow in 13 control and 9 chemodenervated dogs. Each point represents the mean ± standard error. Analysis of variance showed that the four slopes were not significantly different from each other. Point-by-point analysis using Student's *t*-test showed that the minimum difference that was significant was between chemodenervated hypoxic hypoxia and chemodenervated carbon monoxide hypoxia at 8 vol. % ($P = 0.05$) (from Traystman et al., 1978).

with carbon monoxide and hypoxic hypoxia was similar, cerebro-vascular resistance actually decreased more with carbon monoxide hypoxia. This represents the effect of the carotid chemoreceptors on systemic blood pressure during each type of hypoxia. When the carotid chemoreceptors were denervated, cerebrovascular resistance decreased to the same level as with carbon monoxide hypoxia. The authors concluded that the carotid chemoreceptors do not play an important role in the global cerebral vasodilator response to either carbon monoxide or hypoxic hypoxia.

Both human and animal histopathology studies have suggested that there are regional differences in tissue injury following severe carbon monoxide exposure. One potential source of these differences is regional differences in the cerebral blood flow response to carbon monoxide exposure. Two logical comparisons of the regional cerebral blood flow response to carbon monoxide hypoxia are (1) anatomical (i.e., rostral to caudal [cortex to brain-stem] comparison) and

(2) physiological (i.e., brain areas with a functional blood–brain barrier versus brain areas without an intact blood–brain barrier). Koehler et al. (1984) observed interesting regional cerebral blood flow responses to hypoxic and carbon monoxide hypoxia in newborn lambs and adult sheep. In adults, regions with high normoxic blood flows such as the caudate nucleus and midbrain showed a large response to hypoxia, whereas lower blood flow regions with large white matter tracts, such as the cervical spinal cord, pons, diencephalon and piriform lobe, showed a relatively lower response. Other brain regions were essentially homogeneous in their responses. Carbon monoxide hypoxia increased cerebral blood flow to a greater extent than hypoxic hypoxia in all brain regions, but the overall pattern of regional cerebral blood flow was similar for the two types of hypoxia in the adults.

In the newborn, regional responses differed for each type of hypoxia. With hypoxic hypoxia in the newborn, the brain-stem regions had a significantly greater response than all other regions except the caudate nucleus, whereas all cerebral lobes responded significantly less than all other regions. With carbon monoxide hypoxia, the difference between brain-stem responses and those of other regions was less marked. In the adults, in contrast, there was no significant interactive effect between the type of hypoxia and the pattern of regional response. With both types of hypoxia, the caudate nucleus had a significantly greater response than all other regions, and the cervical spinal cord responded significantly less than all other regions.

Ludbrook et al. (1992) exposed rabbits to 1% carbon monoxide (11 000 mg/m^3 [10 000 ppm]) or to hypoxic hypoxia. Both conditions caused a significant increase in cerebral blood flow of up to 300%, such that oxygen delivery to the brain was unchanged. Despite the maintenance of oxygen delivery to the brain during and after the carbon monoxide exposure, the cortical somatosensory evoked response voltages were halved during the exposure and recovered to only about 80% of baseline subsequently. The authors concluded that the primary toxicity of carbon monoxide to the brain in rabbits is not due to a reduction in oxygen delivery.

Okeda et al. (1987) also demonstrated carbon monoxide-induced regional cerebral blood flow differences in cats. This group attempted to demonstrate that there is a selective vulnerability of the pallidum and cerebral white matter and showed low cerebral blood flow values for these brain areas. In newborn lambs (Koehler et al., 1984), unlike adult sheep, the patterns of regional cerebral blood flow responses were not similar with the two types of hypoxia. Brain-stem regions, especially the medulla, had marked responses relative to the cerebrum during hypoxic hypoxia. Peeters et al. (1979) also made this observation in unanaesthetized fetal lambs.

There is at least one area of the brain that does not respond to alterations in arterial oxygen content and partial pressure as do other brain areas — the neurohypophysis. The neurohypophysis is an ana-tomically unique region of the brain, and the regulation of blood flow to this area appears to be different from that to other areas of the brain. Hanley et al. (1986) demonstrated that when arterial oxygen content was reduced equivalently with hypoxic hypoxia and carbon monoxide hypoxia, global cerebral blood flow increased by 239% and 300%, respectively. Regional cerebral blood flow also showed similar responses for all brain areas except the neurohypophysis. With hypox-ic hypoxia, neurohypophyseal blood flow increased markedly (320%), but it was unchanged with carbon monoxide. These blood flow responses of the neurohypophysis occur independently of alterations in blood pressure.

Wilson et al. (1987) determined the role of the chemoreceptors in the neurohypophyseal response to hypoxia and found that chemo-receptor denervation completely inhibited the increase in neuro-hypophyseal blood flow associated with hypoxia. The response to carbon monoxide was unaltered. These data (Hanley et al., 1986; Wilson et al., 1987) demonstrated that the mechanism responsible for the increase in neurohypophyseal blood flow with hypoxia is unique compared with other brain regions. The only animals in which neuro-hypophyseal blood flow did not respond were the denervated animals and those given carbon monoxide. Both of these conditions are ones in which the chemoreceptors have been shown to be inactive (Comroe, 1974; Traystman & Fitzgerald, 1981). Although the blood flow response to hypoxia does not involve the peripheral chemoreceptors for most brain regions, this is not true for the neurohypophysis. Thus, here is an example of one regional brain area that does not respond to carbon monoxide hypoxia (i.e., a change in arterial oxygen content) but does respond to a change in arterial oxygen tension. This suggests that the chemoreceptor represents the mechanism involved in the neurohypophyseal response to hypoxic hypoxia and that local changes in arterial oxygen content are not involved in this response, because the neurohypophysis does not respond to carbon monoxide. It is unclear whether other regional brain areas have similar responses to hypoxic and carbon monoxide hypoxia.

Carbon monoxide hypoxia results in an increase in cerebral blood flow, and this has been demonstrated by a number of investigators (Sjostrand, 1948b; Haggendal & Norback, 1966; Paulson et al., 1973; Traystman et al., 1978; Traystman & Fitzgerald, 1981).

Traystman (1978) examined the cerebral blood flow responses to carbon monoxide hypoxia in anaesthetized dogs, particularly in the range of carboxyhaemoglobin less than 20% (Fig. 10). A carboxy-haemoglobin level as low as 2.5% resulted in a small, but significant, increase in cerebral blood flow to 102% of control. With reductions in oxygen carrying capacity of 5, 10, 20 and 30% (5, 10, 20 and 30% carboxyhaemoglobin), cerebral blood flow increased to approximately 105, 110, 120 and 130% of control, respectively. At each of these levels, cerebral oxygen consumption remained unchanged. At car-boxyhaemoglobin levels above 30%, cerebral blood flow increased out of proportion to the decrease in oxygen carrying capacity, but the brain could no longer maintain cerebral oxygen consumption constant. At a carboxyhaemoglobin level of 50%, cerebral blood flow increased to about 200% of control. These findings are in general agreement with those of MacMillan (1975), who demonstrated that as carboxy-haemoglobin increased to 20, 50 and 65%, cerebral blood flow increased to 200, 300 and then 400%, respectively, in cats. These cerebral blood flow increases at 20% carboxyhaemoglobin are higher than those reported in Traystman's (1978) study, but the reason for this is not known. Haggendal & Norback (1966) demonstrated a 50–150% of control increase in cerebral blood flow with carboxy-haemoglobin levels of 30–70%, and Paulson et al. (1973) showed a 26% increase in cerebral blood flow with a carboxyhaemoglobin level of 20%. These findings also indicate that cerebral blood flow increases progressively with increasing carboxyhaemoglobin concentrations and that cerebral oxygen consumption is maintained constant even at a carboxyhaemoglobin level of 30%. This has important implications regarding the behavioural and electrophysiological consequences of carbon monoxide exposure. These findings are also consistent with those of Dyer & Annau (1978), who found that superior colliculus evoked potential latencies are not affected by carboxyhaemoglobin

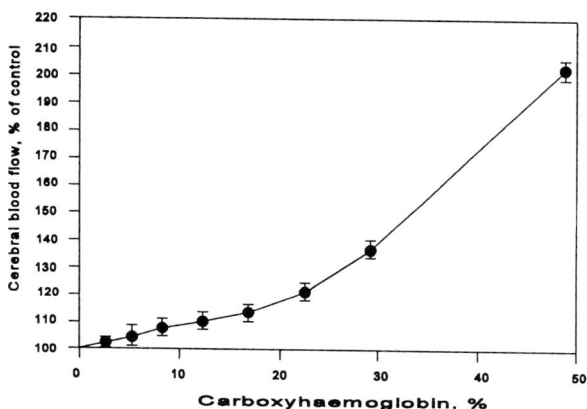

Fig. 10. Effect of increasing carboxyhaemoglobin levels on cerebral blood flow, with special reference to low-level administration (below 20% carboxyhaemo-globin). Each point represents the mean ± standard error of 10 dog preparations (from Traystman, 1978).

levels up to 40%. At levels above this, the brain cannot increase blood flow enough to compensate for decreased tissue oxygen delivery. At these high carboxyhaemoglobin levels, then, behavioural and neuro-physiological abnormalities should be quite evident. At lower car-boxyhaemoglobin levels, these abnormalities should not be seen, because the brain can increase its blood flow or oxygen extraction to maintain a constant cerebral oxygen consumption. These data lead to the suggestion that as long as the brain can compensate for the decrease in oxygen availability by increasing its blood flow or oxygen extraction to maintain a constant cerebral oxygen consumption, there should be no detrimental effects of hypoxia at these levels (i.e., up to a carboxyhaemoglobin level of 30% or more). However, when these compensatory mechanisms fail, detrimental effects on behavioural or electrophysiological aspects should be observed. This is a completely different hypothesis from the one that suggests there are behavioural or neurophysiological effects at carboxyhaemoglobin levels of less than 10%, or even 5%.

Lanston et al. (1996) examined cerebral blood flow in sheep during carbon monoxide exposure (65% carboxyhaemoglobin). Mean arterial blood pressure was generally unchanged. Oxygen delivery to the brain was sustained throughout the administration of carbon monoxide because of a significant increase in cerebral blood flow. Despite lack of evidence of metabolic acidosis or brain hypoxia, oxygen consumption by the brain fell progressively, and the sheep showed behavioural changes that varied from agitation to sedation and narcosis. The authors hypothesized that the mechanism of these changes was probably unrelated to hypoxia but may have been due to raised intracranial pressure or a direct effect of carbon monoxide on brain function.

Brian et al. (1994) examined the hypothesis that carbon monoxide directly relaxes cerebral blood vessels. The aorta and basilar and middle cerebral arteries were used for tension recordings *in vitro*. After pre-contraction, cumulative relaxation concentration–response curves for carbon monoxide, nitric oxide and other substances were obtained. Carbon monoxide (1×10^{-6} to 3×10^{-4} mol/litre) did not affect tension in rabbit or dog cerebral arteries. In rabbit aorta, carbon monoxide induced $29 \pm 4\%$ (mean $\pm$ standard error of the mean) relaxation at the highest concentration used (3×10^{-4} mol/litre). In contrast, nitric oxide produced 80–100% relaxation of all arteries. The authors suggested that carbon monoxide does not have a significant effect on tone in cerebral arteries, but at high concentrations it produces concentration-dependent relaxation in rabbit aorta. Factors that account for this regional heterogeneity are not clear.

7.4.1.2 Effects on brain metabolism

Ischiropoulos et al. (1996) studied nitric oxide production in the brain, documenting by electron paramagnetic resonance spectroscopy that nitric oxide is increased ninefold immediately after carbon monoxide poisoning. Evidence that peroxynitrite was generated was sought by looking for nitrotyrosine in the brains of carbon monoxide-poisoned rats. Nitrotyrosine was found deposited in vascular walls and also diffusely throughout the parenchyma in immunocytochemical studies. A 10-fold increase in nitrotyrosine was found in the brains of carbon monoxide-poisoned rats. When rats were pretreated with the nitric oxide synthase inhibitor L-nitroarginine methyl ester, formation of both nitric oxide and nitrotyrosine in response to carbon monoxide poisoning was abolished, as well as leukocyte sequestration in the microvasculature, endothelial xanthine dehydrogenase conversion to xanthine oxidase and brain lipid peroxidation. The authors concluded that perivascular reactions mediated by peroxynitrite are important in the cascade of events that lead to brain oxidative stress in carbon monoxide poisoning.

Song and collaborators (1990) exposed rat pups to carbon monoxide acutely and subacutely and made measurements of brain monoamines (norepinephrine [NE], dopamine [DA] and 5-hydroxy-tryptamine [5-HT]) and 5-hydroxytryptamine metabolite, 5-hydroxy-indoleacetic acid (5-HIAA), using fluorospectroscopy. They found that 57 mg carbon monoxide/m^3 (50 ppm) was the minimal concentration that produced significant changes in monoamine levels and also showed a definite dose–effect–time relationship. The authors suggested that monoamines could be useful biochemical indices for toxicity of carbon monoxide in the central nervous system.

In a review article, Penney and his group (1993) described the results of a series of studies on acute severe carbon monoxide poisoning using a defined animal model. Rats were exposed to high levels of carbon monoxide (1700–2900 mg/m^3 [1500–2500 ppm]) for 90 min. The common carotid artery and jugular vein serving one side of the brain of the rats were occluded by indwelling catheter. The usual changes noted were hypotension, bradycardia, hypothermia, altered blood glucose, unconsciousness, cerebral oedema and behavioural evidence of central nervous system damage. The authors found that both greatly depressed and greatly elevated glucose during and/or after carbon monoxide exposure increased morbidity and mortality. Added hypothermia induced during carbon monoxide exposure was found to reduce morbidity/enhance survival, whereas rapid rewarming was found to be the best strategy to reduce morbidity/enhance survival post-carbon monoxide exposure. Infusion of ethanol during carbon monoxide exposure was found to significantly decrease morbidity and mortality. The use of the *N*-methyl-D-aspartic acid receptor blocker ketamine had a significant beneficial effect in reducing morbidity and mortality (Penney & Chen, 1996). Cerebral cortical oedema and changes in brain phosphagens were assessed following carbon monoxide poisoning. Through the use of magnetic resonance imaging and spectroscopic techniques, cyanide poisoning, in contrast to carbon monoxide exposure, was found to produce a different pattern of changes in blood glucose and lactate and failed to slow cardiac AV conduction and ventricular repolarization.

7.4.1.3   Evidence for a reoxygenation injury in carbon monoxide poisoning

The fact that cytochrome a_3 may remain inhibited when a patient is reoxygenated leads to consideration of the possibility of free oxygen radical formation and then occurrence of reoxygenation injury. It has long been recognized that, with regard to pathological aspects of brain lesions, similarities are striking between carbon monoxide poisoning and post-ischaemic reperfusion injury, and a common mechanism has been hypothesized for these different forms of brain injury (Okeda et al., 1981, 1982).

More recently, Thom (1990) showed evidence for the occurrence of lipid peroxidation in brain of carbon monoxide-poisoned rats. An increase of conjugated diene and malonyldialdehyde concentrations appears only after a 90-min period of normal air breathing following carbon monoxide exposure. Thom (1992) was able to demonstrate that blocking xanthine oxidase by allopurinol or depleting animals of this enzyme by feeding them a tungsten-supplemented diet decreases the magnitude of brain lipid peroxidation. This offers further evidence that, at least in part, a common mechanism exists between carbon monoxide poisoning and reperfusion injury.

Brown & Piantadosi (1992) reported further evidence for oxygen free radical generation during the reoxygenation phase after carbon monoxide exposure by showing a decrease in brain catalase activity that demonstrates hydrogen peroxide production. There was also a decrease in the ratio of reduced/oxidized glutathione with an increase in salicylate hydroxylation products and a decrease in intracellular pH and energetic compounds. This was rapidly corrected by hyperbaric oxygen as opposed to normobaric oxygen, under which it continued to decrease during the first 45 min.

7.4.2   Effects on schedule-controlled behaviour

Because of the high levels of carboxyhaemoglobin that are employed in studies of laboratory animals using schedule-controlled behaviour, effects of carboxyhaemoglobin are reported in all articles on the subject. There are a number of problems with the published literature. Only a few investigators measured carboxyhaemoglobin;

instead, most simply specified the exposure parameters. Another problem is that of hypothermia, which occurs in rats when carboxy-haemoglobin levels rise (Annau & Dyer, 1977; Mullin & Krivanek, 1982). If hypothermia develops as a consequence of carboxyhaemo-globin elevation, behavioural effects may be secondary to the hypo-thermia, rather than directly attributable to the carboxyhaemoglobin. None of the experimenters attempted to control for hypothermia effects. Thus, behavioural effects of carboxyhaemoglobin may be overestimated in the rat with respect to humans who do not exhibit hypothermia from elevated carboxyhaemoglobin.

The level of carboxyhaemoglobin may be estimated for studies in which it was not given by use of normative curves published in Montgomery & Rubin (1971). Schrot & Thomas (1986) and Schrot et al. (1984) published corroborating curves. For all rat studies in which carboxyhaemoglobin was not measured, carboxyhaemoglobin levels were estimated from exposure parameters.

With one exception (Mullin & Krivanek, 1982), effects of carboxyhaemoglobin on schedule-controlled behaviour did not occur until carboxyhaemoglobin exceeded approximately 20%. In some studies, no effect was observed until even higher levels. It is possible, however, that a number of the studies were insensitive because of the small numbers of subjects employed. In the study by Mullin & Krivanek (1982), it was reported that conditioned avoidance behaviour was affected at carboxyhaemoglobin levels as low as 12.2%. The carboxyhaemoglobin level reported in this study is, however, about half of the value that would be estimated from the Montgomery & Rubin (1971) data. It seems likely that either exposure or carboxy-haemoglobin values were erroneous in the report. If the exposure data were correct, the effects threshold would fit the other data in the literature. It thus appears that carboxyhaemoglobin does not affect schedule-controlled behaviour in laboratory animals until levels exceed 20%.

Miyagawa et al. (1995) examined the acute effects of carbon monoxide exposure on a steady-state operant behaviour involving bar pressing. This was done in (1) rats exposed to various concentrations of carbon monoxide (570, 1100, 1700 and 2300 mg/m^3 [500, 1000, 1500 and 2000 ppm]) for 1 h and (2) rats exposed to 1700 mg/m^3 (1500 ppm) for different periods (1, 2 and 4 h). Abrupt cessation of the response was produced by exposure to 1100 mg/m^3 (1000 ppm) or higher concentrations of carbon monoxide. Recovery from the effects of carbon monoxide exposure was observed as a sudden resumption of responding during the post-exposure period. The duration of exposure required to produce response inhibition was closely correlated with the exposure concentration. The post-exposure interval required for response recovery was also correlated with the exposure concentration. This post-exposure response recovery interval, however, was constant and independent of the duration of exposure when the concentration was fixed at 1700 mg/m^3 (1500 ppm). It was found that carboxyhaemoglobin levels were within the range 33–43% when response recovery occurred. This suggested the existence of a threshold carboxyhaemoglobin level associated with the drastic behavioural change.

When there were frank effects on schedule-controlled behaviour, they all seemed to be in the direction of a slowing of rate or speed of response. Schedule-produced patterns of behaviour were not disrupted, in general. Thus, it appears that the effect of elevated carboxyhaemoglobin is on some general aspect of behavioural control having to do with the rate of processing.

Table 14 summarizes health effects, other than developmental effects, of carbon monoxide on laboratory animals.

7.5 Adaptation

The earliest study on adaptation to carbon monoxide exposure in laboratory animals was performed by Killick (1937). Mice were exposed to successively higher concentrations of carbon monoxide, which, over a period of 6–15 weeks, reached 2300–3300 mg/m^3 (2000–2850 ppm) and produced 60–70% carboxyhaemoglobin. The non-adapted mice exhibited much more extreme symptoms when exposed to such levels. A control group was used to partially rule out effects of selection of carbon monoxide-resistant animals.

Clark & Otis (1952) exposed mice to gradually increasing carbon monoxide levels over a period of 14 days until a level of 1400 mg/m^3 (1200 ppm) was reached. When exposed to a simulated altitude of 10 400 m, survival of the carbon monoxide-adapted groups was much greater than that of controls. Similarly, when mice were acclimatized

Table 14. Health effects, other than developmental effects, of carbon monoxide on laboratory animals[a]

Species	LOEL[b] mg/m³	ppm	COHb (%)	Duration	Health effects	References
Disturbances in cardiac rhythm						
Dog	57 and 110	50 and 100	2.6–12.0	Intermittently or continuously, 6 weeks	Abnormal electrocardiograms	Preziosi et al., 1970
Dog	110	100	6.3–6.5	2 h	Ventricular fibrillation threshold reduced	Aronow et al., 1978, 1979
Dog, anaesthetized	570	500	20	90 min	Enhanced sensitivity to digitalis-induced ventricular tachycardia	Kaul et al., 1974
Dog, anaesthetized	5700	5000	4.9	5 sequential exposures	Myocardial ischaemia increased 1 h after coronary artery ligation	Becker & Haak, 1979
Dog	3400, then 150	3000, then 130	13–15	15 min, then 1 h	Carbon monoxide inhaled prior to coronary artery ligation increased severity and extent of ischaemic injury and magnitude of ST-segment elevation more than ligation alone	Sekiya et al., 1983
Monkey	110–120	96–102	12.4	Continuously (23 h/day for 24 weeks)	Higher P-wave amplitudes in both infarcted and non-infarcted subjects; higher incidence of T-wave inversion in infarcted subjects	DeBias et al., 1973

Table 14 (contd).

Species	LOEL mg/m³	LOEL ppm	COHb (%)	Duration	Health effects	References
Monkey	110	100	9.3	16 h	Reduction of the threshold for ventricular fibrillation induced by an electrical stimulus applied to the myocardium during the final stage of ventricular repolarization	DeBias et al., 1976
Haemodynamic studies						
Rat, anaesthetized	180	160	NA[c]	20 min, repeatedly	Brain P_{O_2} decreased	Weiss & Cohen, 1974
Rat, anaesthetized	170	150	7.5	0.5–2 h	Heart rate, cardiac output, cardiac index, time derivative of maximal force, stroke volume increased, and mean arterial pressure, total peripheral resistance, left ventricular systolic pressure decreased	Kanten et al., 1983
Rat, anaesthetized	570	500	24	1 h	Vessel inside diameter increased (36–40%) and vessel flow rate increased (38–54%)	Gannon et al., 1988
Rat	1700	1500	NA	90 min	Heart rate decreased	Penney et al., 1993
Rabbit, anaesthetized	11 000	10 000	21–28	3 min	Regional blood flow to the myocardium increased	Kleinert et al., 1980
Rabbit	9200	8000	63	15 min	Mean blood pressure and arterial pH decreased	Fein et al., 1980

Table 14 (contd).

Dog, anaesthetized	1700	1500	30 min	23.1	Heart rate increased, myocardial oxygen consumption decreased	Adams et al., 1973
Dog	1700	1500	NA	15	Heart rate increased with increasing carboxy-haemoglobin, at rest and during exercise in both resistant and susceptible subjects	Farber et al., 1990
Dog, anaesthetized	11 000 then 1100	10 000 then 1000	15–20 min (total)	61–67	Cardiac output and stroke volume increased, mean arterial pressure and total peripheral resistance decreased	Sylvester et al., 1979
Cardiomegaly						
Rat	230	200	1–42 days	15.8	Heart weight increased	Penney et al., 1974b
Rat	460	400	6 weeks	35	Myocardial lactate dehydrogenase M subunits elevated 5–6%	Styka & Penney, 1978
Rat	570	500	38–47 days	38–40	Left ventricular apex-to-base length and left ventricular outside diameter increased 6.4% and 7.3%, respectively	Penney et al., 1984
Rat	230	200	Prenatally until 29 days of age	NA	Cardiomegaly induced; haemoglobin, haematocrit and red blood cell counts decreased	Penney et al., 1980
Rat	570	500	Newborns, continuous, 30 days	NA	Number and size of arteries and veins in heart increased	Penney et al., 1993

Table 14 (contd).

Species	LOEL		COHb (%)	Duration	Health effects	References
	mg/m³	ppm				
Rabbit	210	180	16.7	2 weeks	Ultrastructural changes in the myocardium	Kjeldsen et al., 1972
Haematology studies						
Rat	69	60	NA	21 days	Fetal haematocrit increased	Prigge & Hochrainer, 1977
Rat	230	200	27.8	18 days before birth	Fetal haemoglobin concentration decreased; fetal haematocrit decreased	Penney et al., 1980
Rat	180	157	21.8	17 days before birth	Fetal red blood cell count decreased	Penney et al., 1983
Rat	110	100	9.3	1–42 days	Haemoglobin concentration increased	Penney et al., 1974b
Rat	110	96	4.9–12.7	Intermittently or continuously, for up to 90 days	Haemoglobin and haematocrit increased	Jones et al., 1971
Guinea-pig	230	200	9.4–12.0	Intermittently or continuously, for up to 90 days	Haemoglobin concentration and haematocrit increased	Jones et al., 1971

Table 14 (contd).

Monkey	110	96	4.9–12.7	Intermittently or continuously, for up to 90 days	Haematocrit increased	Jones et al., 1971
Monkey	230	200	9.4–12.0	Intermittently or continuously, for up to 90 days	Haemoglobin concentration and haematocrit increased	Jones et al., 1971
Dog	220	195	30	72 h	Haemoglobin concentration and haematocrit increased	Syvertsen & Harris, 1973
Dog	57	50	7.3	3 months	Haemoglobin concentration, haematocrit and red blood cell count slightly increased	Musselman et al., 1959
Atherosclerosis and thrombosis						
Rabbit	190–400	170–350	17–33	10 weeks, cholesterol added to diet	Atheromatosis and increased cholesterol accumulation in aorta	Astrup et al., 1967
Rabbit	230	200	17	6 weeks continuously and intermittently	Cholesterol concentration in aortic arch increased	Stender et al., 1977
Rabbit	290	250	20	Intermittent, 10 weeks, cholesterol added to diet	Coronary artery atherosclerosis	Davies et al., 1976

Table 14 (contd).

Species	LOEL		COHb (%)	Duration	Health effects	References
	mg/m³	ppm				
Rabbit	170	150	NA	45 days, cholesterol added to diet	Low-density lipoprotein particles richer in cholesterol and with a higher cholesterol-to-phospholipid molar ratio	Alcindor et al., 1984
Rabbit	460	400	20	7–10 days	Increased PGI$_2$ production; significant in the ventricular myocardium	Effeney, 1987
Pigeon	170	150	10	6 h/day, 5 days/week for 52 weeks, cholesterol added to diet	Development of coronary artery lesions enhanced	Turner et al., 1979
Mini-pig	180 & 210	160 & 185	NA	4 h/day for 1–16 days	Adhesion of platelets to arterial endothelium and to fossae of degenerated endothelial cells; platelet aggregation, haematocrit and blood viscosity increased	Marshall & Hess, 1981
Squirrel monkey	110–340	100–300	9–26	Intermittent, 7 months, cholesterol added to diet	Atherosclerosis in coronary arteries enhanced	Webster et al., 1970

Table 14 (contd).

Pulmonary morphology

Mouse	57	50	<10	1–5 days	Cristae increased in the mitochondria in non-ciliated bronchiolar (Clara) cell; smooth endoplasmic reticulum dilated in the non-ciliated bronchiolar (Clara) cell	Niden, 1971
Rat	5700–11 000	5000–10 000	NA	15–45 min	Oedema of capillary endothelium and alveolar epithelium	Niden & Schulz, 1965
Rat	290	250	NA	7.5 weeks	Increase in gross lung weight	Penney et al., 1988
Rat	570	500	41–42	Birth through 33 days of age	Phagocytic ability of alveolar macrophages enhanced	Chen et al., 1982
Rabbit	5700–11 000	5000–10 000	NA	15–45 min	Oedema of capillary endothelium and alveolar epithelium	Fein et al., 1980
Rabbit	9200	8000	63	15 min	Capillary endothelial and alveolar epithelial swelling and oedema, permeability to ^{51}Cr-labelled ethylenediaminetetraacetic acid from alveoli to arterial blood increased, detachment of the endothelium from the basement membrane, mitochondrial disintegration, and alveolar Type 2 cells depleted of lamellar bodies	Fein et al., 1980

Table 14 (contd).

Species	LOEL		COHb (%)	Duration	Health effects	References
	mg/m³	ppm				
Lung function						
Rat, anaesthetized	33 000	28 400	>60	4 min	Total pulmonary resistance increased	Mordelet-Dambrine et al., 1978; Stupfel et al., 1981
Guinea-pig, anaesthetized	33 000	28 400	>60	4 min	Total pulmonary resistance increased	Mordelet-Dambrine et al., 1978; Stupfel et al., 1981
Rabbit	9200	8000	63	15 min	Dynamic lung compliance decreased, and airway resistance increased	Fein et al., 1980
Cat, anaesthetized	11 000	10 000	NA	NA	Ventilation increased	Lahiri & Delaney, 1976
Cat	1700	1500	NA	60 min	Ventilation decreased	Gautier & Bonora, 1983

Table 14 (contd).

Effects on cerebral blood flow and brain metabolism

					Effect	Reference
Rat	57	50	NA	Pups exposed acutely and subacutely	Monoamine ([NE, DA, & 5-HT] and 5-HT metabolite, 5-HIAA) levels changed	Song et al., 1990
Rat	1700	1500	NA	90 min	Blood glucose altered, unconsciousness induced, cerebral oedema, central nervous system damage and hypothermia	Penney et al., 1993
Rabbit	11 000	10 000	NA	NA	Increase in cerebral blood flow; cortical somatosensory evoked response voltages halved	Ludbrook et al., 1992
Dog, anaesthetized	?	?	2.5	?	Cerebral blood flow increased	Traystman, 1978
Cat	?	?	20	?	Cerebral blood flow increased 200%	MacMillan, 1975

Effects on schedule-controlled behaviour

					Effect	Reference
Rat	1100	1000	NA	1–4 h	Bar pressing ceased	Miyagawa et al., 1995

[a] Experiments involving cigarette smoke have not been included.
[b] LOEL = lowest-observed-effect level.
[c] NA = not available.

to a simulated altitude of 5500 m, Clark & Otis (1952) showed that these altitude-adapted mice survived 2900 mg carbon monoxide/m^3 (2500 ppm) better than controls. Wilks et al. (1959) reported similar effects in dogs.

Chronic exposure of rats to carbon monoxide increases haemoglobin concentration, haematocrit and erythrocyte counts via erythropoietin production. Penney et al. (1974a) concluded that the threshold for the erythropoietin response was 110 mg/m^3 (100 ppm; 9.26% carboxyhaemoglobin). Cardiac enlargement, involving the entire heart during carbon monoxide exposure (compared with right ventricular hypertrophy with high-altitude exposure), is induced when ambient carbon monoxide is near 230 mg/m^3 (200 ppm), producing a carboxyhaemoglobin level of 15.8% (Penney et al., 1974b). Blood volume of the rats exposed for 7.5 weeks to carbon monoxide exposures peaking at 1500 mg/m^3 (1300 ppm) nearly doubled, and erythrocyte mass more than tripled (Penney et al., 1988). After 42 days of continuous exposure to 570 mg/m^3 (500 ppm), rat blood volume almost doubled, primarily as a consequence of increases in numbers of erythrocytes (Davidson & Penney, 1988). It should be noted that all the demonstrated effects on tissues and fluids are induced by long-term exposures to high carbon monoxide concentrations. McGrath (1989) exposed rats for 6 weeks to altitudes ranging from 1000 m (ambient) to 5500 m and to concentrations of carbon monoxide ranging from 0 to 570 mg/m^3 (0 to 500 ppm). At 10 and 40 mg carbon monoxide/m^3 (9 and 35 ppm), where carboxyhaemoglobin levels ranged from 0.9% to 3.3%, there were no significant changes in body weight, right ventricular weight, haematocrit or haemoglobin. Small but non-significant changes in these variables were measured when the carbon monoxide concentration was 110 mg/m^3 (100 ppm) and carboxyhaemoglobin levels ranged from 9.4% to 10.2%. This is consistent with the observations noted above (Penney et al., 1974a) — that the threshold for erythropoietic effects was 110 mg/m^3 (100 ppm).

Besides the level of exposure, the time course of exposure to carbon monoxide is also important. As discussed above, haemoglobin increases in laboratory animals exposed to carbon monoxide after about 48 h and continues to increase in the course of continued exposure until about 30 days, depending perhaps upon exposure level. This haematopoietic response to long-term carbon monoxide exposure is similar to that shown for long-term hypoxic hypoxia, except that it is slower to start and tends to offset carbon monoxide hypoxic effects.

Most investigators have at least implied that increased haemoglobin level is the mechanism by which adaptation occurs. Certainly this explanation is reasonable for the studies showing increased survival in groups adapted for several days. Little has been done, however, to elucidate the extent to which such increases offset the deleterious effects of carbon monoxide. The probability that some adaptation occurs is supported theoretically by haemoglobin increases and empirically in the findings of laboratory animal studies measuring survival time. However, adaptation has not been demonstrated for specific health effects other than survival time.

7.6 Developmental toxicity

7.6.1 Introduction

Developmental toxicity has been described as including death of the developing organism, structural abnormalities, altered growth and functional deficits resulting from toxic exposures that occur pre- and postnatally up to the subject's attaining sexual maturity. The appearance of toxic effects may occur at any time throughout life. Concern for special vulnerability of immature organisms to toxic compounds focuses on the possibilities that (1) a toxic exposure that is not sufficient to produce maternal toxicity or toxicity in the adult organism will adversely affect the fetus or neonate or (2) at a level of exposure that does produce a toxic consequence in the adult or mother, the fetus or neonate suffers a qualitatively different toxic response. Toxic responses that occur early in life but are not permanent may or may not be a cause of concern. In some cases, they truly may be transient events with no persisting consequences. However, in other cases, such results may have their own consequences for development of the organism or may reappear under conditions of ill health produced by other toxic exposures, environmental stresses or exposure to pathogenic agents.

There are theoretical reasons and supporting experimental data that suggest that the fetus and developing organism are especially vulnerable to carbon monoxide. One reason for approaching the fetus as a separate entity for purposes of regulation is that the fetus is likely to experience a different carbon monoxide exposure from the adult given identical concentrations of the gas in air. This is due to differences in uptake and elimination of carbon monoxide from fetal haemoglobin, which are documented below. Further, differences appear to exist among species in the relative affinity of fetal and adult haemoglobin for carbon monoxide. These data are reviewed by Longo (1970).

Less studied is the possibility that tissue hypoxia may differ between the fetus and adult even at equivalent carboxyhaemoglobin concentrations as a result of differences in perfusion of critical organs, differences in maturation of adaptive cardiovascular responses to hypoxia and differences in tissue requirements for oxygen. Inferences concerning these factors are obtained principally from experimentation performed in laboratory animals in which the immature organism does show enhanced toxicity relative to the adult. Concern must also be expressed for the development of sensitive and appropriate animal models. It is necessary to bear in mind the relative state of development of the human and laboratory animal in question at the time of birth in developing useful animal models. For example, the neonatal rat is very immature relative to the neonatal human at birth with respect to development of the central nervous system (Fechter et al., 1986); thus, a combined prenatal and neonatal exposure model may be more accurate in predicting consequences of prenatal exposure in the human.

Hill et al. (1977) aptly described mathematical models for predicting fetal exposure to carbon monoxide based upon placental transport and the differences between maternal and fetal haemoglobin affinity for carbon monoxide and oxygen. They predicted that for any maternal carbon monoxide exposure of moderate duration, fetal carboxyhaemoglobin levels would lag behind maternal carboxyhaemoglobin levels for several hours, but would, given sufficient time, surpass maternal carboxyhaemoglobin levels by as much as 10 percentage points (in humans) owing to the higher carbon monoxide affinity of fetal haemoglobin compared with adult haemoglobin. Moreover, they predicted a far longer wash-out period for the fetal circulation to eliminate carbon monoxide following termination of exposure than that found in the adult. Data accumulated in both laboratory animal and human studies support these conclusions.

Table 15 summarizes developmental effects of carbon monoxide on laboratory animals.

7.6.2 Evidence for elevated fetal carboxyhaemoglobin relative to maternal haemoglobin

A fairly wide range of neonatal and maternal carboxyhaemoglobin levels has been published for humans, probably due to wide differences in cigarette smoking patterns prior to and during labour. In one recent study, measurement of fetal cord blood in the offspring of cigarette smokers who smoked during labour showed fetal carboxyhaemoglobin levels 2.55 times higher than in maternal blood. Cord blood averaged 10.1% carboxyhaemoglobin at delivery, whereas maternal blood averaged 5.6% carboxyhaemoglobin on the mother's arrival at the hospital and 4.1% carboxyhaemoglobin at delivery (Bureau et al., 1982). These values for fetal carboxyhaemoglobin are fairly high relative to other published sources (Longo, 1977). However, higher fetal carboxyhaemoglobin levels have been found in laboratory studies across a broad range of animal species when sufficient time was allowed for carboxyhaemoglobin to equilibrate in the fetal compartment. Christensen et al. (1986) ultimately observed higher carbon monoxide levels in fetal guinea-pigs than in their dams following carbon monoxide exposure given near term. Immediately following maternal exposure, at gestational age 62–65 days, to a bolus of carbon monoxide gas (5 ml given over 65 s through a tracheal catheter), these investigators reported a faster elevation in maternal carboxyhaemoglobin levels than in fetal levels, a finding consistent with the models of Hill et al. (1977).

Anders & Sunram (1982) exposed gravid rats to 25 mg carbon monoxide/m^3 (22 ppm) for 1 h on day 21 of gestation and reported that fetal carboxyhaemoglobin levels averaged 12% higher than levels taken at the same time period in the dam. These results are consistent with those of Garvey & Longo (1978), whose study involved chronic carbon monoxide exposures in rats. Dominick & Carson (1983) exposed pregnant sows to carbon monoxide concentrations of 170–460 mg/m^3 (150–400 ppm) for 48–96 h between gestation days 108 and 110. They reported fetal carboxyhaemoglobin levels that exceeded maternal levels by 3–22% using a CO-Oximeter.

Studies with control and protein-deficient CD-1 mice exposed to 0 (endogenous), 74 or 140 mg carbon monoxide/m^3 (0, 65 or 125

Table 15. Developmental effects of carbon monoxide exposure on laboratory animals

Species	CO concentration		Duration	Maternal COHb (%)	Effects	References
	mg/m³	ppm				
Mice	290	250	7 and 24 h/day on GDs 6–15	10–11	Increase in resorptions at 7 h/day	Schwetz et al., 1979
CD-1 mice	74, 140, 290, 570	65, 125, 250, 500	24 h/day on GDs 6–17	NA	Increase in embryolethality; dose-dependent decrease in live litter size from 140 mg/m³ (125 ppm) up; decrease in mean fetal weights at 290 and 570 mg/m³ (250 and 500 ppm)	Singh & Scott, 1984
ICR mice	1) 1700, 2900, 4000 2) 570	1) 1500, 2500, 3500 2) 500	1) 10 min on one of GDs 5, 11 or 16 2) 1 h/day on GDs 0–6, 7–13, 14–20	NA	Dose-related increases in micronuclei (maternal bone marrow and fetal blood) and sister chromatid exchanges (maternal bone marrow and fetal cell suspension) in all dose groups (values for fetuses somewhat lower than in dams)	Kwak et al., 1986
Rats	230	200	24 h/day during the last 18 days of gestation	28	Decrease in litter size and birth weights	Penney et al., 1980, 1983
Rats	860	750	3 h/day on GDs 7, 8 or 9	NA	Increase in fetal resorptions and stillbirths; decrease in body length; skeletal anomalies	Choi & Oh, 1975
Rabbit	210	180	continuously throughout gestation	16–18	Increase in fetal mortality; malformations	Astrup et al., 1972

Table 15 (contd).

Rabbit	3100–6200	2700–5400	12 puffs; short periods daily from GDs 6 to 18	16	Increase in fetal mortality	Rosen-krantz et al., 1986
Sows	170–460	150–400	48–96 h between GDs 108 and 110	23 (290 mg/m³ [250 ppm])	Increase in the number of stillbirths at 290 mg/m³ (250 ppm)	Dominick & Carson, 1983
CF-1 mice	290	250	24 h/day on GDs 6–15	NA	Minor skeletal variants	Schwetz et al., 1979
Pigmented rats	170	150	continuously throughout gestation	NA	Decrease in birth weights	Fechter & Annau, 1980b
Rats	86, 170, 340	75, 150, 300	continuously throughout gestation until PD 10	NA	Dose-dependent decrease in body weights (from 86 mg/m³ [75 ppm] up)	Storm et al., 1986
Rats	110	100	GDs 0–21 or 9–21	10–14	Increase in placental weight and decrease in fetal weights in both treatment groups	Lynch & Bruce, 1989
Rats	1300–1400	1100–1200	2 h/day throughout gestation	NA	~20% increase in maternal haematocrit; decrease in placental and fetal weights	Leichter, 1993

Table 15 (contd).

Species	CO concentration		Duration	Maternal COHb (%)	Effects	References
	mg/m³	ppm				
Rats	69, 230	60, 200	chronic; no other information available	NA	Increase in heart weight (69 mg/m³ [60 ppm]); increase in the number of cardiac muscle fibres (230 mg/m³ [200 ppm])	Prigge & Hochrainer, 1977
Rats	170	150	prenatal exposure	NA	Increase in wet heart weights; slight decrease in body weights	Fechter et al., 1980
Rats	230	200	fetal and neonatal exposure	NA	Increase in right ventricular weight; increase in left ventricular weight	Clubb et al., 1986
Rats	1) 230 2) 230 + 570	1) 200 2) 200 + 500	1) continuously throughout gestation 2) continuously throughout gestation (200) + neonatally until the age of 29 days (500)	28	Cardiomegaly, decrease in haemoglobin, haematocrit and red blood cells; heart weight/body weight was still elevated in young adult rats receiving carbon monoxide both prenatally and neonatally	Penney et al., 1980
CD-1 mice	74, 140	65, 125	GD 7–18	NA	Impairment of the righting reflex on PD 1 and negative geotaxis on PD 10 at 140 mg/m³ (125 ppm); impairment of aerial righting (at 74 or 140 mg/m³ [65 or 125 ppm])	Singh, 1986

Table 15 (contd).

Rats	170	150	24 h/day continuously throughout gestation	Decrease in splenic macrophage phagocytosis, killing and oxygen release in 15-day-old rats; decrease in leukocyte common antigen in 21-day-old rats	Giustino et al., 1993, 1994
Rats	170	150	continuously throughout gestation	Delays in the development of negative geotaxis and homing behaviour	Fechter & Annau, 1980a,b
Rats	170	150	continuously throughout gestation	Impairment of acquisition and retention of a learned active avoidance task in rats 30–31 days of age	Mactutus & Fechter, 1984,1985
Rats	170	150	24 h/day continuously throughout gestation	Reduction in the minimum frequency of ultrasonic calls emitted by pups removed from their nest; decrease in the responsiveness (rate of calling) to a challenge dose of diazepam; impairment of acquisition of an active avoidance task	Di Giovanni et al., 1993
Rats	170	150	chronic, prenatal	Decrease in cerebellar wet weight; increase in norepinephrine level between ages 14 and 42 days	Storm & Fechter, 1985a
Rats	170	150	24 h/day continuously throughout gestation	Impairment of acquisition of a two-way active avoidance task in 90-day-old rats; impairment of acquisition and reacquisition of a two-way active avoidance task in 18-month-old rats	De Salvia et al., 1995

Table 15 (contd).

Species	CO concentration		Duration	Maternal COHb (%)	Effects	References
	mg/m³	ppm				
Rats	86, 170, 340	75, 150, 300	From GD 0 to PD 10	11.5, 18.5 and 26.8	Decrease in cerebellar weights and total GABA levels at 170 and 340 mg/m³ (150 and 300 ppm) on both PD 10 and PD 21; reduction in total GABA uptake	Storm et al., 1986
Rats	86, 170, 340	75, 150, 300	chronic prenatal exposure	NA	Dose-dependent decrease in norepinephrine and serotonin concentration in the pons/medulla at 21 days of age; dose-dependent increase in norepinephrine concentration in the neocortex at 42 days of age	Storm & Fechter, 1985b
Rats	170	150	24 h/day continuously throughout gestation	14	Decrease in resting chloride conductance and delay in the developmental reduction of resting potassium conductance in skeletal muscle within the first 60 days of postnatal life	De Luca et al., 1996
Rats	86, 170	75, 150	24 h/day on GD 0–20	14 (170 mg/m³ [150 ppm])	Increase in time constant of sodium current inactivation (on PD 40) and negative shift of equilibrium potential (on PD 40 and 270) in myelinated nerve fibres; depression of the rate of myelin formation	Carratù et al., 1993, 1995
Pigs	230, 290	200, 250	from GD until birth	20 and 22	Impairment of negative geotaxis behaviour and open field activity 24 h after birth at 290 mg/m³ (250 ppm); impairment of open field activity 48 h after birth at both 230 and 290 mg/m³ (200 and 250 ppm)	Morris et al., 1985a

Table 15 (contd).

Pigs	340	300	96 h	NA	Multifocal haemorrages and vacuolation of the neuropile throughout the cortical white matter and brain-stem; cerebellar oedema with swollen oligodendrocytes and astrocytes	Dominick & Carson, 1983

GD = gestational day; PD = postnatal day; GABA = γ-aminobutyric acid; NA = not available.

ppm) from days 8 to 18 of gestation revealed that maternal carboxy-haemoglobin levels were related to carbon monoxide concentration and were inversely related to dietary protein intake. The authors suggested that protein deficiency may enhance placental carboxy-haemoglobin levels, which may exacerbate fetal hypoxia (Singh et al., 1992).

Carbon monoxide uptake and elimination were studied in maternal and fetal sheep. When pregnant sheep in the last quarter of gestation were exposed to 34 mg carbon monoxide/m^3 (30 ppm), the maternal carboxyhaemoglobin increased from a baseline level of 1.1% to around 4.6% over 8–10 h. The fetal carboxyhaemoglobin increased more slowly from a baseline value of around 1.8% to a steady-state value of 7.4% by 36–48 h. At 57 mg/m^3 (50 ppm), the steady-state carboxyhaemoglobin in the dams was 12.2% and in the fetus 19.8%. The increase in the maternal carboxyhaemoglobin resembled a simple exponential process with a half-time of 1.5 h; the half-time for fetal carboxyhaemoglobin was 5 h. The decay curve for carbon monoxide elimination showed a similar relationship, with elimination occurring more slowly in the fetus than in the mother (Longo & Hill, 1977).

Fetal carboxyhaemoglobin kinetics may not be static throughout pregnancy. Bissonnette & Wickham (1977) studied transplacental carbon monoxide uptake in guinea-pigs at approximate gestational ages 45–68 days. They reported that placental diffusing capacity increases significantly with increased gestational age and appears to be correlated with fetal weight rather than placental weight. Longo & Ching (1977) also showed increases in carbon monoxide diffusion rates across the placenta of the ewe during the last trimester of pregnancy. However, they did not find a consistent increase when diffusion rate was corrected for fetal weight (i.e., when diffusing capacity was expressed on a per kilogram fetal weight basis).

7.6.3 *Effect of maternal carboxyhaemoglobin on placental oxygen transport*

Gurtner et al. (1982) studied the transport of oxygen across the placenta in the presence of carbon monoxide by cannulating both the maternal and fetal vessels of anaesthetized sheep. They measured the transport of oxygen across the placenta compared with transport of argon, urea and tritiated water when carbon monoxide was introduced. They showed a reduction in oxygen diffusing capacity relative to argon, which appeared to be related to the level of maternal carboxy-haemoglobin. Reduction of oxygen transport was observed below 10% carboxyhaemoglobin, and oxygen transport approached zero at car-boxyhaemoglobin values of 40–50%. Gurtner et al. (1982) interpreted these data as supporting the role of carrier-mediated transport for oxygen and suggested that carbon monoxide competitively binds to this carrier. An alternative explanation is that the introduction of carbon monoxide simply reduces the amount of fetal haemoglobin available to bind oxygen. Moreover, Longo & Ching (1977), for example, were unable to alter carbon monoxide diffusing capacity across the placenta by administration of a series of drugs that bind to cytochrome P-450.

Christensen et al. (1986) suggested that maternal carbon monox-ide exposure may independently impair oxygen diffusion across the placenta as a result of the enhanced affinity of maternal haemoglobin for oxygen in the presence of carboxyhaemoglobin (the Haldane effect). Using the guinea-pig, these authors demonstrated an initial, almost instantaneous, fall in fetal P_{O_2} in arterial blood and an increase in fetal partial pressure of carbon dioxide, which subsequently was followed by an increase in fetal carboxyhaemoglobin between approx-imately 5 and 10 min (the last time point studied, but a time when fetal carboxyhaemoglobin values were still far below maternal values). They calculated that the decrease in fetal oxygen transfer was due mostly to a decrease in maternal oxygen carrying capacity, but also, perhaps up to one-third, to the increased affinity of haemoglobin for oxygen in the presence of carbon monoxide. This model assumes that uterine perfusion remains constant under the experimental conditions used. Longo (1976) also showed a significant dose-related drop in fetal oxygen levels in blood taken from the fetal descending artery and fetal inferior vena cava after pregnant ewes were exposed to variable levels of carbon monoxide for durations sufficient to yield carboxy-haemoglobin equilibration in both the fetal and maternal compart-ments.

To summarize, it has been demonstrated that the presence of maternal carboxyhaemoglobin over a range of values results in depressed oxygen levels in fetal blood. The simplest explanations for the inverse relationship between maternal carboxyhaemoglobin and fetal oxygen levels are reduced maternal oxygen carrying capacity, impaired dissociation of oxygen from maternal haemoglobin (the

Haldane effect) and reduced availability of free fetal haemoglobin that is able to bind oxygen.

7.6.3.1 Measurement of carboxyhaemoglobin content in fetal blood

Zwart et al. (1981a) and Huch et al. (1983) called into question the accuracy of spectrophotometric measurements of carboxyhaemoglobin in fetal blood using the IL 182 and IL 282 CO-Oximeter. The CO-Oximeter is effectively a spectrophotometer preset to read samples at specific wavelengths that correspond to absorbance maxima for oxyhaemoglobin, carboxyhaemoglobin and methaemoglobin determined using adult blood samples. Different plug-in modules (IL 182) or programmed absorbance values (IL 282) can be used to correct for species differences in the absorbance spectrum of rat, human, dog and cow. Some investigators have used these instruments for estimating carboxyhaemoglobin levels in species for which the instrument has not been calibrated, such as the pig and guinea-pig. Typically, individual investigators have calibrated the CO-Oximeter using blood standards fully saturated with carbon monoxide and with oxygen. The adequacy of such a procedure is not certain. Further, the correspondence of absorbance maxima between adult and fetal haemoglobin for species upon which the CO-Oximeter is calibrated at the factory is an empirical question for which few data are published. Noting the finding of higher apparent carboxyhaemoglobin levels in the venous cord blood of humans than in the uterine artery, Huch et al. (1983) examined the possibility that oxyhaemoglobin in fetal blood might interfere with accurate measurement of carboxyhaemoglobin levels in the fetus, presumably because of different absorbance maxima for fetal than for adult blood. Working *in vitro*, Huch et al. (1983) deoxygenated fetal and maternal blood by flushing a tonometer with nitrogen and 5% carbon dioxide. They introduced a "small volume" of carbon monoxide gas, measured the blood gases using the IL 282 CO-Oximeter and then studied the effect of stepwise addition of oxygen to the apparent carboxyhaemoglobin levels. They showed little influence of oxygen saturation upon maternal carboxyhaemoglobin but indicated that oxygen saturation did affect readings of fetal carboxyhaemoglobin so as to overestimate carboxyhaemoglobin. This error is particularly likely at high oxyhaemoglobin concentrations. Zwart et al. (1981b) suggested an apparent elevation of carboxyhaemoglobin levels of approximately 2% with 40% oxyhaemoglobin saturation and of approximately 6% with oxyhaemoglobin levels of 90–95%. Such errors do not invalidate the finding that fetal carboxyhaemoglobin exceeds maternal values, but they do bring into question the magnitude of this difference. Whether similar errors also occur in measuring fetal carboxyhaemoglobin levels in animal blood is uncertain and should be subjected to experimental test. The calibration of spectrophotometers based upon fetal haemoglobin absorbance spectra rather than automated analysis based upon adult absorbance spectra is recommended to achieve greater accuracy in determining absolute levels of carbon monoxide in fetal blood. Vreman et al. (1984) described a gas chromatographic method for measuring carboxyhaemoglobin that has been applied to human neonates. Because of the very small volume of blood needed to make these measurements and because they eliminate the problem of absorbance spectra of fetal haemoglobin, this may be considered a useful means of accurately assessing carboxyhaemoglobin in developing organisms. There is also a new model of the CO-Oximeter (No. 482) that apparently allows for use of absorbance spectra based on calibration of fetal blood.

6.4 Consequences of carbon monoxide in development

This section presents the evidence that carbon monoxide exposure during early development has the potential of producing untoward effects. The four types of toxic outcomes — fetotoxicity, gross teratogenicity, altered growth and functional deficiencies in sensitive organ systems — are considered in order. As is the case in adult organisms, the nervous and cardiovascular systems appear to be most sensitive to carbon monoxide exposure.

4.1 Fetotoxic and teratogenic consequences of prenatal carbon monoxide exposure

There is clear evidence from laboratory animal studies that very high levels of carbon monoxide exposure may be fetotoxic. However, there exists some question concerning the level of exposure that causes fetal death, because both the duration and concentration of maternal exposure are critical values in determining fetal exposure. The data that suggest that prenatal carbon monoxide exposure produces terata are extremely limited and, again, come largely from studies using quite high exposure levels.

1)   Fetotoxicity

CD-1 mice (17 per group) were exposed to 0, 74, 140, 290 or 570 mg carbon monoxide/m^3 (0, 65, 125, 250 or 500 ppm), 24 h per day, from days 6 to 17 of gestation and were sacrificed on day 18. No signs of maternal toxicity were observed at any dose. Embryo-fetal toxicity became apparent at 140 mg/m^3 (125 ppm), resulting in a dose-dependent decrease in live litter size, an increase in embryolethality and decreased mean fetal weights. No malformations were detected. Although a significant decrease in fetal weights was found at the two highest doses, skeletal abnormalities (reduced ossification) were reported as not being dose dependent (Singh & Scott, 1984).

Schwetz et al. (1979) also exposed mice to 290 mg carbon monoxide/m^3 (250 ppm) for 7 and 24 h per day on gestation days 6–15. They found no effect on number of implantation sites or number of live fetuses per litter, but they did find a significant elevation in resorptions with the 7-h exposure (10–11% carboxyhaemoglobin), but not with 24 h per day exposures.

Rates of micronuclei and sister chromatid exchanges were analysed in ICR mice from mothers and fetuses on day 20 of pregnancy after exposure to high acute doses (0, 1700, 2900 or 4000 mg/m^3 [0, 1500, 2500 or 3500 ppm] for 10 min) of carbon monoxide on one of day 5, day 11 or day 16 of pregnancy. In a more chronic exposure regime, pregnant mice were treated with 570 mg carbon monoxide/m^3 (500 ppm) for 1 h per day on days 0–6, 7–13 or 14–20. Dose-related increases in both micronuclei and sister chromatid exchanges were observed in all dose groups, values for fetuses being somewhat lower than for dams. This may relate to the different sensitivity of the tissues examined (maternal bone marrow and fetal blood for the micronucleus test, maternal bone marrow and fetal cell suspension for sister chromatid exchanges). No relevant differences were found between maternal and fetal values or between exposure on different days or time periods of gestation (Kwak et al., 1986).

Dominick & Carson (1983) exposed pregnant sows to carbon monoxide concentrations of 170–460 mg/m^3 (150–400 ppm) for 48–96 h between gestation days 108 and 110 (average gestation was 114 days). They showed a significant linear increase in the number of stillbirths as a function of increasing carbon monoxide exposure. Stillbirths were significantly elevated above control levels when the maternal carboxyhaemoglobin levels exceeded 23% saturation. These saturation levels were obtained at approximately 290 mg carbon monoxide/m^3 (250 ppm). Carboxyhaemoglobin levels were measured using an IL 182 CO-Oximeter equipped with a human blood board; pig blood samples fully saturated with carbon monoxide and with oxygen were run each day to calibrate the instrument. There was very large variability among litters at a given concentration level in the percentage of stillbirths that occurred. Penney et al. (1980) found evidence of reduced litter size in rats exposed for the last 18 days of gestation to 230 mg carbon monoxide/m^3 (200 ppm; maternal carboxyhaemoglobin levels averaged 28%). However, Fechter et al. (1987) did not observe similar effects on litter size in rats exposed to levels of carbon monoxide as high as 340 mg/m^3 (300 ppm; maternal carboxyhaemoglobin levels of 24%).

2) Teratogenicity

There are limited data (Astrup et al., 1972) suggesting increased fetal mortality and malformations among rabbits exposed to 210 mg carbon monoxide/m^3 (180 ppm; 16–18% carboxyhaemoglobin) throughout gestation. The frequency of malformations reported was small and the historical rate of such anomalies in the laboratory was undocumented, so these results require replication by other workers before they can be considered as the basis for regulation. Rosenkrantz et al. (1986) exposed rabbits to high doses (12 puffs of 3100–6200 mg/m^3 [2700–5400 ppm]) of carbon monoxide for short periods daily from gestation days 6 to 18. Carboxyhaemoglobin levels were estimated at 16%, although animals had not equilibrated with the inhaled mixture. Despite a large number of fetal deaths, there was no evidence of terata in the carbon monoxide-exposed animals.

Choi & Oh (1975) reported skeletal anomalies in rats exposed to 860 mg carbon monoxide/m^3 (750 ppm) for 3 h per day on gestation day 7, 8 or 9. They also reported an excess in fetal absorptions and stillbirths and a decrease in body length. Schwetz et al. (1979) reported no teratogenic effects but did report an increase in minor skeletal variants in CF-1 mice exposed to 290 mg carbon monoxide/m^3 (250 ppm) for 24 h per day from gestation days 6 to 15.

One of the best-studied and possibly one of the most sensitive measures of early carbon monoxide exposure is a depression in birth weight. The effect seen in animals following fetal carbon monoxide exposure is generally transitory and occurs despite the fact that maternal body weight growth through pregnancy does not appear to be adversely affected. Inasmuch as the depressed birth weight observed is a transient event, its significance is not clear. However, in humans, low-birth-weight babies may be at particular risk for many other developmental disorders, so the effect cannot be disregarded casually. Moreover, in humans, there is a strong correlation between maternal cigarette smoking and reduced birth weight (see chapter 8). Whether the causative agent here is carbon monoxide, nicotine or a combination of these or other agents is uncertain.

Fechter & Annau (1980b) replicated earlier data from their laboratory showing significantly depressed birth weights in pigmented rats exposed throughout gestation to 170 mg carbon monoxide/m^3 (150 ppm). Penney et al. (1980) also found a significant depression in birth weight among rats exposed for the last 18 days of gestation to 230 mg carbon monoxide/m^3 (200 ppm). Penney et al. (1983) showed a trend towards divergence in body weight among fetuses exposed to 230 mg carbon monoxide/m^3 (200 ppm), which developed progressively during the last 17 days of parturition, suggesting that late gestational exposure to carbon monoxide may be essential to observe the effect. Storm et al. (1986) reported that following carbon monoxide exposure from the beginning of gestation through postnatal day 10, body weight was depressed in a dose-dependent fashion at 86, 170 and 340 mg carbon monoxide/m^3 (75, 150 and 300 ppm). Moreover, these values were all significantly lower than air-control subjects. By age 21 days, no significant body weight differences were seen among the test groups. At no time did Fechter and colleagues observe evidence of maternal toxicity as identified by death, reduced maternal weight gain or gross physical appearance.

Exposure of Wistar rats (11–17 per group) to 110 mg carbon monoxide/m^3 (100 ppm) during pregnancy, with maternal carboxyhaemoglobin levels estimated to be in the order of 10–14%, resulted in fetal weight reductions of 8% and 6% in groups treated over days 0–21 or 9–21 of gestation, respectively. No effects on fetal weight were observed with exposure on days 0–15, 3–11 or 17–21 of gestation. Placental weight increased significantly in all groups for which carbon monoxide exposure continued to term but was unaffected when treatment ceased before term (Lynch & Bruce, 1989).

Leichter (1993) exposed groups of 7–8 pregnant Sprague-Dawley rats to 1300–1400 mg carbon monoxide/m³ (1100–1200 ppm) for 2 h per day throughout gestation. As food intake was reduced in carbon monoxide-exposed animals, a pair-fed control group was introduced in addition to a control group allowed food *ad libitum*. When the animals were killed on day 20 of pregnancy, maternal haematocrit values in the carbon monoxide-exposed group were increased by about 20% compared with both control group values. Fetal and placental weights were significantly decreased in the carbon monoxide-exposed group compared with the other two groups.

Morris et al. (1985b) exposed neonatal piglets chronically to carbon monoxide for 21 days starting at approximately 28 days of age (230 and 340 mg/m³ [200 and 300 ppm]; carboxyhaemoglobin levels averaged 16% and 21%, respectively). They reported a significant impairment in weight gain in pigs exposed to 340 mg carbon monoxide/m³ (300 ppm), but no effect in pigs exposed to 230 mg/m³ (200 ppm).

.4.2 Alteration in cardiovascular development following early carbon monoxide exposure

It is known that a variety of cardiovascular and haematopoietic changes can accompany hypoxia in neonates and adult subjects, including elevation in haemoglobin, haematocrit and heart weight. Data gathered in adult laboratory animals suggest that these changes may be related. Cardiomegaly resulting from hypoxia reflects the amount of work performed to extract an adequate supply of oxygen. Whether or not the same processes occur in prenatal and neonatal carbon monoxide-induced hypoxia has been the subject of several reports. For prenatal exposure, the accumulated laboratory animal data suggest that carbon monoxide-induced cardiomegaly may be proportionately greater than in adult animals at a given maternal carbon monoxide exposure level. Whether or not this is due to higher fetal carboxyhaemoglobin levels, as a consequence of fetal haemoglobin's higher affinity for carbon monoxide, is not clear. Although the cardiomegaly may resolve when the neonatal subject is placed in a normal air environment, there is evidence for a persisting increase in the number of muscle fibres. The functional significance of these changes, if any, is uncertain. The lowest-observed-effect level (LOEL) for fetal cardiomegaly has not been well determined. One experiment has shown significant elevation of heart weight following carbon monoxide exposures as low as 69 mg/m^3 (60 ppm) (Prigge & Hochrainer, 1977), and there are no published dose–response experiments that provide a no-observed-effect level (NOEL). Chronic prenatal carbon monoxide exposure at approximately 230 mg/m^3 (200 ppm) results in a significant increase in the number of muscle fibres in the heart. The NOEL for this change has not been determined.

Fechter et al. (1980) measured wet and dry heart weight and protein and nucleic acid levels at several time points between birth and weaning in rats prenatally exposed to 170 mg carbon monoxide/m^3 (150 ppm) or to air. They reported that neonates had significantly elevated wet heart weights despite a slightly reduced body weight at birth. Groups did not differ at birth in dry heart weight, total protein or RNA or DNA levels in whole heart. No significant differences between groups on any measure were present at postnatal day 4 or subsequent ages studied. The data were interpreted as evidence for cardiac oedema rather than a change in heart muscle mass itself. The finding of a heavier heart at birth replicated the finding of Prigge & Hochrainer (1977), who exposed rats prenatally to carbon monoxide at levels as low as 69 mg/m^3 (60 ppm) and observed a similar increase in heart weight. Clubb et al. (1986) observed increases in right ventricular weight due to fetal carbon monoxide exposure and increases in left ventricular weight following neonatal exposure to 230 mg carbon monoxide/m^3 (200 ppm). As in the case of Fechter et al. (1980), they showed a gradual return to normal heart weight when prenatally exposed neonates were placed in an air environment neonatally.

These results were not verified by Penney et al. (1983) in offspring of pregnant rats exposed to 180, 190 or 230 mg carbon monoxide/m^3 (157, 166 or 200 ppm; 21.8–33.5% carboxyhaemoglobin) for the last 17 of 22 gestation days. These workers observed that wet and dry heart weights increase proportionately and concluded that cardiomegaly, present at birth, is not due to elevated myocardial water content. They also determined that cardiac lactate dehydrogenase M subunit composition and myoglobin concentration were elevated at 230 mg carbon monoxide/m^3 (200 ppm). They concluded that maternal carbon monoxide inhalation exerts significant effects on fetal body and placental weights, heart weight, enzyme constituents and enzyme composition. Moreover, in newborn rats inhaling 570 mg carbon monoxide/m^3 (500 ppm; 38–42% carboxyhaemoglobin) for

32 days and then developing in air, Penney et al. (1982) observed that heart weight to body weight ratios increased sharply after birth, peaked at 14 days of age and then fell progressively; they remained higher in rats exposed prenatally to carbon monoxide than in control rats for up to 107 days of age. The persistent cardiomegaly could not be explained by changes in DNA or hydroxyproline.

Male offspring of Wistar rats (number not stated) exposed to 0, 86 or 170 mg carbon monoxide/m^3 (0, 75 or 150 ppm) from day 0 to day 20 of gestation were evaluated for the effects of carbon monoxide exposure on splenic macrophage function (macrophage phagocytosis and killing ability, macrophage respiratory bursts) on day 15, 21 or 60 postpartum. Literature values and a pilot study showed 15% carboxy-haemoglobin saturation in pregnant rats exposed to 170 mg carbon monoxide/m^3 (150 ppm). No differences in maternal mortality or weight gain were observed in exposed rats. Litters were reduced to six male pups where possible within 24 h of birth and fostered to untreated mothers. Splenic macrophage phagocytosis and oxygen release were significantly decreased in 15- and 21-day-old males exposed to 170 mg carbon monoxide/m^3 (150 ppm) during pregnancy, and splenic macrophage killing was significantly reduced in 15-day-old males prenatally exposed to 86 or 170 mg carbon monoxide/m^3 (75 or 150 ppm). No carbon monoxide-induced alterations in splenic macrophage function were observed in 60-day-old rats (Giustino et al., 1993).

Female Wistar rats were exposed to 0, 86 or 170 mg carbon monoxide/m^3 (0, 75 or 150 ppm) from day 0 to day 20 of gestation, and the frequency of splenic immunocompetent cells was evaluated in young (15- to 21-day-old) or aged (18 months) male offspring. A significant decrease in leukocyte common antigen (LCA +) cells in 21-day-old animals prenatally exposed to 170 mg carbon monoxide/m^3 (150 ppm) was found, but other cell populations (macrophages, major histocompatibility [MHC] II cells, T and B lymphocytes) were not significantly lower than control values, although there was a trend towards a reduction. No changes were observed at the lower dose of 86 mg/m^3 (75 ppm) at 15 days or 18 months of age (Giustino et al., 1994).

To summarize, there is good evidence for the development of severe cardiomegaly following early-life carbon monoxide exposure at doses between 69 and 230 mg/m^3 (60 and 200 ppm). These effects are transitory if exposure is prenatal, and it is not clear whether they alter cardiac function or produce latent cardiovascular effects that may become overt later in life. There are many published reports suggesting increases in heart weight associated with neonatal carbon monoxide exposures up to 570 mg/m^3 (500 ppm), maintained over the first 30–60 days of life (see section 7.2.3). Persisting elevation in heart weight results from combined prenatal carbon monoxide exposure at 230 mg/m^3 (200 ppm) and neonatal exposure at 570 mg/m^3 (500 ppm). The LOEL for this effect has not been determined.

7.6.4.3   *Neurobehavioural consequences of perinatal carbon monoxide exposure*

Fechter & Annau (1980a,b) reported delays in the development of negative geotaxis and homing in rats exposed prenatally to 170 mg carbon monoxide/m^3 (150 ppm) (maternal carboxyhaemoglobin levels were not reported in this paper, but levels previously reported in this laboratory under that exposure regimen are 15–17%; Fechter & Annau, 1977). These data were replicated by Singh (1986) using CD-1 mice exposed to 0, 74 or 140 mg carbon monoxide/m^3 (0, 65 or 125 ppm) from gestation days 7 to 18. He found that exposure at 140 mg/m^3 (125 ppm) significantly impaired the righting reflex on postnatal day 1 and negative geotaxis on postnatal day 10. He also reported impaired aerial righting among subjects exposed prenatally to 74 or 140 mg carbon monoxide/m^3 (65 or 125 ppm). Morris et al. (1985a) studied the consequences of moderate carbon monoxide exposure given very late in gestation. They exposed pigs to 230 or 290 mg carbon monoxide/m^3 (200 or 250 ppm; carboxyhaemoglobin levels of 20% and 22%) from gestation day 109 until birth. They found impairment of negative geotaxis behaviour and open field activity 24 h after birth in pigs exposed to 290 mg/m^3 (250 ppm). Activity in the open field was significantly reduced in subjects exposed to both 230 and 290 mg/m^3 (200 and 250 ppm) 48 h after birth. The significance of these behavioural dysfunctions is that they point to delays in behavioural development that may themselves contribute to impairments in the way in which the individual interacts with its environment.

There are also reports of impaired cognitive function produced by prenatal carbon monoxide exposure that may be related to permanent neurological damage. Mactutus & Fechter (1984) showed poorer acquisition and retention of a learned active avoidance task in rats 30–31 days of age that had received continuous prenatal exposures of 170 mg/m³ (150 ppm). This study is noteworthy because very careful efforts were made to distinguish cognitive deficits such as motivational and emotional factors and performance deficits such as motor factors. These findings were replicated and extended by Mactutus & Fechter (1985), who studied the effects of prenatal exposures to 170 mg carbon monoxide/m³ (150 ppm; 16% maternal carboxyhaemoglobin) on learning and retention in weanling, young adult and aging (1-year-old) rats. They found that both the weanling and young adult rats showed significant retention deficits, whereas impairments were found in both learning and retention in aging rats relative to control subjects. They interpreted these results to mean that there are permanent neurological sequelae of prenatal carbon monoxide exposure. They raised the important issue that sensitivity of tests for consequences of early toxic exposure may reflect the developmental status of the test subject and complexity of the task. In this case, a learning impairment not observed in the early adult period was detected by working with aged subjects. No systematic attempts have been made to replicate these effects using lower levels of carbon monoxide.

Di Giovanni et al. (1993) exposed Wistar rats to 0, 86 or 170 mg carbon monoxide/m³ (0, 75 or 150 ppm) 24 h per day continuously throughout gestation. The high dose (170 mg/m³ [150 ppm]) produced 14% carboxyhaemoglobin saturation. The exposure regimen did not affect maternal body weight gain, duration of pregnancy, live litter size, pup mortality or pup weight and pup body weight gain. Litters were culled to six males within 24 h of birth and fostered to untreated dams. One pup from each litter was tested in a single behavioural test. Ultrasonic vocalization recorded on postnatal day 5 showed a decrease in the minimal frequency (kHz) of calling in pups exposed prenatally to 170 mg carbon monoxide/m³ (150 ppm). On day 10, a challenge dose of diazepam (0.25 mg/kg body weight) reduced the calling rate by about 90% in the control and low-dose group; in the high-dose group, the reduction was less than 50%. Locomotor activity with and without an amphetamine challenge was unaffected by maternal carbon monoxide exposure when rats were tested in an open field at 14 and 21 days of age. At 90 days of age, rats exposed prenatally to 170 mg/m³ (150 ppm) showed impaired acquisition of the correct response in a two-way active avoidance task. However, prenatal exposure to carbon monoxide (170 mg/m^3 [150 ppm]) significantly impaired the acquisition of a two-way active avoidance task in 3-month-old male rats as well as the acquisition and reacquisition of this schedule in 18-month-old rats subjected to six daily 20-trial sessions (De Salvia et al., 1995). These findings show that gestational exposure to carbon monoxide induces permanent learning and memory impairment in rat offspring.

7.6.4.4   *Neurochemical effects of prenatal and perinatal carbon monoxide exposure*

A significant number of studies have appeared concerning the effects of acute and chronic prenatal and perinatal carbon monoxide exposure upon a variety of neurochemical parameters. These experiments are important because the transmission of information between nerve cells is based upon neurochemical processes. Labelled neurotransmitters can sometimes act as markers for the development of specific neurons in the brain, thereby serving as a sensitive alternative to histopathological investigation, particularly when the toxic agent selectively affects neurons based upon a biochemical target. The absence of a specific cell group may have important consequences for subsequent brain development, because the absence of growth targets for synapse formation can have additional consequences on brain development. Altered neurochemical development has been observed at carbon monoxide exposure levels lower than those necessary to produce signs of maternal toxicity or gross teratogenicity in neonates. Chronic prenatal and perinatal exposures to 170–340 mg carbon monoxide/m^3 (150–300 ppm) have been shown to yield persisting alterations in norepinephrine, serotonin and γ-aminobutyric acid levels and in γ-aminobutyric acid uptake in rats. There are also a substantial number of acute exposure studies that have demonstrated neurochemical effects of carbon monoxide. However, these generally have been conducted at life-threatening levels and are not particularly relevant to setting ambient air standards for carbon monoxide.

Storm & Fechter (1985a) reported that chronic prenatal carbon monoxide exposures of 170 mg/m^3 (150 ppm; approximately 16–18% carboxyhaemoglobin based upon other research in this laboratory) decreased cerebellar wet weight but increased norepinephrine levels in this structure when expressed in terms of either concentration (nanograms per milligram wet weight) or total cerebellar content above control values between the ages of 14 and 42 days. This period represented the duration of the experiment. Although this persisting elevation in norepinephrine cannot be considered permanent, it is the case that rats do obtain normal adult values of monoamine neurotransmitters at about the age of 40–45 days. There was no effect of carbon monoxide treatment on norepinephrine levels in the cerebral cortex. Because noradrenergic neurons have their cell bodies outside of the cerebellum and project axons that terminate on cell bodies in this structure, Storm & Fechter's (1985a) data may reflect an effect of increased noradrenergic innervation secondary to toxic injury to target neurons in the cerebellum. Consistent with this hypothesis, Storm et al. (1986) reported deficits in cerebellar weight and, more importantly, deficits in markers of γ-aminobutyric acid-ergic activity in the cerebellum following prenatal and perinatal carbon monoxide exposures. γ-Aminobutyric acid is thought to be an inhibitory neurotransmitter present in several neuronal cell types that are intrinsic to the cerebellum. Subjects in this experiment received 0, 86, 170 or 340 mg carbon monoxide/m^3 (0, 75, 150 or 300 ppm; corresponding maternal carboxyhaemoglobin levels were 2.5, 11.5, 18.5 and 26.8%) from the beginning of gestation until postnatal day 10. Neurochemical measurements were made either on postnatal day 10 or on postnatal day 21. They showed reduced total γ-aminobutyric acid levels in the cerebellum following either 170 or 340 mg carbon monoxide/m^3 (150 or 300 ppm) exposure at both measurement times. They also reported a significant reduction in total γ-aminobutyric acid uptake, but not glutamate uptake, in synaptosomes prepared from cerebella of 21-day-old neonates. Glutamate is an excitatory neurotransmitter found within the cerebellum. Histological markers of cerebellar toxicity were also obtained that were compatible with the neurochemical data, and these are described below in section 7.6.4.5.

In a related paper, Storm & Fechter (1985b) evaluated norepinephrine and serotonin levels at postnatal days 21 and 42 in four different brain regions (pons/medulla, neocortex, hippocampus and cerebellum) of rats following chronic prenatal exposures of rats to 86, 170 or 340 mg carbon monoxide/m^3 (75, 150 or 300 ppm). They reported that norepinephrine and serotonin concentrations decreased linearly with dose in the pons/medulla at 21, but not at 42, days of age (i.e., evidence of a transient effect); the LOEL was 170 mg/m^3 (150 ppm). Norepinephrine increased linearly with carbon monoxide dose in the neocortex at 42, but not at 21, days of age. The authors also showed that cerebellar weight was significantly reduced for the rats exposed at 170 and 340 mg/m^3 (150 and 300 ppm) when measured on postnatal 21 and for the rats exposed at 340 mg/m^3 (300 ppm) at 42 days of age.

7.6.4.5   Morphological consequences of acute prenatal carbon monoxide exposure

Profound acute carbon monoxide exposures do result in obvious neurological pathology that can be predicted to be inconsistent with life or with normal neurological development. These data are not as relevant to setting standards for ambient air quality as they are in demonstrating the danger of accidental high-level carbon monoxide exposures and in providing possible insight into the susceptibility of the developing brain to toxic exposure (Daughtrey & Norton, 1982; Okeda et al., 1986; Storm et al., 1986). The one possible exception is a study conducted in fetal pigs exposed via the mother to 340 mg carbon monoxide/m^3 (300 ppm) for 96 h (Dominick & Carson, 1983). The authors reported quite marked sensitivity to the carbon monoxide as reflected in fetotoxicity, but they also identified multifocal haemorrhages and vacuolation of the neuropile throughout the cortical white matter and brain-stem. They also observed cerebellar oedema with swollen oligodendrocytes and astrocytes, two non-neuronal cell types that have important roles in supporting neural function.

Full characterization of the histopathological effects of very low, subchronic carbon monoxide exposure on development is impeded by the absence of additional research in the published literature.

7.6.4.6   Neuromuscular effects of prenatal carbon monoxide exposure

Prenatal exposure of pregnant rats to 170 mg carbon monoxide/m^3 (150 ppm) but not 86 mg/m^3 (75 ppm) (carboxyhaemoglobin levels not measured) throughout gestation produced a depression of potassium and chloride conductances in gastrocnemius muscle fibres during the first 60 days of postnatal life, reaching control values at 80 days. These effects were responsible for a decrease in excitability of skeletal muscle fibres (De Luca et al., 1996).

A voltage clamp analysis of ionic currents recorded from peripheral (sciatic) nerve fibres isolated from prenatally exposed male Wistar rats (five litters per group exposed to 0, 86 or 170 mg carbon monoxide/m^3 [0, 75 or 150 ppm] on days 0–20 of gestation) was performed in animals aged 40 and 270 days. In 40-day-old carbon monoxide-exposed rats from both dose groups, changes in voltage and time-dependent properties of sodium channels were observed, resulting in slower inactivation of activity. These changes in membrane excitability were still present to a lesser degree at 270 days (Carratú et al., 1993). Morphometric examination of the sciatic nerve in the experiments reported by Carratú et al. (1995) showed that the rate of myelin formation in rats exposed to 170 mg carbon monoxide/m³ (150 ppm) (14% carboxyhaemoglobin) was depressed between 28 and 40 days of age, but it was not studied at later stages.

4.7 Summary

The data reviewed in this section provide strong evidence that prenatal carbon monoxide exposures of 170–230 mg/m³ (150–200 ppm; ~15–25% maternal carboxyhaemoglobin levels) produce reductions in birth weight, cardiomegaly, delays in behavioural development and disruption in cognitive function in laboratory animals of several species. Isolated experiments suggest that some of these effects may be present at carbon monoxide concentrations as low as 69–74 mg/m³ (60–65 ppm; ~6–11% carboxyhaemoglobin) maintained throughout gestation.

7.7 Other systemic effects

Several studies suggest that enzyme metabolism of xenobiotic compounds may be affected by carbon monoxide exposure (Montgomery & Rubin, 1971; Kustov et al., 1972; Pankow & Ponsold, 1972, 1974; Martynjuk & Dacenko, 1973; Swiecicki, 1973; Pankow et al., 1974; Roth & Rubin, 1976a,b). Most of the authors have concluded, however, that effects on metabolism at low carboxyhaemoglobin levels (≤15%) are attributable entirely to tissue hypoxia produced by increased levels of carboxyhaemoglobin, because they are no greater than the effects produced by comparable levels of hypoxic hypoxia. At higher levels of exposure, at which carboxyhaemoglobin concentrations exceed 15–20%, there may be direct inhibitory effects of carbon monoxide on the activity of mixed-function oxidases, but more basic research is needed. The decreases in xenobiotic metabolism shown with carbon monoxide exposure might be important to individuals receiving treatment with drugs (see section 8.1.6).

The effects of carbon monoxide on tissue metabolism noted above may partially explain the body weight changes associated with carbon monoxide. Short-term exposure to 290–1100 mg/m³ (250–1000 ppm) for 24 h was reported previously to cause weight loss in laboratory rats (Koob et al., 1974), but no significant body weight effects were reported in long-term exposure studies in laboratory animals at carbon monoxide concentrations ranging from 57 mg/m³ (50 ppm) for 3 months to 3400 mg/m³ (3000 ppm) for 300 days (Campbell, 1934; Musselman et al., 1959; Stupfel & Bouley, 1970; Theodore et al., 1971). It is quite probable that the initial hypoxic stress resulted in decreased weight gain followed by compensation for the hypoxia with continued exposure by adaptive changes in the blood and circulatory system. It is known, however, that carbon monoxide-induced hypoxia during gestation will cause a reduction in the birth weight of laboratory animals. Although a similar effect has been difficult to demonstrate in humans exposed to carbon monoxide alone, there is a strong correlation between maternal cigarette smoking and reduced birth weight (see section 8.2.7.2).

In experiments with adult rats exposed to 570 mg carbon monoxide/m³ (500 ppm) whose food and water consumption and body weight were measured on a daily basis, it was noted that after an initial period of a few days in which appetite was reduced, food and water consumption renormalized and body weight gain progressed (Penney, 1984). During weeks 2–5 of the study, body weight gain of the carbon monoxide-exposed rats was significantly greater than that of the matched air controls. It was suggested by the authors that carbon monoxide may have suppressed basal metabolic rate without curtailing food intake, thus allowing added body mass.

Inhalation of high levels of carbon monoxide, leading to carboxy-haemoglobin concentrations greater than 10–15%, has been reported to cause a number of systemic effects in laboratory animals as well as effects in humans suffering from acute carbon monoxide poisoning (see section 8.3). Tissues of highly active oxygen metabolism, such as heart, brain, liver, kidney and muscle, may be particularly sensitive to carbon monoxide poisoning. Other systemic effects of carbon monoxide poisoning are not as well known and are, therefore, less certain. There are reports in the literature of effects on liver (Katsumata et al., 1980), kidney (Kuska et al., 1980) and bone (Zebro et al., 1983). Results from one additional study in adult guinea-pigs suggest that immune capacity in the lung and spleen was affected by intermittent exposure to high levels of carbon monoxide for 3–4 weeks (Snella & Rylander, 1979). It is generally agreed that these systemic effects are caused by the severe tissue damage occurring during acute carbon monoxide poisoning due to (1) ischaemia resulting from the formation of carboxyhaemoglobin, (2) inhibition of oxygen release from oxyhaemoglobin, (3) inhibition of cellular cytochrome function (e.g., cytochrome oxidases) and (4) metabolic acidosis.

Gong & Wang (1995) investigated the effects of carbon monoxide at low concentration on lipid peroxidation and glutathione peroxidase in the rat. It was shown that lipid peroxide increased significantly in plasma when the rats were exposed to 500 mg carbon monoxide/m^3 (440 ppm) for 20 days, and also in heart when exposed for 30 days. No changes in lipid peroxide were caused by 25 and 50 mg carbon monoxide/m^3 (22 and 44 ppm), but 25 mg carbon monoxide/m^3 (22 ppm) resulted in changes in glutathione peroxidase. These findings indicate that carbon monoxide can induce free radicals and thus form lipid peroxide. Glutathione peroxidase may serve as an early sensitive indicator for the effects of carbon monoxide.

Shinomiya and collaborators (1994) investigated the relationship between degree of carboxyhaemoglobin saturation and the amount of carbon monoxide in various organs. Rats inhaled air containing 0.195% carbon monoxide (2200 mg/m^3 [1950 ppm]) for varying periods of time to give five experimental groups. After sacrifice, blood was collected for measurement of carboxyhaemoglobin, giving carboxyhaemoglobin values (%) of 12.2 ± 1.16, 31.6 ± 2.38, 42.4 ± 2.11, 52.3 ± 1.81 and 73.6 ± 3.01. The carbon monoxide content in the rat organs decreased in the following order: blood > spleen > liver > lung > kidney/myocardium > encephalon/pectoral muscles. There was a good positive correlation between the degree of carboxyhaemoglobin saturation and carbon monoxide content of the organ examined.

7.8 Interactions

7.8.1 Combinations with psychoactive drugs

8.1.1 Alcohol

There have been a number of animal studies of combinations of alcohol and carbon monoxide. Although there is some evidence that alcohol metabolism can be reduced in rat liver *in situ* by a carboxy-haemoglobin level of 20% (Topping et al., 1981), an *in vivo* study in mice found no effects of carbon monoxide exposure on alcohol metabolism (Kim & Carlson, 1983). On the other hand, Pankow et al. (1974) provided some evidence that high doses of carbon monoxide decreased blood alcohol levels in rats 30 min after a large dose of alcohol. They also reported that this dose of alcohol significantly lowered carboxyhaemoglobin levels associated with a large sub-cutaneous dose of carbon monoxide. These high-dose combinations were also associated with additive effects on enzyme markers of hepatotoxicity, but no interactions were observed when lower doses of carbon monoxide were given.

In contrast to the inconsistent metabolic effects seen with combinations of carbon monoxide and alcohol, results of two behavioural studies in animals have shown substantial interaction effects. Mitchell et al. (1978) studied the interaction of inhaled carbon monoxide with two doses of alcohol in rats using two behavioural measures. Sensorimotor incapacitation was assessed by failure to remain on a rotating rod. An additional measure of motor effects was the inability to withdraw the leg from a source of electric shock. The length of exposure to approximately 2300 mg carbon monoxide/m³ (2000 ppm) before the animals failed in these performances was decreased in a dose-dependent manner by alcohol. Carboxyhaemo-globin determinations made at the time of behavioural incapacitation were inversely related to alcohol dose.

Knisely et al. (1989) reported a large interaction of carbon monoxide exposure and alcohol administration on operant behaviour in animals. Mice, trained to lever press for water reinforcement, were tested with 1.1 g alcohol/kg body weight and various doses of carbon monoxide, alone and in combination. An unusual feature of this study was that both alcohol and carbon monoxide were administered by intraperitoneal injection. A dose of alcohol that had little effect on rates of lever pressing when given alone resulted in large rate-decreasing effects when given in combination with doses of carbon monoxide that also had no effects when given alone. Typically, behavioural effects of carbon monoxide alone were not seen under these test conditions until carboxyhaemoglobin saturations greater than 40–50% were obtained (Knisely et al., 1987). Thus, alcohol about doubled the acute toxicity of carbon monoxide in this study.

1.2   Barbiturates

There has been some interest in the interaction of carbon monoxide with barbiturates, because prolongation of barbiturate effects can reflect effects of toxicants on drug metabolism. In an early evaluation of the functional significance of the binding of carbon monoxide to cytochrome P-450, Montgomery & Rubin (1971) examined the effects of carbon monoxide exposure on the duration of action of hexobarbital and the skeletal muscle relaxant zoxazolamine in rats. Both drugs are largely deactivated by the hepatic mixed-function oxidase system. Although carbon monoxide was found to dose-dependently enhance both hexobarbital sleeping time and zoxazolamine paralysis, subsequent research indicated that this was probably not due to a specific inhibition of the mixed-function oxidase system by carbon monoxide, but rather was a non-specific effect of hypoxia, because even greater effects could be produced at a similar level of arterial oxygen produced by hypoxic hypoxia (Montgomery & Rubin, 1973; Roth & Rubin, 1976c). In support of the lack of effects of carbon monoxide on drug metabolism, Kim & Carlson (1983) found no effect of carbon monoxide exposure on the plasma half-life for either hexobarbital or zoxazolamine in mice.

There have been two studies of the interaction of carbon monoxide and pentobarbital using operant behaviour in laboratory animals. McMillan & Miller (1974) found that exposure of pigeons to 440 mg carbon monoxide/m^3 (380 ppm), a concentration that had little effect on behaviour when given alone, reduced the response rate, thereby increasing the effects of an intermediate dose of pentobarbital. On the other hand, the disruptive effects of all doses of pentobarbital on the temporal patterning of fixed-interval responding were enhanced markedly by 1200 mg carbon monoxide/m^3 (1030 ppm). This concentration of carbon monoxide by itself did not alter response patterning, but did lower overall rates of responding. In the study, Knisely et al. (1989) found generally additive effects of intraperitoneal carbon monoxide administration with the effects of pentobarbital in mice responding under a fixed-ratio schedule. In that study, the interaction of carbon monoxide with pentobarbital was not as evident as the interaction with alcohol, suggesting that general conclusions about carbon monoxide interactions with central nervous system depressant drugs may not be possible.

7.8.1.3   Other psychoactive drugs

Even more limited data are available on interactions of carbon monoxide exposure with other psychoactive drugs. In the study by Knisely et al. (1989) of interactions of intraperitoneal carbon monoxide administration with psychoactive drugs on operant behaviour of mice, *d*-amphetamine, chlorpromazine, nicotine, diazepam and morphine were studied in addition to alcohol and pentobarbital. As with alcohol, a suggestion of greater than additive effects was obtained from combinations of carbon monoxide with both *d*-amphetamine and chlorpromazine; however, in these cases, the differences from additivity did not reach statistical significance. Effects of carbon monoxide in combination with nicotine, caffeine and morphine were additive. McMillan & Miller (1974) also found evidence for an interaction of carbon monoxide and *d*-amphetamine on operant behaviour in pigeons. In this study, carbon monoxide concentrations as low as 560 and 1100 mg/m^3 (490 and 930 ppm) were able to modify the behavioural effects of *d*-amphetamine.

7.8.2   Combinations with other air pollutants and environmental factors

Many of the data concerning the combined effects of carbon monoxide and other pollutants found in ambient air are based on animal experiments; they are briefly reviewed here. The few human studies available are reviewed in chapter 8.

7.8.2.1   Other air pollutants

Groll-Knapp et al. (1988) reported that combined exposure of rats to carbon monoxide plus nitric oxide for 3 h caused a significant ($P <$ 0.01) increase in mean methaemoglobin levels when compared with methaemoglobin levels in rats exposed to nitric oxide alone. No significant changes were observed in blood carboxyhaemoglobin levels compared with exposure to carbon monoxide alone or to carbon monoxide plus nitric oxide. Combined exposure also caused significant behavioural changes. Hugod (1981) reported that combined exposure to carbon monoxide plus nitric oxide plus hydrogen cyanide for 2 weeks produced no morphological changes in the lungs, pulmonary arteries, coronary arteries or aortas of rabbits.

In a 1-year inhalation toxicity study, no adverse toxic effects were seen in groups of rats exposed to relatively low levels of carbon monoxide plus nitrogen dioxide or carbon monoxide plus sulfur dioxide compared with rats exposed to one of these pollutants alone (Busey, 1972). Murray et al. (1978) observed no teratogenic effects in offspring of mice or rabbits exposed to carbon monoxide plus sulfur dioxide for 7 h per day during gestation days 6–15 or 6–18, respectively.

Male Sprague-Dawley rats were given 3 or 6 mmol dibromomethane/kg body weight or breathed 260 mg carbon monoxide/m³ (225 ppm) for 120 min (Fozo & Penney, 1993). Peak carboxyhaemoglobin levels were 16% after 8 h with 3 mmol dibromomethane/kg body weight, 18% after 12 h with 6 mmol dibromomethane/kg body weight and 17% in carbon monoxide-exposed rats. Systolic blood pressure dropped 1.6 kPa (12 mmHg) in rats receiving carbon monoxide, but there was no significant change in rats receiving dibromomethane. Body temperature dropped approximately 1 °C in the rats receiving carbon monoxide or dibromomethane. There was no significant change in heart rate or in blood glucose or lactate in any of the rats.

8.2.2   Combustion products

A common condition in an atmosphere produced by a fire is the presence of a rapidly changing combination of potentially toxic gases (primarily carbon monoxide, carbon dioxide and hydrogen cyanide), reduced oxygen levels (hypoxic hypoxia) and high temperatures. Combined exposure to these gases occurs during smoke inhalation under conditions of hypoxic hypoxia. In addition, both carbon monoxide and carbon dioxide are common products of carbon-containing materials; consequently, accidental exposure to high levels of carbon monoxide will rarely occur without simultaneous exposure to carbon dioxide. Exposure to carbon monoxide and hydrogen cyanide is of concern because both carbon monoxide and hydrogen cyanide produce effects by influencing tissue oxygen delivery. Increased carboxyhaemoglobin reduces oxygen carrying capacity and may interfere with tissue oxygen release, whereas hydrogen cyanide inhibits tissue respiration. Studies were conducted to determine the toxicological interactions of the combustion products with and without reduced oxygen.

Several studies have investigated the effects resulting from combined exposure to carbon monoxide and combustion products from fires. Rodkey & Collison (1979) reported a significant ($P < 0.02$) decrease in mean survival time in mice jointly exposed until death to carbon monoxide plus carbon dioxide compared with mice exposed to carbon monoxide alone. In contrast, Crane (1985) observed no differences in the times to incapacitation or times to death in rats exposed until death to various concentrations of carbon monoxide plus carbon dioxide. In a recent study, Levin et al. (1987a) demonstrated a synergistic effect between carbon monoxide and carbon dioxide in rats exposed to various concentrations of carbon monoxide plus carbon dioxide. Simultaneous exposure to non-lethal levels of carbon dioxide (1.7–17.3%) and to sublethal levels of carbon monoxide (2900–4600 mg/m³ [2500–4000 ppm]) caused deaths in rats both during and following (up to 24 h) a 30-min exposure. Although the equilibrium levels of carboxyhaemoglobin were not changed by the presence of carbon dioxide, the rate of carboxyhaemoglobin formation was 1.5 times greater in rats exposed to carbon monoxide plus carbon dioxide than in rats exposed to carbon monoxide alone. The synergistic effects of carbon dioxide on carbon monoxide toxicity were also observed at other exposure times (Levin et al., 1988a).

Combined exposure to carbon monoxide plus hydrogen cyanide had an additive effect in rats, as evidenced by increases in mortality rate (Levin et al., 1987b, 1988a). Results from this series of experiments showed that the exposed animals died at lower carbon monoxide concentrations as the levels of hydrogen cyanide increased. In the presence of hydrogen cyanide, carboxyhaemoglobin at equilibrium was less than that measured in the absence of hydrogen cyanide; however, the initial rate of carboxyhaemoglobin formation was the same. This apparent depressive effect of hydrogen cyanide on carboxyhaemoglobin formation may explain the reason for the low carboxyhaemoglobin levels (<50%) seen in some people who died in a fire (Levin et al., 1987b). However, Hugod (1981) found no morphological changes in the lung, pulmonary and coronary arteries or aorta of rabbits exposed to carbon monoxide plus nitric oxide plus hydrogen cyanide (see section 7.8.2.1).

Combined exposures to carbon monoxide plus potassium cyanide have produced conflicting results. Norris et al. (1986) reported that the dose that was lethal to 50% of test subjects (LD_{50}) was significantly lower in mice pretreated with carbon monoxide prior to intraperitoneal injection of potassium cyanide. Sublethal doses of potassium cyanide produced a synergistic effect on mortality. On the other hand, Winston & Roberts (1975) observed no alterations in lethality in mice pre-treated with carbon monoxide and then treated with intraperitoneal injections of potassium cyanide.

The effects of carbon monoxide and potassium cyanide on the electrocardiogram were examined in female rats (1) treated with 1700 or 2700 mg carbon monoxide/m³ (1500 or 2400 ppm) for 90 min, (2) given 4 mg potassium cyanide/kg body weight alone or (3) given 1700 mg carbon monoxide/m³ (1500 ppm) plus 4 mg potassium cyanide/kg body weight (Katzman & Penney, 1993). Carbon monoxide significantly increased PR interval by 4.5–17.0 ms, with or without combination with potassium cyanide. Administration of potassium cyanide alone produced minimal change in the PR interval. QT interval was increased up to 20 ms by exposure to carbon monoxide, with or without combination with potassium cyanide. Potassium cyanide alone produced no change in the QT interval. T-wave duration was increased up to 22.5 ms by exposure to 1700 mg carbon monoxide/m³ (1500 ppm), with or without combination with potassium cyanide. Potassium cyanide alone produced minimal changes in T-wave duration. There were no changes in duration of the QRS complex or of the R wave. No correlation was observed between body temperature and QT interval. The results indicate that carbon monoxide at the concentrations used has major effects on the electrocardiogram in slowing AV conduction and ventricular repolarization (Katzman & Penney, 1993).

Levine-prepared female rats were exposed to 2700 mg carbon monoxide/m³ (2400 ppm), 1700 mg carbon monoxide/m³ (1500 ppm), 4 mg sodium cyanide/kg body weight or both 1700 mg carbon monoxide/m³ (1500 ppm) and 4 mg sodium cyanide/kg body weight (Dodds et al., 1992). The carbon monoxide exposures lasted 90 min, followed by recovery in room air. Rats exposed to the highest carbon monoxide concentration experienced significant bradycardia. All groups exhibited an initial hypotension, which was either maintained or exaggerated during exposure in all but the rats receiving cyanide and which returned to pre-exposure values by 90 min. All groups experienced a significant hypothermia during the exposure period, with those in the 1700 mg carbon monoxide/m³ (1500 ppm) group or the cyanide group returning to initial values over the recovery period. During exposure, all groups experienced an initial surge in glucose concentration, which was maintained in all but rats exposed to 2700 mg carbon monoxide/m³ (2400 ppm). The greatest hyper-glycaemic response resulted from the combination of carbon monoxide and cyanide, whereas 2700 mg carbon monoxide/m³ (2400 ppm) produced the smallest. Cyanide alone produced no significant rise in lactate concentration. However, lactate concentration in all other groups was significantly elevated during the exposure period, returning to initial values by 4 h of recovery. Lactate concentrations and neurological deficit in rats exposed to 1700 mg carbon monoxide/m³ (1500 ppm), when added to those rats treated with cyanide, closely approximated the lactate and neurological deficit of the combination treatment. Neurological deficit was greatest in rats exposed to 2700 mg carbon monoxide/m³ (2400 ppm). While in most cases the responses of the rats to carbon monoxide and cyanide differed whether the substances were administered alone or in com-bination, a synergistic relationship is not suggested. An additive or less than additive relationship is more likely.

A number of studies examined the effects of carbon monoxide administered under conditions of hypoxic hypoxia. Rodkey & Collison (1979) observed a lower mean survival time and a higher level of carboxyhaemoglobin in mice exposed to carbon monoxide in the presence of low oxygen (14%) than in those exposed to an ambient oxygen (21%) level. Winston & Roberts (1975) showed that pre-exposure of mice to carbon monoxide, followed by exposure to 7% oxygen 24 h later, had no effect on lethality compared with controls exposed to 7% oxygen only. Thus, pre-exposure to carbon monoxide had no protective effect against hypoxic hypoxia. However, pre-exposure to 10% oxygen caused a significant decrease in lethality in mice exposed 24 h later to carbon monoxide. Alterations in colethality were not associated with alterations in carboxyhaemoglobin levels. In a behavioural study in mice, Cagliostro & Islas (1982) showed that reaction times gradually increased with a decrease in oxygen levels to 10%. At <10% oxygen, reaction time increased dramatically. At reduced oxygen levels and in the presence of carbon monoxide, the decreases in performance were even greater than those observed in mice exposed to reduced oxygen levels alone.

Three and four gas combinations of combustion products were also examined (Levin et al., 1988b). The combinations tested included carbon monoxide, carbon dioxide, hydrogen cyanide and reduced oxygen. Carbon dioxide showed synergistic effects when tested with any of the other gases. The other gases were additive with carbon monoxide.

.2.3 Environmental factors

Yang et al. (1988) studied the combined effects of high temperature and carbon monoxide exposure in laboratory mice and rats. They were exposed 1 h per day for 23 consecutive days to environmental chamber temperatures of 25 and 35 °C at carbon monoxide concentrations ranging from 660 to 700 mg/m^3 (580 to 607 ppm). Carboxyhaemoglobin levels after 1 h of exposure ranged from 31.5% to 46.5%. The toxicity of carbon monoxide to mice, based on the concentration that is lethal to 50% of test subjects (LC$_{50}$) and survival time, was found to be 3 times higher at 35 °C. High temperature was also found to enhance the effects of carbon monoxide on the function of oxidative phosphorylation of liver mitochondria in rats. Body temperature regulation and heat tolerance were also affected by carbon monoxide exposure. The authors speculated that these effects of combined exposure to carbon monoxide and high temperature are due to the production of higher carboxyhaemoglobin, possibly as a result of hyperventilation. [Contrary to this notion, the Task Group noted that it is likely that the same equilibrium carboxyhaemoglobin saturation would only be reached sooner owing to increased ventilation caused by heat.]

Fechter et al. (1987, 1988) and Fechter (1988), using laboratory rats exposed to high levels of carbon monoxide (290–1400 mg/m^3 [250–1200 ppm] for 3.5 h) with and without broad-band noise (105 dB for 120 min), showed that carbon monoxide acts in a dose-dependent manner to potentiate noise-induced auditory dysfunction. Although carbon monoxide or noise alone did not have an effect, carbon monoxide combined with noise produced a more severe loss of hair cells at the basal end of the cochlea, especially at the high-frequency tones. A previous pilot study by Young et al. (1987) conducted at 1400 mg carbon monoxide/m^3 (1200 ppm) also showed that combined exposure to noise and carbon monoxide produced high-frequency shifts of greater magnitude than those produced by exposure to noise alone. The carbon monoxide levels used in these studies, however, are much greater than those encountered in the typical ambient environment, or even in the typical occupational environment.

There have been few studies of the long-term effects of carbon monoxide at altitude conducted in various laboratory animal species. James et al. (1979) studied cardiac function in six unsedated goats that were chronically instrumented and exposed to 180–230 mg carbon monoxide/m^3 (160–200 ppm; 20% carboxyhaemoglobin) for 6 weeks at 1500 m (5000 ft). Cardiac index and stroke volume were unchanged during and after the exposure. Heart rate and contractility of the left ventricular myocardium were unchanged during exposure to carbon monoxide, but both were depressed during the first week after removal of the carbon monoxide. The authors concluded that if there was a decrease in intrinsic myocardial function during the carbon monoxide exposure, it may have been masked by increased sympathetic activity.

McGrath (1988, 1989) studied cardiovascular, body and organ weight changes in rats exposed continuously for 6 weeks to (1) ambient altitude, (2) ambient altitude plus carbon monoxide, (3) simulated high altitude and (4) carbon monoxide at high altitude. Altitudes ranged from 1000 m (3300 ft) to 5500 m (18 000 ft), and carbon monoxide concentrations ranged from 0 to 570 mg/m^3 (0 to 500 ppm).

Carbon monoxide had no effect on body weight at any altitude. There was a tendency for haematocrit to increase even at the lowest concentration of carbon monoxide (10 mg/m^3 [9 ppm]), but the increase did not become significant until 110 mg/m^3 (100 ppm). At 3000 m (10 000 ft), there was a tendency for the total heart weight to increase in rats inhaling 110 mg carbon monoxide/m^3 (100 ppm). Although its effects on the heart at high altitude are complex, carbon monoxide had little effect on the right ventricle in concentrations of 570 mg/m^3 (500 ppm) or less; it did not exacerbate any effects due to altitude. There was a tendency for the left ventricle weight to increase with exposure to 40 mg carbon monoxide/m^3 (35 ppm) at high altitude, but the increase was not significant until 110 mg carbon monoxide/m^3 (100 ppm). Heart rate, blood pressure, cardiac output and peripheral resistance were unaffected by exposure to 40 mg carbon monoxide/m^3 (35 ppm) or 3000-m (10 000-ft) altitude, singly or in combination. The author concluded that 6 weeks of exposure to 40 mg carbon monoxide/m^3 (35 ppm) does not produce measurable effects in the healthy laboratory rat, nor does it exacerbate the effects produced by exposure to 3000-m (10 000-ft) altitude.

The data reported by McGrath (1988, 1989) are generally in agreement with findings reported by other investigators. The carboxy-haemoglobin concentrations at the end of the 6 weeks of exposure to carbon monoxide at 1000 m (3300 ft) were 0.6, 0.9, 2.4, 3.7 and 8.5% for ambient carbon monoxide levels of 0, 10, 40 and 110 mg/m^3 (0, 9, 35, 50 and 100 ppm), respectively. This relationship can be expressed as:

$$\% \ COHb = 0.115 + 0.08x \qquad (7\text{-}1)$$

where x is the carbon monoxide exposure in parts per million. The correlation coefficient (r) for this relationship was 0.99. The data from other altitudes were not sufficient to calculate the rate of increase. Exposure of rats to 570 mg/m^3 (500 ppm) and altitudes up to 5500 m (18 000 ft) resulted in carboxyhaemoglobin levels of 40–42%.

An interesting, but not unexpected, finding in this study was that high-altitude residence in the absence of exogenous carbon monoxide resulted in increased basal carboxyhaemoglobin concentrations. These values were 0.6, 1.3, 1.7 and 1.9% for altitudes from 1000 to 5500 m (3300 to 18 000 ft). These increases can be expressed as:

$$\% \ COHb = 0.0000914 + 0.26687x \qquad (7\text{-}2)$$

where x is altitude, in feet. The correlation coefficient (r) for this relationship was 0.99. Whether or not similar increases in basal carboxyhaemoglobin concentrations would be observed in humans adapted to altitude needs to be determined. The author concluded that although there was a tendency for haematocrit ratios, spleen weights and total heart weights to be elevated by combined carbon monoxide–altitude exposure, the results were not significant, and, in general, the effects produced by 4600-m (15 000-ft) altitude were not intensified by exposure to 110 mg carbon monoxide/m^3 (100 ppm).

McDonagh et al. (1986) studied cardiac hypertrophy and ventri-cular capillarity in rats exposed to 5500 m (18 000 ft) and 57, 110 and 570 mg carbon monoxide/m^3 (50, 100 and 500 ppm). Coronary capillarity increased after exposure to 5500 m (18 000 ft) for 6 weeks, but this response was blocked by carbon monoxide. Right ventricular thickness was increased by altitude but was not increased further by carbon monoxide. At 570 mg carbon monoxide/m^3 (500 ppm), the right ventricular hypertrophy was attenuated, but the results are uncertain owing to the high mortality in this group. Left ventricular thickness was also increased at 5500 m (18 000 ft) and was increased further by carbon monoxide. The authors concluded that because the ventricular thickness is increased while capillarity is reduced, it is possible that the myocardium can be underperfused in the altitude plus carbon monoxide group.

Cooper et al. (1985) evaluated the effects of carbon monoxide at altitude on electrocardiograms and cardiac weights in rats exposed for 6 weeks to (1) ambient, (2) ambient + 570 mg carbon monoxide/m^3 (500 ppm), (3) 4600 m (15 000 ft) and (4) 4600 m (15 000 ft) + 570 mg carbon monoxide/m^3 (500 ppm) conditions. Carboxyhaemoglobin values were 36.2% and 34.1% in the ambient + carbon monoxide and altitude + carbon monoxide groups, respectively. Haematocrits were $54 \pm 1\%$, $77 \pm 1\%$, $68 \pm 1\%$ and $82 \pm 1\%$ in the ambient, ambient + carbon monoxide, altitude and altitude + carbon monoxide groups, respectively. In the ambient + carbon monoxide, altitude and altitude + carbon monoxide groups, respectively, the mean electrical axis shifted 33.2° left, 30° right and 116.4° right. Heart weight to body weight ratios were 2.6×10^{-3}, 3.2×10^{-3}, 3.2×10^{-3} and 4.0×10^{-3} in the ambient, ambient + carbon monoxide, altitude and altitude + carbon monoxide groups, respectively. Whereas carbon monoxide increased left ventricular weight and altitude increased right ventricular weight, altitude + carbon monoxide increased both. Changes in electrocardiogram were consistent with changes in cardiac weight.

These results indicate that whereas carbon monoxide inhaled at ambient altitude causes a left electrical axis deviation, carbon monoxide inhaled at 4600 m (15 000 ft) exacerbates the well-known phenomenon of right electrical axis deviation. Thus, the results from chronic animal studies indicate that there is little effect of carbon monoxide on the cardiovascular system of rats exposed to carbon monoxide concentrations of 110 mg/m^3 (100 ppm) or less and altitudes up to 3000 m (10 000 ft).

7.8.2.4   *Summary*

In animal studies, no interaction was observed following combined exposure of carbon monoxide and ambient air pollutants such as nitrogen dioxide or sulfur dioxide (Busey, 1972; Murray et al., 1978; Hugod, 1981). However, an additive effect was observed following combined exposure of high levels of carbon monoxide plus nitric oxide (Groll-Knapp et al., 1988), and a synergistic effect was observed after combined exposure to carbon monoxide and ozone (Murphy, 1964).

Toxicological interactions of combustion products, carbon monoxide, carbon dioxide and hydrogen cyanide, at levels typically produced by indoor and outdoor fires, have shown a synergistic effect following carbon monoxide plus carbon dioxide exposure (Rodkey & Collison, 1979; Levin et al., 1987a) and an additive effect with carbon monoxide plus hydrogen cyanide (Levin et al., 1987b). Additive effects were also observed when carbon monoxide, hydrogen cyanide and low oxygen were combined; adding carbon dioxide to this combination was synergistic (Levin et al., 1988b). Additional studies are needed, however, to evaluate the effects of carbon monoxide under conditions of hypoxic hypoxia.

Finally, laboratory animal studies (Young et al., 1987; Fechter et al., 1988; Yang et al., 1988) suggest that the combination of environmental factors such as heat stress and noise with exposure to carbon monoxide may be important determinants of health effects. Of the effects described, the one potentially most relevant to typical human exposures is a greater decrement in exercise performance seen when heat stress is combined with 57 mg carbon monoxide/m^3 (50 ppm) (see chapter 8).

8. EFFECTS ON HUMANS

8.1 Healthy subjects

8.1.1 Introduction

Many direct experiments investigating the effects of carbon monoxide on humans have been conducted during the last century. Although many reports describe inadvertent exposures to various levels of carbon monoxide, there are a considerable number of precise and delineated studies utilizing human subjects. Most of these have been conducted by exposing young adult males to concentrations of carbon monoxide equivalent to those frequently or occasionally detected during ambient monitoring. Research on human subjects, however, can be limited by methodological problems that make the data difficult to interpret. These problems include (1) failure to measure blood carboxyhaemoglobin levels; (2) failure to distinguish between the physiological effects from a carbon monoxide dose of high concentration (i.e., bolus effect) and the slow, insidious increment in carboxyhaemoglobin levels over time from lower inhaled carbon monoxide concentrations; (3) failure to distinguish between normal blood flow and blood flow increased in response to hypoxia (compensatory responses); and (4) the use of small numbers of experimental subjects. Other factors involve failure to provide (1) control measures (e.g., double-blind conditions) for experimenter bias and experimenter effects; (2) control periods so that task-learning effects do not mask negative results; (3) homogeneity in the subject pool, particularly in groups labelled "smokers"; (4) control of possible boredom and fatigue effects; and (5) adequate statistical treatment of the data.

8.1.2 Acute pulmonary effects of carbon monoxide

8.1.2.1 Effects on lung morphology

Results from human autopsies have indicated that severe pulmonary congestion and oedema were produced in the lungs of individuals who died from acute smoke inhalation resulting from fires (Fein et al., 1980; Burns et al., 1986). These individuals, however, were exposed to relatively high concentrations of carbon monoxide as well as other combustion components of smoke, such as carbon dioxide, hydrogen cyanide, various aldehydes (e.g., acrolein), hydrochloric acid, phosgene and ammonia. If carbon monoxide, contained in relatively high concentrations in the inhaled smoke, was responsible for the pathological sequelae described in fire victims, then to what extent can oedema be attributed to the primary injury of capillary endothelial or alveolar epithelial cells?

In a study by Parving (1972) on 16 human subjects, transcapillary permeability to ^{131}I-labelled human serum albumin increased from an average 5.6% per hour in controls to 7.5% per hour following exposure to carbon monoxide. The subjects were exposed for 3–5 h to 0.43% (4900 mg/m^3 [4300 ppm]) carbon monoxide, resulting in approximately 23% carboxyhaemoglobin. There were no associated changes in plasma volume, haematocrit or total protein concentration.

The only other relevant permeability studies were conducted with cigarette smoke. Mason et al. (1983) showed rapidly reversible alterations in pulmonary epithelial permeability induced by smoking using radiolabelled diethylenetriaminepentaacetic acid (^{99m}Tc-DTPA) as a marker. This increased permeability reverted to normal fairly rapidly when subjects stopped smoking (Minty et al., 1981). Using a rat model, the permeability changes associated with cigarette smoke were demonstrated later by Minty & Royston (1985) to be due to the particulate matter contained in the smoke. The increase in ^{99m}Tc-DTPA clearance observed after exposure to dilute whole smoke did not occur when the particles were removed, suggesting that the carbon monoxide contained in the gaseous phase does not alter permeability of the alveolar–capillary membrane.

1.2.2   Effects on lung function

Human studies of pulmonary function are mostly devoted to the identification of effects occurring in the lungs of individuals exposed to relatively high concentrations of carbon monoxide. Older studies in the literature describe the effects of brief, controlled experiments with high carbon monoxide–air mixtures. Chevalier et al. (1966) exposed 10 subjects to 5700 mg carbon monoxide/m^3 (5000 ppm) for 2–3 min until carboxyhaemoglobin levels reached 4%. Measurements of pulmonary function and exercise studies were performed before and after exposure. Inspiratory capacity and total lung capacity decreased 7.5% ($P < 0.05$) and 2.1% ($P < 0.02$), respectively, whereas maximum breathing capacity increased 5.7% ($P < 0.05$) following exposure.

Mean resting diffusing capacity of the lungs decreased 7.6% ($P <$ 0.05) compared with air-exposed controls. Fisher et al. (1969) exposed a small number ($n = 4$) of male subjects, aged 23–36 years, to 6% (69 000 mg/m^3 [60 000 ppm]) carbon monoxide for 18 s, resulting in estimated carboxyhaemoglobin concentrations of 17–19%. There were no significant changes in lung volume, mechanics or diffusing capacity. Neither of these studies was definitive, however, and no follow-up studies were reported.

More recent studies in the literature describing the effects of carbon monoxide on pulmonary function have been concerned with exposure to the products of combustion and pyrolysis from such sources as tobacco, fires or gas- and kerosene-fuelled appliances and engines. One group of individuals, representing the largest proportion of the population exposed to carbon monoxide, is tobacco smokers.

A second group evaluated for potential changes in acute ventilatory function includes occupations in which individuals are exposed to variable, and often unknown, concentrations of carbon monoxide in both indoor and outdoor environments. Firefighters, tunnel workers and loggers are typical examples of individuals at possible risk. Unfortunately, these individuals are also exposed to high concentrations of other combustion components of smoke and exhaust. It is very difficult to separate the potential effects of carbon monoxide from those due to other respiratory irritants.

Firefighters have previously been shown to have a greater loss of lung function associated with acute and chronic smoke inhalation (as reviewed by Sparrow et al., 1982). None of these earlier studies, however, characterized the exposure variables, particularly the concentrations of carbon monoxide found in smoke, nor did they report the carboxyhaemoglobin levels found in firefighters after exposure. Most reports of lung function loss associated with other occupational exposures also fail to characterize exposure to carbon monoxide. The following studies have attempted to monitor, or at least estimate, the carbon monoxide and carboxyhaemoglobin levels found in occupational settings where lung function was also measured.

Sheppard et al. (1986) reported that acute decrements in lung function were associated with routine firefighting. Baseline airway responsiveness to methacholine was measured in 29 firefighters from one fire station in San Francisco, California, USA, who were monitored over an 8-week period. Spirometry measurements were taken before and after each 24-h work shift and after each fire. Exhaled gas was sampled 55 times from 21 firefighters immediately after each fire and was analysed for carbon monoxide. Despite the use of personal respiratory protection, exhaled carbon monoxide levels exceeded 110 mg/m^3 (100 ppm) on four occasions, with a maximum of 150 mg/m^3 (132 ppm), corresponding to predicted carboxyhaemoglobin levels of 17–22%. Of the 76 spirometry measurements obtained within 2 h after a fire, 18 showed a greater fall in FEV_1 and/or FVC compared with routine work shifts without fires. Decrements in lung function persisted for as long as 18 h in some of the individuals, but they did not appear to occur selectively in those individuals with pre-existing airway hyperresponsiveness.

Evans et al. (1988) reported on changes in lung function and respiratory symptoms associated with exposure to automobile exhaust among bridge and tunnel officers. Spirometry measurements were obtained in and symptom questionnaires were administered on a voluntary basis to 944 officers of the Triborough Bridge and Tunnel Authority in New York City, New York, USA, over an 11-year period between 1970 and 1981. Regression analyses were performed on 466 individuals (49%) who had been tested at least three times during that period. Carboxyhaemoglobin levels were calculated from expired-air breath samples. Small but significant differences were found between the bridge and tunnel officers. Estimated levels of carboxyhaemoglobin were consistently higher in tunnel workers than in bridge workers for both non-smoking individuals (1.96% and 1.73%, respectively) and smoking individuals (4.47% and 4.25%, respectively). Lung function measures of FEV_1 and FVC were lower, on average, in tunnel workers than in bridge workers. There were no reported differences in respiratory symptoms except for a slightly higher symptom prevalence in tunnel workers who smoked. Because differences in lung function between the two groups were small, it is questionable if the results are clinically significant or if they were even related to carbon monoxide exposure.

Hagberg et al. (1985) evaluated the complaints of 211 loggers reporting dyspnoea and irritative symptoms in their eyes, noses and throats after chain-saw use. Measurements of lung spirometry, carboxyhaemoglobin and exposure to carbon monoxide, hydrocarbons and aldehydes were conducted on 23 loggers over 36 work periods lasting 2 h each. Ventilation levels during tree felling averaged 41 litres/min. Carboxyhaemoglobin levels increased after chain-saw use (P < 0.05) but were weakly although significantly (P < 0.001) correlated (r = 0.63) with mean carbon monoxide concentrations of 19 mg/m^3 (17 ppm) (5–84 mg/m^3 [4–73 ppm] range) in non-smokers. Corresponding carboxyhaemoglobin levels were apparently <2%; unfortunately, the absolute values before and after exposure were not reported. Bronchoconstriction, measured by a decreased FEV_1/FVC (P < 0.03) and forced expiratory flow measured at 25–75% of FVC (P < 0.005), was found after the work periods, but no correlations were obtained between lung function, carboxyhaemoglobin levels and exposure variables. There were no reported significant changes in FEV_1 or FVC.

The potential effects of indoor combustion products of kerosene space heaters on lung function were evaluated by Cooper & Alberti (1984). Carbon monoxide and sulfur dioxide concentrations were monitored in 14 suburban homes in Richmond, Virginia, USA, during January and February of 1983 while modern kerosene heaters were in operation. Spirometry measurements were obtained in 29 subjects over a 2-day period, randomizing exposures between days with and without the heater on. During heater operation, the carbon monoxide concentration was 7.8 ± 6.8 mg/m^3 (6.8 ± 5.9 ppm) (0–16 mg/m^3 [0–14 ppm] range), and the sulfur dioxide concentration was 1.1 ± 1.1 mg/m^3 (0.4 ± 0.4 ppm) (0–2.9 mg/m^3 [0–1 ppm] range). On control days, the indoor carbon monoxide concentration was 0.16 ± 0.61 mg/m^3 (0.14 ± 0.53 ppm), whereas sulfur dioxide was undetectable. Six of the homes had carbon monoxide concentrations exceeding 10 mg/m^3 (9 ppm). Corresponding outdoor carbon monoxide concentrations were 0–3.4 mg/m^3 (0–3 ppm). Carboxyhaemoglobin levels significantly increased from 0.82 ± 0.43% on control days to 1.11 ± 0.52% on days when kerosene heaters were used. Exposure to heater emissions, however, had no effect on FVC, FEV_1 or maximum mid-expiratory flow rate.

Most of the published community population studies on carbon monoxide have investigated the relationship between ambient carbon monoxide levels and hospital admissions, deaths or symptoms attributed to cardiovascular diseases. Little epidemiological information is available on the relationship between carbon monoxide and pulmonary function, symptomatology and disease.

One study by Lutz (1983) attempted to relate levels of ambient pollution to pulmonary diseases seen in a family practice clinic in Salt Lake City, Utah, USA, during the winters of 1980 and 1981, when heavy smog conditions prevailed. Data on patient diagnoses, local climatological conditions and levels of carbon monoxide, ozone and particulate matter were obtained over a 13-week period. Pollutant levels were measured daily and then averaged for each week of the study; absolute values were not reported. For each week, weighted simple linear regression (values not reported) and correlation analyses were performed. Significant correlations ($P = 0.01$) between pollution-related diseases and the environmental variables were found for particulate matter ($r = 0.79$), ozone ($r = -0.67$) and percentage of smoke and fog ($r = 0.79$), but not for carbon monoxide ($r = 0.43$) or percentage of cloud cover ($r = 0.33$). The lack of a significant correlation with carbon monoxide was explained by a small fraction (2%) of diagnoses for ischaemic heart disease compared with a predominance of respiratory tract diseases such as asthma, bronchitis, bronchiolitis and emphysema.

Daily lung function in a large community population exposed to indoor and outdoor air pollution was measured in Tucson, Arizona, USA, by Lebowitz and co-workers (Lebowitz et al., 1983a,b, 1984, 1985, 1987; Lebowitz, 1984; Robertson & Lebowitz, 1984). Subsets of both healthy subjects and subjects with asthma, allergies and airway obstructive disease were drawn from a symptom-stratified, geographic sample of 117 middle-class households. Symptoms, medication use and peak flow measurements were recorded daily over a 2-year period. Indoor/outdoor monitoring was conducted in a random sample of 41 representative houses. Maximum 1-h concentrations of ozone, carbon monoxide and nitrogen dioxide and daily levels of total suspended particulates, allergens and meteorological variables were monitored at central stations within 0.8 km of each population subset. Because gas stoves and tobacco smoking were the predominant indoor sources, indoor pollutant measurements were made for particles and carbon monoxide. Levels of carbon monoxide were low, averaging less than 2.7 mg/m^3 (2.4 ppm) indoors and 4.4–5.6 mg/m^3 (3.8–4.9 ppm) outdoors. Spectral time-series analysis was used to evaluate relationships between environmental exposure and pulmonary effects over time (Robertson & Lebowitz, 1984; Lebowitz et al., 1987). Asthmatics were the most responsive, whereas healthy subjects showed no significant responses. Outdoor ozone, nitrogen dioxide, allergens, meteorology and indoor gas stoves were significantly related to symptoms and peak flow.

8.1.3 Cardiovascular and respiratory response to exercise

The most extensive human studies on the cardiorespiratory effects of carbon monoxide are those involving the measurement of oxygen uptake during exercise. Healthy young individuals were used in most of the studies evaluating the effects of carbon monoxide on exercise performance; healthy older individuals were used in only two studies (Raven et al., 1974a; Aronow & Cassidy, 1975). In all of these studies, oxygen uptake during submaximal exercise for short durations (5–60 min) was not affected by carboxyhaemoglobin levels as high as 15–20%. Under conditions of short-term maximal exercise, however, statistically significant decreases (3–23%) in maximal oxygen uptake ($\dot{V}_{O_2}$ max) were found at carboxyhaemoglobin levels ranging from 5% to 20% (Pirnay et al., 1971; Ekblom & Huot, 1972; Vogel & Gleser, 1972; Stewart et al., 1978; Weiser et al., 1978; Klein et al., 1980). In another study by Horvath et al. (1975), the critical level at which carboxyhaemoglobin marginally influenced $\dot{V}_{O_2}$ max ($P <$ 0.10) was approximately 4.3%. The data obtained by Horvath's group and others are summarized in Fig. 11. There is a linear relationship between decline in $\dot{V}_{O_2}$ max and increase in carboxyhaemoglobin that can be expressed as *% decrease in* $\dot{V}_{O_2}$ *max* $= 0.91$ *(% COHb)* $+ 2.2$ (US EPA, 1979b; Horvath, 1981). Short-term maximal exercise duration has also been shown to be reduced (3–38%) at carboxyhaemoglobin levels ranging from 2.3% to 7% (Ekblom & Huot, 1972; Drinkwater et al., 1974; Raven et al., 1974a,b; Horvath et al., 1975; Weiser et al., 1978).

Numerous studies have demonstrated that an increase in carboxyhaemoglobin is associated with a compensatory increase in brain blood flow. Benignus et al. (1992) conducted two sets of studies on 14 and 12 young healthy men, respectively, measuring brain blood flow by the method of impedance plethysmography. In the first study, subjects were transiently exposed to various concentrations of carbon monoxide. In the second study, the exposure lasted 4 h. The exposures produced carboxyhaemoglobin levels up to 18.4%. The variation of

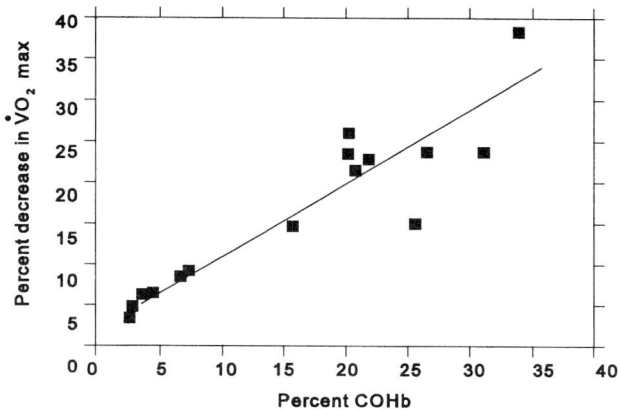

Fig. 11. Relationship between carboxyhaemoglobin level and decrement in maximal oxygen uptake for healthy non-smokers (adapted from US EPA, 1979b; Horvath, 1981).

the brain blood flow response among subjects was large and statistically significant. The authors speculated that changes in carbon monoxide-induced brain blood flow might be related to behavioural effects.

The kinetics of carbon monoxide uptake during the transition phase from rest to exercise was investigated by Kinker et al. (1992). Data from six subjects who switched from rest to constant exercise (at various levels of peak $\dot{V}_{O_2}$) while breathing carbon monoxide (63 mg/m^3 [55 ppm]) show that carbon monoxide uptake increased faster than oxygen uptake at all exercise levels. No significant changes in the diffusing capacity for carbon monoxide were found, suggesting that other factors such as changes in pulmonary blood flow and recruitment of alveoli–capillary surface area might be involved in carbon monoxide uptake.

The work rate-dependent effect of carbon monoxide on minute ventilation ($\dot{V}_E$) during exercise was studied by Koike et al. (1991). Ten healthy subjects were exposed to carbon monoxide to bring the carboxyhaemoglobin levels up to 11% and 20%, respectively. During incremental exercise, $\dot{V}_E$ was not affected by carbon monoxide

257

breathing at work rates below the lactic acidosis threshold. Above the lactic acidosis threshold, however, as the work load and carboxy-haemoglobin concentration increased, the $\dot{V}_E$ increased as well. The increase in $\dot{V}_E$ correlated positively with carboxyhaemoglobin levels ($r \approx 0.83$). Such an increase in exercise $\dot{V}_E$ due to carboxyhaemo-globin might restrict work rates above the lactic acidosis threshold and lead to greater lactic acidosis.

Potential effects of hypoxia on muscle deoxygenation during exercise were studied by Maehara et al. (1997). Seven healthy subjects exercised at two constant work loads under various conditions of hypoxic and carbon monoxide-induced hypoxia (15% carboxyhaemo-globin). They found that progressive muscle deoxygenation was accelerated at exercise levels above the lactic acidosis threshold.

It is well recognized that the same decrease in oxygen carrying capacity of blood due to carbon monoxide-induced hypoxia and anaemia will have different physiological effects. Celsing et al. (1987) found in a series of very carefully performed studies in normal sub-jects that $\dot{V}_{O_2}$ max decreased by 19 ml/min per kilogram per gram per litre change in haemoglobin over a range of haemoglobin concen-trations from 13.7 to 17.0 g/dl. This change represents a 2% decrease in $\dot{V}_{O_2}$ max for every 3% decrease in haemoglobin concentration in a well-trained subject. The decrease also corresponds to the decrease in $\dot{V}_{O_2}$ max reported by Ekblom & Huot (1972) and Horvath et al. (1975). However, Ekblom & Huot (1972) found a much more marked effect on maximal work time (i.e., work on a constant load until exhaustion with a duration of about 6 min). An explanation for the marked decrease in maximal work time could be that carbon monox-ide has a negative effect on the oxidative enzymatic system, whereas the decrease in work time is due to a combination of a decrease in oxygen capacity and a less efficient oxidative enzymatic system. If the data are extrapolated to lower carboxyhaemoglobin values, a 3% level of carboxyhaemoglobin should decrease the maximal work time by about 20%. However, this decrease is more than the 10% average decrease reported by Klausen et al. (1983), who also found more marked effects in less well-trained subjects compared with well-trained subjects. Additional studies need to be performed to resolve the difference in the maximal work time values reported in these studies.

8.1.4 Behavioural changes and work performance

8.4.1 Introduction

Effects of carboxyhaemoglobin elevation above 20% on behaviour have been unambiguously demonstrated in humans. Below this level, results are less consistent. Based on meta-analysis of Benignus (1994), 18–25% carboxyhaemoglobin levels in healthy sedentary persons would be required to produce a 10% decrement in behaviour. Some of the differences among studies of the effect of carboxyhaemoglobin on the behaviour of humans are apparently due to technical problems in the execution of experiments, because single-blind or non-blind experiments tend to yield a much higher rate of significant effects than do double-blind studies. Even when non-double-blind experiments are eliminated from consideration, however, a substantial amount of disparity remains among results of studies. It is possible that such residual disagreement is due to the action of an unsuspected variable that is not being controlled across experiments. Because at present no data are available on the behavioural changes and work performance of individuals with chronically elevated carboxyhaemoglobin, the subsequent sections discuss only findings of acute studies.

8.4.2 Sensory effects

1) Vision

Absolute threshold. In an experiment using four well-trained young subjects, it was demonstrated that visual sensitivity was decreased in a dose-related manner by carboxyhaemoglobin levels of 4.5, 9.4, 15.8 and 19.7% (McFarland et al., 1944). Various aspects of these data were subsequently reported by Halperin et al. (1959) and McFarland (1970). Carboxyhaemoglobin elevations were accomplished by inhalation of boluses of high-concentration carbon monoxide. Visual thresholds were measured repeatedly over a 5-min period at each carboxyhaemoglobin level. Experimenters were not blind to the exposure conditions, and the subjects could have easily deduced the conditions from the experimental design, because no air-only condition was included to control for the effects of the testing scheme itself. Data from only one typical subject were presented. Thresholds were measured at only one level of dark adaptation (0.02 lx).

McFarland (1973), in a scantily documented article, reported that similar threshold shifts occurred at the end of a carbon monoxide exposure period (17% carboxyhaemoglobin) and an air-only session. Thus, it is possible that the effects reported by McFarland et al. (1944) were due to fatigue or some other time-on-task related variable. Von Restorff & Hebisch (1988) found no dark adaptation effects on subjects with carboxyhaemoglobin levels ranging from 9% to 17%. Luria & McKay (1979) found no effect of 9% carboxyhaemoglobin on scotopic visual threshold.

The effect of 17% carboxyhaemoglobin (bolus administration, followed by maintenance carbon monoxide level for 135 min) on the entire dark adaptation curve was studied by Hudnell & Benignus (1989) using 21 young men in a double-blind study. No difference between carbon monoxide and air groups was observed. A power of 0.7 was calculated for the test employed so that the conclusions are reasonably defensible. From the above evidence, it appears that effects on visual sensitivity have not been demonstrated at carboxyhaemoglobin levels up to 17%.

Temporal resolution. The temporal resolution of the visual system has been studied in the form of critical flicker frequency. In the critical flicker frequency paradigm, subjects report the frequency at which light flashes begin to appear as a continuous light.

Seppanen et al. (1977) reported dose-ordinal decreases of critical flicker frequency for carboxyhaemoglobin levels of approximately 4.0, 6.1, 8.4, 10.7 and 12.7%. The experiment was conducted with 22 healthy subjects whose ages ranged from 20 to 62 years. Carboxyhaemoglobin was induced by breathing high concentrations of carbon monoxide from a Douglas bag. Subjects were blind as to the condition, but apparently experimenters were informed. Appropriate controls for fatigue were included, and the exposure levels were randomized.

A study was reported by von Post-Lingen (1964) in which carboxyhaemoglobin levels ranging up to 23% were induced in 100 subjects by breathing carbon monoxide-contaminated air from a spirometer for about 7 min in a single-blind procedure. One group of subjects was given an injection of Evipan (sodium hexobarbitone; see Reynolds & Prasad, 1982), a drug previously shown to have produced decreases in critical flicker frequency only if patients had demonstrable brain damage. In the non-drug group, critical flicker frequency was unaffected until approximately 14% carboxyhaemoglobin. In the drug group, however, effects began at carboxyhaemoglobin levels as low as 6% and were dose proportional up to the highest carboxyhaemoglobin value. When the drug-plus-carbon monoxide study was repeated in a double-blind replication ($n = 15$), no effects were seen. The latter replication study was given only one paragraph in the report, and thus it is not clear exactly what was done.

Beard & Grandstaff (1970) reported significant effects on critical flicker frequency in an earlier study in which four subjects had been exposed to carbon monoxide levels of 57, 170 or 290 mg/m^3 (50, 150 or 250 ppm) for 1 h. Carboxyhaemoglobin was estimated by the authors to have reached 3.0, 5.0 and 7.5%, respectively, by the end of the exposure. Documentation was extremely sparse, and, with only four subjects, power was probably low. Even though the elevated carboxyhaemoglobin groups had decreased critical flicker frequency, the results were not dose ordinal. There is a comparatively large amount of literature published before the Seppanen et al. (1977) article, in which the effect of elevated carboxyhaemoglobin on critical flicker frequency was tested. In none of the earlier studies was critical flicker frequency found to be affected, even though much higher levels of carboxyhaemoglobin were reached. The studies and their maximum carboxyhaemoglobin levels were Lilienthal & Fugitt (1946), 15.4%; Vollmer et al. (1946), 17.5%; Guest et al. (1970), 8.9%; O'Donnell et al. (1971a), 12.7%; Fodor & Winneke (1972), 7.5%; Ramsey (1973), 11.2%; and Winneke (1974), 10.0%. To be sure, there was much variation in size of the subject group, method and experimental design among the above studies, but no pattern emerges as to why the Seppanen et al. (1977), Beard & Grandstaff (1970) and von Post-Lingen (1964) studies found significant effects when the others did not. It is noteworthy that the studies reporting significant effects were all conducted in a single- or non-blind manner.

Miscellaneous visual functions. A number of researchers reported the results of experiments in which visual parameters other than absolute threshold or critical flicker frequency were measured as part of a battery of tests. Many of these experiments studied a large group of subjects.

Beard & Grandstaff (1970) reported a study in which four subjects were exposed to carbon monoxide sufficient to produce estimated (by the authors) carboxyhaemoglobin levels of 3.0, 5.0 and 7.5%. The measurements made were critical flicker frequency (see above), brightness difference thresholds, visual acuity and absolute threshold. Data for the latter variable were unreliable and were not reported. Dose-related impairments in acuity and brightness difference sensitivity were reported. The scant documentation of methods plus the low number of subjects make the results difficult to evaluate.

Five other reports of significant visual function effects by carboxyhaemoglobin elevation are extant. Two of the studies (Bender et al., 1972; Fodor & Winneke, 1972) reported that tachistoscopic pattern detection was impaired by carboxyhaemoglobin levels of 7.3% and 5.3%, respectively. Weir et al. (1973), Ramsey (1972) and Salvatore (1974) reported that brightness discrimination was adversely affected by carboxyhaemoglobin levels of 6–20%.

Tests of visual function after carboxyhaemoglobin elevation conducted by other authors have been uniformly non-significant. Especially noteworthy are studies by Hudnell & Benignus (1989) and Stewart et al. (1975), both of which found no acuity effects as reported by Beard & Grandstaff (1970). Brightness discrimination was similarly not found to be affected (Ramsey, 1973), in contradiction with the reports of others. The latter study is especially interesting in that it represents a failure to replicate an earlier study by the same author (Ramsey, 1972). The earlier study by Ramsey was conducted in a single-blind manner, whereas the later one was double-blind.

The most thorough tests of visual function were performed by Hudnell & Benignus (1989), who tested absolute threshold (see above), acuity and motion detection with carboxyhaemoglobin levels of 17% and found no effects due to carboxyhaemoglobin. The acuity and motion detection were tested at both scotopic and photopic levels.

It would appear that the results of studies on the effects of carboxyhaemoglobin elevation on miscellaneous visual function are not supportive of significant effects. Results that were significant in two studies (Beard & Grandstaff, 1970; Weir et al., 1973) were contradicted by other reports using relatively large groups of subjects under better controlled conditions.

2) Audition

Surprisingly little work has been done concerning the effects of carboxyhaemoglobin on auditory processes. Stewart et al. (1970) reported that the audiogram of subjects exposed to carbon monoxide resulting in up to 12.0% carboxyhaemoglobin was not affected. Haider et al. (1976) exposed subjects to a 105-dB, one-octave bandwidth random noise (centre frequency of 2 kHz) for 15 min while carboxyhaemoglobin level was elevated to 13%. Under continued carboxyhaemoglobin elevation, the temporary threshold shifts were measured after noise cessation. No effects of carboxyhaemoglobin on temporary threshold shift were observed. Guest et al. (1970) tested the effects of elevated carboxyhaemoglobin (8.9%) on auditory flutter fusion and found no significant effect. The flutter fusion test is analogous to critical flicker frequency in vision and was tested by having the subject judge the rate at which an interrupted white noise became apparently continuous. From these data, it would appear that the functioning of the auditory system is not particularly sensitive to carboxyhaemoglobin elevation, but little research has been done.

4.3   *Motor and sensorimotor performance*

1)   Fine motor skills

In a single-blind study, Bender et al. (1972) found that manual dexterity and precision (Purdue pegboard) were impaired by 7% carboxyhaemoglobin. Winneke (1974) reported that hand steadiness was affected by 10% carboxyhaemoglobin, but no supportive statistical test was presented.

Similar motor functions were evaluated by a number of other investigators and were found not to be affected, even at higher carboxyhaemoglobin levels. Vollmer et al. (1946) reported that 20% carboxyhaemoglobin did not affect postural stability. O'Donnell et al. (1971b) used the Pensacola Ataxia Battery to measure various aspects of locomotion and postural stability. Subjects with 6.6% carboxyhaemoglobin were not affected. Stewart et al. (1970, 1975) tested the ability of subjects to manipulate small parts using the Crawford collar and pin test and screw test, the American Automobile Association hand steadiness test and the Flanagan coordination test. Carboxyhaemoglobin levels up to 15% had no effect on any of the measures. Two subjects were taken to 33% and 40% carboxyhaemoglobin, however; in these subjects, the collar and pin performance was impaired and the subjects reported hand fatigue. Manual dexterity (Purdue pegboard), rapid precision movement (Purdue hand precision) and static hand steadiness (pen in hole) and tapping tests were not affected by carboxyhaemoglobin levels of approximately 5.3% (Fodor & Winneke, 1972). Wright et al. (1973) reported that hand steadiness was not affected by carboxyhaemoglobin levels of 5.6%. Weir et al. (1973) found no effects of 14% carboxyhaemoglobin on tapping, star tracing and rail walking. Mihevic et al. (1983) discovered no effect on tapping when the task was performed alone or simultaneously with an arithmetic task. Finally, Seppanen et al. (1977) demonstrated that tapping speed was unaffected by 12.7% carboxyhaemoglobin. Most of the above non-significant studies used a moderate to large number of subjects. The overwhelming evidence in the area of fine motor control indicates that carboxyhaemoglobin levels below approximately 20% (the highest level tested) do not produce effects.

2) Reaction time

Of the 12 different experiments that studied reaction time, only 1 reported a significant result (Weir et al., 1973), and that effect occurred only at 20% carboxyhaemoglobin. A number of the non-significant effects were from studies using a large number of subjects. The consistent finding that carboxyhaemoglobin elevation does not affect reaction time is especially impressive because of the wide range of carboxyhaemoglobin levels employed (5.0–41.0%).

3) Tracking

Tracking is a special form of fine motor behaviour and hand–eye coordination that requires a subject to either follow a moving target or compensate for a moving target's motion by manipulation of a lever, for example. Of the 11 studies on the topic, 4 reported significant effects, and 1 of those found effects only at 20% carboxyhaemoglobin. The matter is more complicated, however, and the literature in the area offers some clues to the reasons for the diversity among the reports.

O'Donnell et al. (1971a,b) used critical instability compensatory tracking in which the task was to keep a meter needle centred. Simultaneous performance of detection tasks was also required in one of the studies. No effects were demonstrated for carboxyhaemoglobin levels as high as 12–13%. The critical instability tracking task was also used by Gliner et al. (1983) in conjunction with peripheral light detection. Carboxyhaemoglobin levels up to 5.8% had no effect on performance. Pursuit rotor tracking was also reported to be unaffected at 5.3% carboxyhaemoglobin (Fodor & Winneke, 1972). Weir et al. (1973) reported that pursuit rotor performance was slightly affected beginning at 20% carboxyhaemoglobin. In a 1988 study, Bunnell & Horvath used a two-dimensional tracking task in which the stimulus was presented on a cathode ray tube and was controlled with a joystick. No effect of carboxyhaemoglobin or exercise or a combination of the two was seen for carboxyhaemoglobin levels up to 10.2%. Schaad et al. (1983) reported that pursuit and compensatory tracking were not affected by a carboxyhaemoglobin level of 20% even during simultaneous performance of monitoring tasks.

In a pair of careful studies of different design, Putz et al. (1976, 1979) studied compensatory tracking by having the subject try to keep a vertically moving spot in the centre of an oscilloscope screen. The tracking was performed while the subject also did a light brightness detection task. In both studies, tracking was significantly affected by carboxyhaemoglobin levels of 5%. The fact that both studies demonstrated significant results despite differences in experimental design lends credibility to the finding. Additional credibility was gained when the Putz et al. (1976) study was replicated with similar results by Benignus et al. (1987).

The consistency of the compensatory tracking results in the Putz et al. (1976) protocol was not continued when Benignus et al. (1990a) attempted to demonstrate a dose–effect relationship using the same experimental design. In the latter study, independent groups were exposed to carbon monoxide sufficient to produce carboxyhaemoglobin levels of control, 5, 12 and 17%. Carbon monoxide was administered via Douglas bag breathing, and then carboxyhaemoglobin was maintained by low-level carbon monoxide in room air. A fifth group was exposed to carbon monoxide in the chamber only, and this group served as a positive control because it was treated in exactly the same ways as the subjects in Putz et al. (1976) and in Benignus et al. (1987). No significant effects on tracking were demonstrated in any group. The means for the tracking error were elevated in a nearly dose-ordinal manner, but not to a statistically significant extent.

At present, there is no apparent reason for the lack of consistency among the reports of tracking performance. The largest study with the widest dose range (Benignus et al., 1990a) appears to be the strongest indicator of no significant effects of carboxyhaemoglobin elevation. However, it is difficult to ignore the several other studies that were well controlled and did demonstrate significant effects. At this point, the best summary seems to be that carboxyhaemoglobin elevation produces small decrements in tracking that are sometimes significant. The possible reasons for such high variability are unclear.

8.1.4.4   Vigilance

A dependent variable that is possibly affected by elevated carboxyhaemoglobin is the performance of extended, low-demand tasks characterized as vigilance tasks. Because of the low-demand characteristic of vigilance tasks, they are usually of a single-task type. Of the eight reports, four reported significant effects. Despite the seemingly greater unanimity in this area, it is noteworthy that for each report of significant effects, there exists a failed attempt at direct replication.

Horvath et al. (1971) reported a significant vigilance effect at 6.6% carboxyhaemoglobin. A second study, conducted in the same laboratory (Christensen et al., 1977), failed to find significant effects of 4.8% carboxyhaemoglobin on the same task. To be sure, the second study used slightly lower carboxyhaemoglobin levels, but the means left no suggestion of an effect. Roche et al. (1981) from the same laboratory reported that performance of the same task after a bolus exposure was used to produce 5% carboxyhaemoglobin was not affected.

Fodor & Winneke (1972) reported a study in which 5.3% carboxyhaemoglobin significantly impaired performance of a vigilance task. When the same task and protocol were tried again in the same laboratory (Winneke, 1974), no significant effects were found, even for carboxyhaemoglobin levels up to 10%.

Groll-Knapp et al. (1972) reported dose-related significant effects of carboxyhaemoglobin levels ranging from estimated values of 3% to 7.6%. Effects were large, but apparently the study was not blind. Haider et al. (1976) reported similar effects at low carboxyhaemo-globin levels, but not at higher levels. The authors have twice mentioned failures to replicate the results (Haider et al., 1976; Groll-Knapp et al., 1978). A similar experiment using a different stimulus failed to produce significant effects at 12% carboxyhaemoglobin (Groll-Knapp et al., 1978).

The fact that all replication attempts for each of the reported significant effects of carboxyhaemoglobin on vigilance have failed to verify the original reports is evidence for some unreliability or the operation of unknown and uncontrolled variables. That the non-verifications were conducted by the original researchers, as well as by others, makes the case for unreliability even more convincing. If vigilance is affected by carboxyhaemoglobin elevation, a convincing demonstration remains to be made. Perhaps a case can be made that behavioural effects of carboxyhaemoglobin levels below 20% are present in the exposed population, but they are probably small and, therefore, difficult to demonstrate reliably (Benignus et al., 1990b).

4.5   Miscellaneous measures of performance

1)   Continuous performance

Continuous performance is a category of behaviour that is related to vigilance. The difference is that many tasks that are performed over a long period of time are more demanding and involve more than simple vigilance. Sometimes the continuous performance tasks are not performed for a sufficiently long period of time to involve decrements in vigilance or are interrupted too frequently.

Putz et al. (1976, 1979) reported that monitoring performed simultaneously with tracking was impaired at carboxyhaemoglobin levels as low as 5%. In a replication attempt of the Putz et al. (1976, 1979) studies, Benignus et al. (1987) failed to find any effects of approximately 8% carboxyhaemoglobin. O'Donnell et al. (1971a) also failed to find effects of carboxyhaemoglobin on a monitoring task performed simultaneously with tracking. Schaad et al. (1983) found no effects on monitoring performed simultaneously with tracking even when carboxyhaemoglobin was 20%. Gliner et al. (1983) reported that signal detection was affected by 5.8% carboxyhaemoglobin when performed singly, but not when performed simultaneously with tracking. The latter results are in conflict with those of Putz et al. (1976, 1979).

Insogna & Warren (1984) reported that the total game score on the performance of a multitask video game was reduced by carboxyhaemoglobin levels of 4.2%. Separate task scores were not collected. Schulte (1963) reported that letter, word and colour detection tasks were dose-ordinally impaired by carboxyhaemoglobin levels as low as 5% and ranging up to 20%. Reported carboxyhaemoglobin levels were at considerable variance with values expected from the exposure parameters (Laties & Merigan, 1979). Benignus et al. (1977) reported that 8.2% carboxyhaemoglobin did not affect a numeric monitoring task.

Again, there is disturbing lack of replicability in the literature. The two most credible studies showing effects of carboxyhaemoglobin on continuous performance (Putz et al., 1976, 1979) were not verified by Benignus et al. (1987). In the latter study, the tracking effects of the Putz et al. (1976, 1979) work were verified. Similar studies of monitoring during tracking (O'Donnell et al., 1971a; Gliner et al., 1983; Schaad et al., 1983) also reported no effects of carboxyhaemoglobin, even with levels as high as 20%. It seems necessary to suspend judgement regarding the continuous performance results until further data and understanding are available. Perhaps the best judgement is to hypothesize small effects.

2)   Time estimation

In 1967, Beard & Wertheim reported that carboxyhaemoglobin produced a dose-related decrement in single-task time estimation accuracy beginning at 2.7%. Various versions of the same task were tested by others with carboxyhaemoglobin levels ranging up to 20% without effects being demonstrated (Stewart et al., 1970, 1973b, 1975; O'Donnell et al., 1971b; Weir et al., 1973; Wright & Shephard, 1978b). An exact replication, which also did not find significant results, was conducted by Otto et al. (1979). It seems safe to assume that time estimation is remarkably impervious to elevated carboxyhaemoglobin.

3)   Cognitive effects

Five of the 11 experiments discussed below that have been reported in the literature found cognitive effects of carboxyhaemoglobin. Bender et al. (1972) reported effects of 7.3% carboxyhaemoglobin on a variety of tasks. Groll-Knapp et al. (1978) reported memory to be affected after exposure to carbon monoxide during sleep (11% carboxyhaemoglobin), but a very similar study performed by the same group later found no effects of 10% carboxyhaemoglobin (Groll-Knapp et al., 1982). Arithmetic performance was affected slightly in a non-dose-ordinal manner when a simultaneous tapping task was performed (Mihevic et al., 1983). Schulte (1963) reported a dose-ordinal effect on arithmetic performance beginning at 5% and ranging to 20% carboxyhaemoglobin. Carboxyhaemoglobin levels in the latter study were considerably different from values expected from the exposure parameters (Laties & Merigan, 1979). Similar variables were tested by others, sometimes at higher levels of carboxyhaemoglobin and with relatively large groups of subjects, without finding effects (O'Donnell et al., 1971a; Stewart et al., 1975; Haider et al., 1976; Groll-Knapp et al., 1978; Schaad et al., 1983). The conclusions are, at best, equivocal.

A study by Bunnell & Horvath (1988) utilized a wide range of cognitive tasks involving short-term memory, Manikin rotation, Stroop word-colour tests, visual search and arithmetic problems (the latter as part of a divided attention task performed simultaneously with tracking). Carboxyhaemoglobin was formed by bag breathing followed by a carbon monoxide level in room air designed to maintain a constant carboxyhaemoglobin level. Subjects were exercised at 0, 35 or 60% of $\dot{V}_{O_2}$ max before cognitive tests were performed. The Stroop test performance was slightly but significantly decreased by both 7% and 10% carboxyhaemoglobin by the same amount, but exercise had no effect. The authors suggested that negative transfer effects (difficulty in reversing instructional sets) were responsible for the decrement. Visual searching improved for carboxyhaemoglobin levels both at rest and at medium exercise. At the high exercise level, however, carboxyhaemoglobin produced dose-ordinal impairments in performance. The authors conjectured that hypoxic depression of cortical function interacted with hypoxic stress and exercise stress to produce the effects.

Most of the data on cognitive effects of carboxyhaemoglobin elevation are not sufficiently consistent to consider. The study by Bunnell & Horvath (1988), however, is suggestive of potentially important effects of interactions of carboxyhaemoglobin and exercise. Before any conclusions may be drawn about the results, the study should be replicated and expanded.

8.1.4.6   Automobile driving

Complex behaviour, in the form of automobile driving, has been tested a number of times for effects of carboxyhaemoglobin elevation. Not only is automobile driving potentially more sensitive to disruption because of its complexity, but it is also an inherently interesting variable because of its direct applicability to non-laboratory situations. The well-practised nature of the behaviour, on the other hand, may make performances more resistant to disruption. The complexity of the behaviour also leads to methodological difficulties. Attempting to exhaustively measure the complex behaviours usually leads investigators to measure many dependent variables. Statistically analysing a large number of variables in a defensible way requires many subjects and leads to greater expense.

In an early study by Forbes et al. (1937), using only five subjects, steering accuracy in a simulator was investigated with carboxyhaemoglobin levels of up to 27.8%. No effects were demonstrated. A sparsely documented experiment by Wright et al. (1973), using 50 subjects with 5.6% carboxyhaemoglobin, tested a number of functions of simulator performance but found no effects. Weir et al. (1973) performed an experiment with actual automobile driving on a highway in which many variables were measured and tested. None of the variables was reliably affected until carboxyhaemoglobin exceeded approximately 20%. Wright & Shephard (1978a) failed to find effects of 7% carboxyhaemoglobin on driving (although they reported effects, they had misapplied the chi-square test). The only effect on driving at a lower carboxyhaemoglobin level (7.6%) was reported by Rummo & Sarlanis (1974), who found that the ability to follow another car at a fixed distance was impaired.

The difference between the experiments of Rummo & Sarlanis (1974) and Weir et al. (1973) is troubling. Both measured following distance, but only the experiment employing the lower-level carboxyhaemoglobin found effects. If automobile driving is affected by carboxyhaemoglobin elevation, it remains to be demonstrated in a conclusive manner.

8.1.4.7   Brain electrical activity

Electrical activity of the brain (see review by Benignus, 1984) offers the possibility of testing the effects of carboxyhaemoglobin without the problem of selecting the most sensitive behavioural dependent variable. It is less dependent upon subject cooperation and effort and may be a more general screening method. The major disadvantage of the measures is the lack of functional interpretability. The area has been plagued with poor quantification and, frequently, a lack of statistical significance testing.

The electroencephalogram is a recording of the continuous voltage fluctuations emitted by the intact brain. The slow-evoked potential originally called the contingent negative variation is computed by averaging over trials and was linked to (among other things) cognitive processes or expectancy (Donchin et al., 1977). The evoked potential is the electrical activity in the brain resulting from sensory stimulation, either auditory or visual. The electroencephalogram, contingent negative variation and evoked potentials have been studied with carboxyhaemoglobin elevation.

Groll-Knapp et al. (1972) reported that the contingent negative variation was decreased in amplitude in a dose-related manner for carboxyhaemoglobin levels ranging from 3% to 7.6%. In a second study, Groll-Knapp et al. (1978) again reported that contingent negative variation amplitude was reduced by 12% carboxyhaemoglobin when subjects missed a signal in a vigilance task. More evidence is required before the functional significance of such an effect can be deduced, but it is a potentially important finding.

Clinical electroencephalograms were analysed by visual inspection by Stewart et al. (1970, 1973a) and Hosko (1970) after exposure to sufficient carbon monoxide to produce carboxyhaemoglobin levels ranging up to 33%. No effects were noticed. Groll-Knapp et al. (1978) reported similar results using spectrum analysis on electroencephalograms from subjects with 12% carboxyhaemoglobin. Haider et al. (1976) reported slight changes in the electroencephalogram spectrum for carboxyhaemoglobin levels of 13%, but no tests of significance were conducted. In view of the above studies, it seems reasonable to assume that no electroencephalogram effects of carboxyhaemoglobin levels below at least 10% should be expected.

O'Donnell et al. (1971b) reported that sleep stages (as determined from the electroencephalogram) were not distributed by carboxyhaemoglobin levels up to 12.4%. Groll-Knapp et al. (1978) and Haider et al. (1976), however, both reported distributed sleep stages at similar carboxyhaemoglobin levels using electroencephalogram spectra. Groll-Knapp et al. (1982) repeated their earlier study and found essentially the same effects.

The visual evoked potential was not affected consistently by carboxyhaemoglobin elevation below approximately 22%, and usually the lowest level for effects was higher. At higher levels, the effects were dose related (Hosko, 1970; Stewart et al., 1970).

Groll-Knapp et al. (1978) found no effect of carboxyhaemoglobin (8.6%) on click auditory evoked potentials during waking but reported increased positive peak amplitudes when subjects were tested during sleep at approximately 11% carboxyhaemoglobin. The finding was verified by Groll-Knapp et al. (1982). The fact that the data were collected during sleep is potentially important.

Putz et al. (1976) conducted a double-blind study in which 30 persons were exposed to 80 mg carbon monoxide/m^3 (70 ppm) for 240 min (5% carboxyhaemoglobin at the end of the session). Among other variables, the auditory evoked potential was measured. The peak-to-peak amplitude of the N_1–P_1 components was increased in a dose-ordinal manner beginning at approximately 3% carboxyhaemo-globin.

Many of the brain electrical activity measures seem to be altered by carboxyhaemoglobin elevation. The functional significance of these changes is not clear. Sometimes an alteration is not an indication of a deleterious effect but merely implies some change in processing. When induced by low levels of carboxyhaemoglobin, however, any change should be viewed as potentially serious.

8.1.5 Adaptation

This section considers whether or not exposure to carbon monoxide will eventually lead to the development of physiological responses that tend to offset some of the deleterious effects. Although there is possibly a temporal continuum in such processes, the term "adaptation" will be used in this review to refer to long-term phenomena, and the term "habituation" will refer to short-term processes. The term "compensatory mechanism" will be used to refer to those physiological responses that tend to ameliorate deleterious effects, whether in the long-term or short-term case.

.5.1   Short-term habituation

Arguments have been made for the possibility that there exist short-term compensatory mechanisms for carbon monoxide exposure. These hypothetical mechanisms have been based upon physiological evidence and have been used to account for certain behavioural findings reported in the literature.

There is physiological evidence for responses that would compensate for the deleterious effects of carbon monoxide in a very short time span. As discussed, carbon monoxide has been demonstrated to produce an increased cerebral blood flow, which is apparently produced by cerebrovascular vasodilation. It has also been shown (Zorn, 1972; Miller and Wood, 1974; Doblar et al., 1977; Traystman, 1978), however, that the tissue P_{O_2} values for various central nervous system sites fall in proportion to carboxyhaemoglobin, despite the increased blood flow. Apparently, the P_{O_2} values would fall considerably more without the increased blood flow. It appears that tissue P_{O_2} falls immediately and continuously as carboxyhaemoglobin rises.

.5.2   Long-term adaptation

Adaptation is an all-inclusive term that incorporates all of the acute or chronic adjustments of an organism to a stressor. It does not indicate (or predict) whether the adjustments are initially or eventually beneficial or detrimental. Acclimatization is an adaptive process that results in reduction of the physiological strain produced by exposure to a stressor. Generally, the main effect of repeated, constant exposure to the stressor is considered to result in an improvement of performance or a reduced physiological cost. Both of these phenomena tend to exploit the reserve potential of the organism.

Whether or not adaptation can occur in individuals chronically exposed to various ambient concentrations of carbon monoxide remains unresolved. Concern for carbon monoxide intoxication in England and Scandinavia led to the speculation that adaptational adjustments could occur in humans (Killick, 1940; Grut, 1949). These concerns were directed to situations where high ambient carbon monoxide concentrations were present. There are only a few available studies conducted in humans.

Killick (1940), using herself as a subject, reported that she developed acclimatization as evidenced by diminished symptoms, slower heart rate and the attainment of a lower carboxyhaemoglobin equilibrium level following exposure to a given inspired carbon monoxide concentration. Interestingly, Haldane & Priestley (1935) had already reported a similar finding as to the attainment of a different carboxyhaemoglobin equilibrium following exposure to a fixed level of carbon monoxide in the ambient air.

Killick (1948) repeated her carbon monoxide exposure studies in an attempt to obtain more precise estimations of the acclimatization effects she had noted previously. The degree of acclimatization was indicated by (1) a diminution in severity of symptoms during successive exposure to the same concentrations of carbon monoxide and (2) a lower carboxyhaemoglobin level after acclimatization than that obtained prior to acclimatization during exposure to the same concentrations of inhaled carbon monoxide.

8.1.6   Carbon monoxide interactions with drugs

There is little direct information on the possible enhancement of carbon monoxide toxicity by concomitant drug use or abuse; however, there are some data suggesting cause for concern. There is evidence that interactions of drug effects with carbon monoxide exposure can occur in both directions; that is, carbon monoxide toxicity may be enhanced by drug use, and the toxic or other effects of drugs may be altered by carbon monoxide exposure.

The effects of combined carbon monoxide exposure and alcohol (ethanol) administration have been the most extensively studied interaction. A study from the Medical College of Wisconsin (1974) found no effects of alcohol doses resulting in blood alcohol levels of about 0.05% and carboxyhaemoglobin levels in the general range of 8–9%, either alone or in combination, on a number of psychomotor behavioural tasks. The lack of sensitivity of these measures to alcohol doses known to affect performance under many other conditions, as well as other problems in the study design, raises the question of the adequacy of this study to detect interactive effects. Rockwell & Weir (1975) studied the interaction of carbon monoxide exposures resulting in nominal 0, 2, 8 and 12% carboxyhaemoglobin levels with alcohol doses resulting in nominal 0.05% blood alcohol levels for effects on actual driving and driving-related performances in young, non-smoking college students. Dose-related effects of carbon monoxide for perceptual narrowing and decreased eye movement were observed. In addition, effects were observed on some measures by this dose of alcohol alone. An effect-addition model was used to evaluate the alcohol–carbon monoxide interaction. In combination, the effects of carbon monoxide and alcohol were often additive, and there was a supra-additive alcohol–carbon monoxide interaction at 12% carboxy-haemoglobin levels. In a retrospective human study, King (1983) noted that the lethal carbon monoxide level was higher in the presence of ethanol, suggesting that alcohol ingestion prior to carbon monoxide exposure may provide some protection.

Because of a concern that persons exposed to carbon monoxide may not be able to detect odours that would indicate a fire or other hazardous condition, especially when consuming alcohol, Engen (1986) conducted a carefully controlled study of combined carbon monoxide–alcohol exposure in human subject volunteers. The detection of a threshold concentration of the smoky odour of quaiacol was evaluated using signal detection analysis. Although not statistically significant, there was a tendency for both alcohol and carbon monoxide to improve odour detection compared with air only. When alcohol and carbon monoxide were combined, the odour detection was significantly poorer than after either treatment alone, but it was not significantly poorer than the air control.

8.1.7 *Combined exposure to carbon monoxide and other air pollutants and environmental factors*

In this section, human effects associated with combined exposure to carbon monoxide and other air pollutants and environmental factors are reviewed. Although a number of studies in the literature have tested exposure to combined pollutants, fewer studies have actually been designed to test specifically for interactions between carbon monoxide and the other exposure components. Therefore, this section emphasizes only those studies providing a combined treatment group where pollutant exposure levels are reported.

8.1.7.1   Exposure to other pollutants in ambient air

Photochemical air pollution is usually associated with two or more pollutants, consisting mainly of carbon monoxide, sulfur oxides, ozone, nitrogen oxides, peroxyacyl nitrates and organic peroxides. The gaseous compounds that constitute tobacco smoke are carbon monoxide, hydrogen cyanide and nitric oxide. As urban living, industrial employment and cigarette smoking bring humans into direct contact with carbon monoxide and other pollutants, it seems appropriate to determine if combined exposure to these pollutants has detrimental health effects.

Several studies have been conducted to determine the effects resulting from combined exposure to carbon monoxide and other pollutants. A study by DeLucia et al. (1983) in adults exposed to carbon monoxide plus ozone during exercise showed no synergistic effects on blood carboxyhaemoglobin levels or pulmonary or cardio-respiratory thresholds. Similarly, simultaneous exposure to carbon monoxide plus ozone plus nitrogen dioxide for 2 h produced no consistent changes (synergistic or additive) in pulmonary function indices and physiological parameters in young male subjects (Hackney et al., 1975a,b).

Combined exposure to carbon monoxide and peroxyacyl nitrates exerted no greater effect on the work capacity of healthy men (young and middle-aged smokers and non-smokers) than did exposure to carbon monoxide alone. Increases in blood carboxyhaemoglobin levels of smokers during the carbon monoxide or carbon monoxide plus peroxyacyl nitrate exposures were observed (Drinkwater et al., 1974; Raven et al., 1974a,b; Gliner et al., 1975).

Halogenated hydrocarbons, such as the dihalomethanes (e.g., methylene bromide, methylene iodide and methylene chloride), are widely used as organic solvents. These chemicals are metabolized in the body to produce carbon monoxide, which is readily bound to haemoglobin. Therefore, any additional exposure to carbon monoxide, producing higher carboxyhaemoglobin levels, could possibly cause greater health effects. For example, up to 80% of inhaled methylene chloride will be metabolized to carbon monoxide. Inhalation of 1800–3500 mg/m^3 (500–1000 ppm), therefore, would result in carboxyhaemoglobin levels of over 14%. Not only can this elevation in carboxyhaemoglobin have a significant effect when combined with carbon monoxide exposure, but the carbon monoxide resulting from metabolism generally requires a longer time to dissipate (Kurppa, 1984).

7.2   Exposure to other environmental factors

1)   Environmental heat

Several of the studies (Drinkwater et al., 1974; Raven et al., 1974a,b; Gliner et al., 1975) describing the effects of carbon monoxide exposure alone and carbon monoxide combined with peroxyacyl nitrates on exercise performance in healthy adult men also evaluated the effects of heat stress. Subjects were exposed to 57 mg carbon monoxide/m^3 (50 ppm) and/or 0.27 ppm peroxyacyl nitrates in environmental exposure chamber conditions of 30% relative humidity at 25 and 30 °C. In these studies, oxygen uptake and exercise duration were assessed during both maximal and submaximal exercise. Heat stress was more effective in reducing maximal exercise performance than exposure to the polluted environments. The combination of heat stress with carbon monoxide exposure was found to be important, however, in producing symptom complaints during submaximal exercise at 35 °C that were not found at 25 °C. Further work in the same laboratory (Bunnell & Horvath, 1989) also demonstrated that subjects experienced significant levels of symptoms, particularly exertion symptoms, associated with elevated carboxyhaemoglobin when exercising in the heat. These studies suggest, therefore, that heat stress may be an important determinant of changes in exercise performance when combined with exposure to carbon monoxide.

2)   Environmental noise

An early epidemiological study by Lumio (1948) of operators of carbon monoxide-fuelled vehicles found significantly greater permanent hearing loss than expected after controlling for possible confounding factors. More recently, Sulkowski & Bojarski (1988) studied age-matched workers with similar length of duty employed in foundry, cast iron and cast steel positions where carbon monoxide and noise exposure varied. The group exposed to the combined effects of 95-dB noise and a mean concentration of 47 mg carbon monoxide/m^3 (41 ppm) did not experience any greater hearing loss than the groups exposed only to noise (96 dB) or carbon monoxide (52 mg/m^3 [45 ppm]). In fact, a permanent threshold shift was significantly larger in workers exposed to noise alone than in those exposed to the combined influence of carbon monoxide and noise.

8.1.8   Exposure to tobacco smoke

8.1.8.1   Environmental tobacco smoke

A common source of carbon monoxide for the general population is tobacco smoke. Exposure to tobacco smoke not only affects the carboxyhaemoglobin level of the smoker, but, under some circumstances, can also affect non-smokers. For example, acute exposure (1–2 h) to smoke-polluted environments has been reported to cause an incremental increase in non-smokers' carboxyhaemoglobin of about 1% (Jarvis, 1987).

The carbon monoxide concentration in tobacco smoke is approximately 4.5% (52 000 mg/m^3 [45 000 ppm]). A smoker may be exposed to 460–570 mg carbon monoxide/m^3 (400–500 ppm) for the approximately 6 min that it takes to smoke a typical cigarette, producing an average baseline carboxyhaemoglobin of 4%, with a typical range of 3–8%. Heavy smokers can have carboxyhaemoglobin levels as high as 15%. In comparison, non-smokers average about 1% carboxyhaemoglobin in their blood. As a result of their higher baseline carboxyhaemoglobin levels, smokers may actually be excreting more carbon monoxide into the air than they are inhaling from the ambient environment. Smokers may even show an adaptive response to the elevated carboxyhaemoglobin levels, as evidenced by increased red blood cell volumes or reduced plasma volumes (Smith & Landaw, 1978a,b).

In addition to being a source of carbon monoxide for smokers as well as for non-smokers, tobacco smoke is also a source of other chemicals with which environmental carbon monoxide could interact. Available data strongly suggest that acute and chronic carbon monoxide exposure attributed to tobacco smoke can affect the cardiopulmonary system, but the potential interaction of carbon monoxide with other products of tobacco smoke confounds the results. In addition, it is not clear if incremental increases in carboxyhaemoglobin caused by environmental exposure would actually be additive to chronically elevated carboxyhaemoglobin levels due to tobacco smoke, because some physiological adaptation may take place.

8.2   Mainstream tobacco smoke

Acute effects of cigarette smoke on maximal exercise performance appear to be similar to those described in subjects exposed to carbon monoxide. Hirsch et al. (1985) studied the acute effect of smoking on cardiorespiratory function during exercise in nine healthy male subjects who were current smokers. They were tested twice — once after smoking three cigarettes per hour for 5 h and once after not having smoked. The exercise tests were done on a bicycle ergometer with analysis of gas exchange and intra-arterial blood gases and pressures. On the smoking day, $\dot{V}_{O_2}$ max was significantly decreased by 4%, and the anaerobic threshold was decreased by 14%. The rate–blood pressure product was a significant 12% higher at comparable work loads of 100 W on the smoking day than on the non-smoking day. There were no changes due to smoking, however, on the duration of exercise or on the mean work rate during maximal exercise testing. The blood carboxyhaemoglobin level before exercise was 1.8% on the non-smoking day and 6.6% on the smoking day. At peak exercise, the carboxyhaemoglobin was 0.9% and 4.8% on the non-smoking and smoking day, respectively. The authors concluded that the main adverse effect of smoking was due to carbon monoxide, although the increase in rate–blood pressure product also might be the result of the simultaneous inhalation of nicotine. They felt that the magnitude of change in performance indicators corresponded well with earlier reports.

It would be interesting, therefore, to determine if smokers and non-smokers had different responses to carbon monoxide exposure. Unfortunately, smokers and non-smokers were not always identified in many of the studies on exercise performance, making it difficult to interpret the available data. Information derived from studies on cigarette smoke is also sparse. As a result, attempts to sort out the acute effects of carbon monoxide from those due to other components of cigarette smoke have been equivocal. Seppanen (1977) reported that the physical work capacities of cigarette smokers decreased at 9.1% carboxyhaemoglobin levels after either breathing boluses of 1300 mg carbon monoxide/m^3 (1100 ppm) or smoking cigarettes. The greatest decrease in maximal work, however, was observed after carbon monoxide inhalation.

Klausen et al. (1983) compared the acute effects of cigarette smoking and inhalation of carbon monoxide on maximal exercise performance. They studied 16 male smokers under three different conditions: after 8 h without smoking (control), after inhalation of the smoke of three cigarettes and after carbon monoxide inhalation. Just before maximal exercise testing, the arterial carboxyhaemoglobin level reached 4.51% and 5.26% after cigarette smoke and carbon monoxide inhalation, respectively, compared with 1.51% for controls. Average $\dot{V}_{O_2}$ max decreased by about 7% with both smoke and carbon monoxide. Exercise time, however, decreased 20% with smoke but only 10% with carbon monoxide, suggesting that nicotine, smoke particles or other components of tobacco smoke may contribute to the observed effects. The authors therefore concluded that a specified carboxyhaemoglobin level induced by either smoke or carbon monoxide decreased maximal work performance to the same degree. Of note is the more marked decrease in work time compared with $\dot{V}_{O_2}$ max induced by carbon monoxide, a finding that agrees with the Ekblom & Huot (1972) results.

8.2 High-risk groups

8.2.1 *Effects in individuals with heart disease*

Aronow et al. (1972) and Aronow & Isbell (1973) demonstrated that patients with angina pectoris, when exposed to low levels of carbon monoxide (2.5–3% carboxyhaemoglobin), experienced reduced time to onset of exercise-induced chest pain as a result of insufficient oxygen supply to the heart muscle. A study by Anderson et al. (1973) reported similar results at mean carboxyhaemoglobin levels of 2.9% and 4.5%.

In 1981, Aronow reported an effect of 2% carboxyhaemoglobin on time to onset of angina in 15 patients. The protocol was similar to that used in previously reported studies, with patients exercising until onset of angina. Only 8 of the 15 subjects developed 1 mm or greater ischaemic ST-segment depression at the onset of angina during the control periods. This was not significantly affected by carbon monoxide. One millimetre or greater ST-segment depression is the commonly accepted criterion for exercise-induced ischaemia. It is questionable, therefore, as to whether the remaining patients truly met adequate criteria for ischaemia despite angiographically documented cardiac disease. After breathing 57 mg carbon monoxide/m³ (50 ppm) for 1 h, the patients' times to onset of angina significantly decreased from a mean of 321.7 ± 96 s to a mean of 289.2 ± 88 s.

The cardiovascular studies by Aronow et al. were severely criticized because the subjective measure of symptoms (angina) was affected by lack of adequate double-blind experimental conditions. There was also a lack of significant objective findings.

In an attempt to improve upon these earlier preliminary studies, the more recent studies placed greater emphasis on electrocardiogram changes as objective measures of ischaemia. Another consideration in the conduct of the newer studies on angina was to better establish the dose–response relationships for low levels of carbon monoxide exposure. Although carboxyhaemoglobin level is accepted as the best measure of the effective dose of carbon monoxide, the reporting of low-level effects is problematic because of inconsistencies in the rigour with which the devices for measuring carboxyhaemoglobin have been validated. The most frequently used technique for measuring carboxyhaemoglobin has been the optical method found in the IL series of CO-Oximeters. Not only is there a lot of individual variability in these machines, but recent comparisons with the gas chromatographic technique of measuring carboxyhaemoglobin have suggested that the optical method may not be a suitable reference technique for measuring low levels of carboxyhaemoglobin. Several additional studies have appeared in the literature to help define the precise carboxyhaemoglobin levels at which cardiovascular effects occur in angina patients. Because the range of carboxyhaemoglobin values obtained with the optical method of analysis may be different from that obtained by gas chromatography, the method used to measure carboxyhaemoglobin will be indicated in parentheses for each of these studies.

Sheps et al. (1987) studied 30 patients aged 38–75 years with ischaemic heart disease and assessed not only symptoms during exercise, but also radionuclide evidence of ischaemia (left ventricular ejection fraction changes). Patients were non-smokers with ischaemia, defined by exercise-induced ST-segment depression, angina or abnormal ejection fraction response (i.e., all patients had documented evidence of ischaemia).

Patients were exposed to carbon monoxide (110 mg/m^3 [100 ppm]) or air during a 3-day, randomized double-blind protocol to achieve a post-exposure level of 4% carboxyhaemoglobin (CO-Oximeter measurement). Resting pre-exposure levels were 1.7%, post-exposure levels were 4.1% and post-exercise levels were 3.6% on the carbon monoxide exposure day; thus, the study examined acute elevation of carboxyhaemoglobin levels from 1.7% to an average of 3.8%, or an average increase of 2.2% carboxyhaemoglobin from resting values. Comparing exposure to carbon monoxide with exposure to air, there was no significant difference in time to onset of angina, maximal exercise time, maximal ST-segment depression (1.5 mm for both) or time to significant ST-segment depression. The conclusion of this study was that 3.8% carboxyhaemoglobin produces no clinically significant effects on this patient population.

Interestingly, further analysis of the time to onset of angina data in this paper demonstrated that 3 of the 30 patients experienced angina on the carbon monoxide exposure day but not on the air-control day. These patients had to be deleted from the classical analysis of differences between time to onset of angina that was reported in the publication. However, actuarial analysis of time to onset of angina including these patients revealed a statistically significant ($P < 0.05$) difference in time to onset of angina favouring an earlier time under the carbon monoxide exposure conditions (Bissette et al., 1986). None of the patients had angina only on the air exposure day.

Subsequent work from these same investigators (Adams et al., 1988) focused on repeating the study at 6% carboxyhaemoglobin (CO-Oximeter measurement). Thirty subjects with obstructive coronary artery disease and evidence of exercise-induced ischaemia were exposed to air or carbon monoxide on successive days in a randomized double-blind crossover fashion. Post-exposure carboxyhaemoglobin levels averaged 5.9 ± 0.1% compared with 1.6 ± 0.1% after air exposure, representing an increase of 4.3% carboxyhaemoglobin. The mean duration of exercise was significantly longer after air exposure than after carbon monoxide exposure (626 ± 50 s for air versus 585 ± 49 s for carbon monoxide, $P < 0.05$). Actuarial methods suggested that subjects experienced angina earlier during exercise on the day of carbon monoxide exposure ($P < 0.05$). In addition, this study showed that, at a slightly higher level of carbon monoxide exposure, both the level and change in ejection fraction at submaximal exercise were greater on the air day than on the carbon monoxide day. The peak exercise left ventricular ejection fraction, however, was not different for the two exposures.

These results demonstrated earlier onset of ventricular dysfunction and angina and poorer exercise performance in patients with ischaemic heart disease after acute carbon monoxide exposure sufficient to increase carboxyhaemoglobin to 6%. It is of interest that in both the 4% study and the 6% study reported by this group, seven of the patients experienced angina on the carbon monoxide exposure day, but not on the air exposure day. There were no patients who experienced angina in the reverse sequence, providing further support for a significant effect of carbon monoxide exposure on angina occurrence.

Kleinman & Whittenberger (1985) and Kleinman et al. (1989) studied non-smoking male subjects with a history of stable angina pectoris and positive exercise tests. All but 2 of the 26 subjects had additional confirmation of ischaemic heart disease, such as previous myocardial infarction, positive angiogram, positive thallium scan, prior angioplasty or prior bypass surgery. Subjects were exposed for 1 h in a randomized double-blind crossover fashion to either 110 mg carbon monoxide/m^3 (100 ppm) or clean air on 2 separate days. Subjects performed an incremental exercise test on a cycle ergometer to the point at which they noticed the onset of angina. For the study group, the 1-h exposure to 110 mg carbon monoxide/m^3 (100 ppm) resulted in an increase in carboxyhaemoglobin from 1.4% after clean air to 3% (CO-Oximeter measurement) after carbon monoxide. For the entire study group ($n = 26$), the 1-h exposure to 110 mg/m^3 (100 ppm) resulted in a decrease of the time to onset of angina by 6.9% from 6.5 to 6.05 min (Kleinman & Whittenberger, 1985). This difference was significant in a one-tailed paired t-test ($P = 0.03$). When using a two-tailed test, the difference loses statistical significance at the $P = 0.05$ level.

In the published version of results from this study (Kleinman et al., 1989), the two subjects with inconsistencies in their medical records and histories were dropped from the analysis. For this study group ($n = 24$), the 1-h exposure to 110 mg carbon monoxide/m^3 (100 ppm) (3% carboxyhaemoglobin by CO-Oximeter measurement) resulted in a significant decrease of time to onset of angina by 5.9% using a one-tailed, two-factor analysis of variance ($P = 0.046$). There was no significant effect on the duration of angina, but oxygen uptake at angina point was reduced 2.7% ($P = 0.04$). Only eight of the subjects exhibited depression in the ST-segment of their electro-cardiogram traces during exercise. For this subgroup, there was a 10% reduction ($P < 0.036$) in time to onset of angina and a 19% reduction ($P < 0.044$) in the time to onset of 1-mm ST-segment depression.

A multicentre study of effects of low levels of carboxyhaemo-globin has been conducted in three cities on a relatively large sample ($n = 63$) of individuals with coronary artery disease (Allred et al., 1989a,b, 1991). The purpose of this study was to determine the effects of carbon monoxide exposures producing 2.0% and 3.9% carboxy-haemoglobin (gas chromatographic measurement) on time to onset of significant ischaemia during a standard treadmill exercise test. Significant ischaemia was measured subjectively by the duration of exercise required for the development of angina (time to onset of angina) and objectively by the time required to demonstrate a 1-mm change in the ST-segment of the electrocardiogram (time to ST). The time to onset of ST-segment changes was measured to the nearest second, rather than to the nearest minute as in the other studies on angina, a strength of this study. Male subjects aged 41–75 (mean = 62.1 years) with stable exertional angina pectoris and a positive stress test, as measured by a greater than 1-mm ST-segment change, were studied. Further evidence that these subjects had coronary artery disease was provided by the presence of at least one of the following criteria: angiographic evidence of narrowing ($\geq 70\%$) of at least one coronary artery, documented prior myocardial infarc-tion or a positive stress thallium test demonstrating an unequivocal perfusion defect. Thus, as opposed to some previous studies reported, this study critically identified patients with documented coronary artery disease.

The protocol for this study was similar to that used in the Aronow studies, because two exercise tests were performed on the same day. The two tests were separated by a recovery period and a double-blind exposure period. On each of the 3 exposure days, the subject per-formed a symptom-limited exercise test on a treadmill, then was exposed for 50–70 min to carbon monoxide concentrations that were experimentally determined to produce end-exposure carboxyhaemo-globin levels of 2% and 4%. The mean exposure levels and ranges for the test environment were clean air (0 mg carbon monoxide/m^3

[0 ppm]), 134 mg carbon monoxide/m^3 (117 ppm; range 48–230 mg/m^3 [42–202 ppm]) and 290 mg carbon monoxide/m^3 (253 ppm; range 160–410 mg/m^3 [143–357 ppm]). The subject then performed a second symptom-limited exercise test. The time to onset of angina and the time to 1-mm ST-segment change were determined for each test. The percent changes following exposure at both 2.0% and 3.9% carboxyhaemoglobin (gas chromatographic measurement) were then compared with the same subject's response to the randomized exposure to room air (less than 2.3 mg carbon monoxide/m^3 [2 ppm]).

When potential exacerbation of the exercise-induced ischaemia by exposure to carbon monoxide was tested using the objective measure of time to 1-mm ST-segment change, exposure to carbon monoxide levels producing 2.0% carboxyhaemoglobin resulted in a 5.1% decrease ($P = 0.01$) in the time to attain this level of ischaemia. At 3.9% carboxyhaemoglobin, the decrease in time to the ST criterion was 12.1% ($P \leq 0.0001$) relative to the air-day results; this reduction in time to ST-segment depression was accompanied by a significant ($P = 0.03$) reduction in the heart rate–blood pressure product (double product), an index of myocardial work. The maximal amplitude of the ST-segment change was also significantly affected by the carbon monoxide exposures: at 2% carboxyhaemoglobin, the increase was 11% ($P = 0.002$), and at 3.9% carboxyhaemoglobin, the increase was 17% relative to the air day ($P \leq 0.0001$).

When the individual centre data in the Allred et al. (1989a,b, 1991) study were analysed for covariates that may have influenced the results of this study, only the absolute level of carboxyhaemoglobin was found to have had a significant effect. This finding is not surprising, given the dose–response relationship between carbon monoxide and time to 1-mm ST-segment change. This analysis compared the slopes for each individual subject. The three times to 1-mm ST-segment change were plotted against the three actual carboxyhaemoglobin levels. The 62 individual slopes were then combined to yield a significant ($P < 0.005$) regression model: *Change in time to 1-mm ST-segment change* $= (-3.85 \pm 0.63)$ *(% COHb)* $+ (8.01\% \pm 2.48\%)$. This dose–response relationship indicates that there is a 3.9% decrease in the time to ST criterion for every 1% increase in carboxyhaemoglobin.

The time to onset of angina was also significantly reduced in these subjects. At 2.0% carboxyhaemoglobin, the time to angina was reduced by 4.2% ($P = 0.027$), and at 3.9% carboxyhaemoglobin, the time was reduced by 7.1% ($P = 0.002$). There were no significant changes in the double products at the time of onset of angina in either exposure condition. The regression analysis for the time to angina data also resulted in a significant relationship ($P < 0.025$). The average regression was *Time to angina* = $(-1.89\% \pm 0.81\%)$ (% *COHb*) + $(1.00\% \pm 2.11\%)$. The lower level of significance and the larger error terms for the angina regression relative to the ST analysis indicate that the angina end-point is more variable. This may be due to the subjective nature of this end-point or the variability in the ability of subjects to clearly recognize the onset of the pain.

The two end-points (time to angina and time to ST change) in the Allred study were also correlated, with a Spearman rank correlation coefficient of 0.49 ($P \le 0.0001$). The conclusion of all of the analyses from this multicentre study is that the response of the myocardium in these patients with coronary artery disease is consistent, although the effects are relatively small.

The analysis of the covariates in this multicentre study also provides answers to ancillary questions that have been raised elsewhere in this document. The medication being used by these subjects did not significantly influence the results (i.e., there does not appear to be any drug interaction with the effects of carbon monoxide). The major medications being used in this group were β-blockers (used by 38 of the 63 subjects), nitrates (used by 36 of the 63 subjects) and calcium channel blockers (used by 40 of the 63 subjects). The other major concern was the influence of the severity of the disease. The simplest approach to this was to evaluate the influence of the duration of the exercise, because the subjects with more severe disease were limited in their exercise performance. No significant correlation was found between duration of exercise and the percent change in time to angina or ST criterion. There was also no relationship between the average time of exercise until the onset of angina and either of the end-points. Not was there a relationship between the presence of a previous myocardial infarction and the study end-points.

The duration of exercise was significantly shortened at 3.9% carboxyhaemoglobin but not at 2.0%. This finding must be used cautiously, because these subjects were not exercised to their maximum capacity in the usual sense. The major reason for termination of the exercise was the progression of the angina (306 of 376 exercise tests). The subjects were to grade their angina on a four-point scale; when the exercise progressed beyond level two, they were stopped. Therefore, this significant decrease in exercise time of 40 s at the 3.9% carboxyhaemoglobin level is undoubtedly due to the earlier onset of angina followed by the normal rate of progression of the severity of the angina.

The individual centre data provide insight into the interpretation of other studies that have been conducted in this area. Each of the centres enrolled the numbers of subjects that have been reported by other investigators. The findings reported above were not substantiated in all instances at each centre. When one considers the responses of the group to even 3.9% carboxyhaemoglobin, it is clear why one might not find significance in one parameter or another. For the decrease in ST segment at 3.9%, only 49 of 62 subjects demonstrated this effect on the day tested. The potential for finding significance at this effect rate with a smaller sample size is reduced. Random sampling of this population with a smaller sample could easily provide subjects that would not show significant effects of these low levels of carbon monoxide on the test day.

The recent reports (Allred et al., 1989b, 1991) of the multicentre study, organized and supported by the Health Effects Institute, discuss some reasons for differences between the results of the studies cited above (see Table 16). The studies have different designs, types of exercise tests, inclusion criteria (and, therefore, patient populations), exposure conditions and means of measuring carboxyhaemoglobin. All of the studies have shown an effect of carboxyhaemoglobin elevation on time to onset of angina (see Fig. 12). Results from the Kleinman et al. (1989) study showed a 6% decrease in exercise time to angina at 3.0% carboxyhaemoglobin (CO-Oximeter measurement) measured at the end of exposure. Allred et al. (1989a,b) reported a 5% and 7% decrease in time to onset of angina after increasing carboxy-haemoglobin levels to 3.2% and 5.6% (CO-Oximeter measurement), respectively, at the end of exposure. Although the Sheps et al. (1987) and Adams et al. (1988) studies did not observe statistically significant

Table 16. Comparison of subjects in studies of the effect of carbon monoxide exposure on occurrence of angina during exercise[a]

Study	Number of subjects	Gender	Medication	Smoking history	Description of disease	Age (years)
Anderson et al., 1973	10	male	1 subject took digitalis; drug therapy basis for exclusion	5 smokers (refrained for 12 h prior to exposure)	Stable angina pectoris, positive exercise test (ST changes); reproducible angina on treadmill	mean = 49.9
Kleinman et al., 1989	24	male	14 on β-blockers; 19 on nitrates; 8 on calcium channel blockers	No current smokers	Ischaemic heart disease, stable exertional angina pectoris	49–66 (mean = 59)
Allred et al., 1989a,b, 1991	63	male	38 on β-blockers; 36 on nitrates; 40 on calcium antagonists	No current smokers	Stable exertional angina and positive exercise test (ST changes) plus one or more of the following: (1) ≥70% lesion by angiography in one or more major vessels, (2) prior myocardial infarction, (3) positive exercise thallium test	41–75 (mean = 62.1)
Sheps et al., 1987	30 (23 with angina)	25 male 5 female	26 subjects on medication; 19 on β-blockers; 11 on calcium channel blockers; 1 on long-acting nitrates	No current smokers	Ischaemia during exercise (ST changes or abnormal ejection fraction response) and one or more of the following: (1) angiographically proven coronary artery disease, (2) prior myocardial infarction, (3) typical angina	36–75 (mean = 58.2)

Table 16 (contd).

| Adams et al., 1988 | 30 (25 with angina) | 22 male 8 female | 25 subjects on medication; 13 on β-blockers + calcium channel blockers; 6 on β-blockers; 5 on calcium channel blockers; 1 on long-acting nitrates | No current smokers | Ischaemia during exercise (ST changes or abnormal ejection fraction response) and one or more of the following: (1) angiographically proven coronary artery disease, (2) prior myocardial infarction, (3) typical angina | 36–75 (mean = 58) |

[a] Adapted from Allred et al. (1989b, 1991).

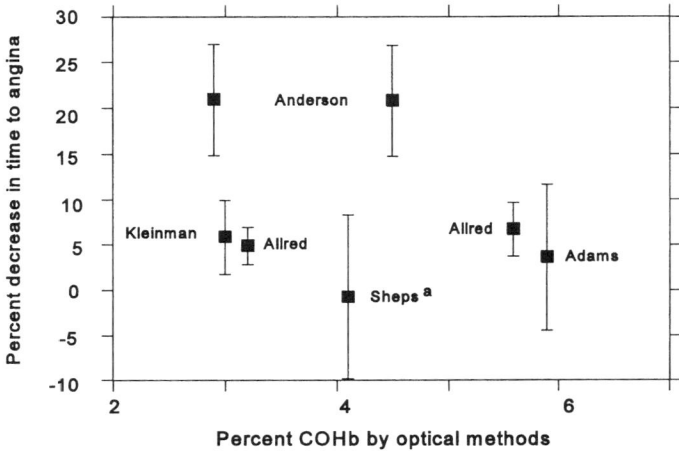

ᵃ Alternative statistical analyses of the Sheps data (Bissette et al., 1986) indicate a significant decrease in time to onset of angina at 4.1% carboxyhaemoglobin if subjects that did not experience exercise-induced angina during air exposure are also included in the analyses.

Fig. 12. The effect of carbon monoxide exposure on time to onset of angina. For comparison across studies, data are presented as mean percent differences between air and carbon monoxide exposure days for individual subjects calculated from each study. Bars indicate calculated standard errors of the mean. Carboxyhaemoglobin levels were measured at the end of exposure; however, because of protocol differences among studies and lack of precision in optical measurements of carboxyhaemoglobin, comparisons must be interpreted with caution (see text and Table 16 for more details) (adapted from Allred et al., 1989b, 1991).

changes in time to onset of angina using conventional statistical procedures, the results of these studies are not incompatible with the rest of the studies reporting an effect of carbon monoxide. Both studies reported a significant decrease in time to onset of angina on days when carboxyhaemoglobin levels at the end of exposure were 4.1% and 5.9% (CO-Oximeter measurement), respectively, if the data analysis by actuarial method included subjects who experienced angina on the carbon monoxide day but not on the air day. In addition, the Adams et al. (1988) study reported that left ventricular performance, assessed by radionuclide measurement of the ejection fraction, was reduced during submaximal exercise after carbon monoxide exposure compared with air exposure.

Of particular importance in this group of studies was the fact that the multicentre study (Allred et al., 1989a,b, 1991) demonstrated a dose–response effect of carboxyhaemoglobin on time to onset of angina. The only other single study that investigated more than a single target level of carboxyhaemoglobin was that by Anderson et al. (1973), and their results, based on a smaller number of subjects, did not show a dose–response relationship for angina.

The time to onset of significant electrocardiogram ST-segment changes, which are indicative of myocardial ischaemia in patients with documented coronary artery disease, is a more objective indicator of ischaemia than is angina. Allred et al. (1989a,b, 1991) reported a 5.1% and 12.1% decrease in time to ST-segment depression at carboxyhaemoglobin levels of 2.0% and 3.9% (gas chromatographic measurement), respectively, measured at the end of exercise. An additional measurement of the ST change was made by Allred et al. (1989b) to confirm this response — all the leads showing ST-segment changes were summed. This summed ST score was also significantly affected by both levels of carboxyhaemoglobin. The significant finding for the summed ST score indicates that the effect reported for time to 1-mm ST-segment change was not dependent upon changes observed in a single electrocardiogram lead.

The differences between the results of these five studies on exercise-induced angina can largely be explained by differences in experimental methodology and analysis of data and, to some extent, by differences in subject populations and sample size. For example, the Kleinman et al. (1989) study and the Allred et al. (1989a,b, 1991) study used one-tailed P values, whereas the Sheps et al. (1987) and Adams et al. (1988) studies used two-tailed P values. The Allred et al. (1989a,b, 1991) study also used trimmed means (with the two highest and two lowest values deleted) to guard against outliers. If a two-sided P value was utilized on the time to onset of angina variable observed at the lowest carboxyhaemoglobin level in the Allred et al. (1989a,b, 1991) study, it would become 0.054 rather than 0.027, a result that would be considered borderline significant. If a two-sided P value were used in the Kleinman et al. (1989) study, the difference in time to onset of angina would lose significance at the $P = 0.05$ level.

The entry criteria in the Allred et al. (1989a,b, 1991) study were more rigorous than in the other studies. All subjects were required to have stable exertional angina and *reproducible* exercise-induced ST depression and angina. Besides these criteria, all subjects were required to have a previous myocardial infarction, angiographic disease or a positive thallium stress test. In addition, only men were studied. These strict entry criteria were helpful in allowing the investigators to more precisely measure an adverse effect of carbon monoxide exposure. The protocol for the multicentre study, however, was slightly different from some of the protocols previously reported. On each test day, the subject performed a symptom-limited exercise test on a treadmill, then was exposed for approximately 1 h to air or one of two levels of carbon monoxide in air before undergoing a *second* exercise test. Time to onset of ischaemic electrocardiogram changes and time to onset of angina were determined for each exercise test. The percent difference for these end-points from the pre- and post-exposure test was then determined. The results on the 2% target day and then the 4% target day were compared with those on the control day.

The statistical significance reported at the low-level carbon monoxide exposure is present only when the *differences* between pre- and post-exposure exercise tests are analysed. Analysis of only the post-exposure test results in a loss of statistical significance for the 2% carboxyhaemoglobin level. Some of the differences between the results of this multicentre study and previous studies may be related to the fact that the exposure was conducted shortly after patients exercised to angina. The length of time required for resolution of exercise-induced ischaemia is not known. However, exercise treadmill testing of patients with coronary artery disease has been shown to induce regional wall motion abnormalities of the left ventricle that persist for over 30–45 min after exercise when chest pain and electro-cardiogram abnormalities are usually resolved (Kloner et al., 1991). In addition, radionuclide studies in these patients have shown meta-bolic effects of ischaemia to last for more than 1 h after exercise (Camici et al., 1986). Because the effects of ischaemia may have a variable duration, differences between pre- and post-exposure tests may have been due to effects of carbon monoxide exposure on *recovery* from a previous episode of exercise-induced ischaemia rather than detrimental effects only during exercise.

In conclusion, five key studies have investigated the potential for carbon monoxide exposure to enhance the development of myocardial ischaemia during progressive exercise tests. Despite differences between them, it is impressive that all of the studies identified in Fig. 12 show a decrease in time to onset of angina at post-exposure carboxyhaemoglobin levels ranging from 2.9% to 5.9%. This represents incremental increases of 1.5–4.4% carboxyhaemoglobin from pre-exposure baseline levels. Therefore, there are clearly demonstrable effects of low-level carbon monoxide exposure in patients with ischaemic heart disease. The adverse health consequences of these types of effects, however, are very difficult to predict in the at-risk population of individuals with heart disease. There exists a distribution of professional judgements on the clinical significance of small performance decrements occurring with the levels of exertion and carbon monoxide exposure defined in these five studies. The decrements in performance that have been described at the lowest levels (≤3% carboxyhaemoglobin) are in the range of reproducibility of the test and may not be alarming to some physicians. On the other hand, the consistency of the responses in time to onset of angina across the studies and the dose–response relationship described by Allred et al. (1989a,b, 1991) between carboxyhaemoglobin and time to ST-segment changes would strengthen the argument in the minds of other physicians that, although small, the effects could limit the activity of these individuals and affect the quality of their life. In addition, it has been argued by Bassan (1990) that 58% of cardiologists believe that recurrent episodes of exertional angina are associated with a substantial risk of precipitating a myocardial infarction, a fatal arrhythmia or slight but cumulative myocardial damage.

8.2.2 Effects in individuals with chronic obstructive lung disease

Aronow et al. (1977) studied the effects of a 1-h exposure to 110 mg carbon monoxide/m^3 (100 ppm) on exercise performance in 10 men, aged 53–67 years, with chronic obstructive lung disease. The resting mean carboxyhaemoglobin levels increased from 1.4% baseline levels to 4.1% after breathing carbon monoxide. The mean exercise time until marked dyspnoea significantly decreased (33%) from 218 s in the air-control period to 147 s after breathing carbon monoxide. The authors speculated that the reduction in exercise performance was due to a cardiovascular limitation rather than respiratory impairment.

Only one other study in the literature, by Calverley et al. (1981), looked at the effects of carbon monoxide on exercise performance in older subjects with chronic lung disease. They evaluated 15 patients with severe reversible airway obstruction due to chronic bronchitis and emphysema. Six of the patients were current smokers, but they were asked to stop smoking for 12 h before each study. The distance walked within 12 min was measured before and after each subject breathed 0.02% (230 mg/m^3 [200 ppm]) carbon monoxide in air from a mouthpiece for 20–30 min until carboxyhaemoglobin levels were 8–12% above their initial levels. A significant decrease in walking distance was reported when the mean carboxyhaemoglobin concentration reached 12.3%, a level much higher than most of those reported in studies on healthy subjects.

Thus, although it is possible that individuals with hypoxia due to chronic lung diseases such as bronchitis and emphysema may be susceptible to carbon monoxide during submaximal exercise typically found during normal daily exercise, these effects have not been studied adequately at relevant carboxyhaemoglobin concentrations of less than 5%.

8.2.3 Effects in individuals with chronic anaemia

An additional study by Aronow et al. (1984) on the effect of carbon monoxide on exercise performance in anaemic subjects found a highly significant decrease in work time (16%) induced by an increase of 1.24% carboxyhaemoglobin. The magnitude of change seems to be very unlikely, however, even considering the report by Ekblom & Huot (1972). The study was double-blind and randomized, but with only 10 subjects. The exercise tests were done on a bicycle in the upright position with an increase in workload of 25 W every 3 min. However, no measure of maximal performance such as blood lactate was used. The mean maximal heart rate was only 139–146 beats per minute compared with a predicted maximal heart rate of 170 beats per minute for the mean age of the subjects. A subject repeating a test within the same day, which was the case in the Aronow et al. (1984) study, will often remember the time and work load and try to do the same in the second test. Normally, however, some subjects will increase while others will decrease the time. This situation was apparent on the air-control day, with an increase demonstrated in 6 out of 10 subjects, despite the high reproducibility for such a soft, subjective end-point. Also, comparing the control tests on the air day with the carbon monoxide day, 7 out of 10 subjects increased their work time. After carbon monoxide exposure, however, all subjects decreased their time between 29 and 65 s. These data appear to be implausible given the soft end-point used, when 2–3 of the subjects would be expected to increase their time even if there were a true effect of carbon monoxide.

.2.4 Arrhythmogenic effects

Until recent years, the literature has been confusing with regard to potential arrhythmogenic effects of carbon monoxide.

Davies & Smith (1980) studied the effects of moderate carbon monoxide exposure on healthy individuals. Six matched groups of human subjects lived in a closed, environmental exposure chamber for 18 days and were exposed to varying levels of carbon monoxide. Standard 12-lead electrocardiograms were recorded during five control, eight exposure and five recovery days. P-wave changes of at least 0.1 mV were seen in the electrocardiograms during the carbon monoxide exposure period in 3 of 15 subjects at 2.4% carboxyhaemoglobin and in 6 of 15 subjects at 7.1% carboxyhaemoglobin compared with 0 of 14 at 0.5% carboxyhaemoglobin. The authors felt that carbon monoxide had a specific toxic effect on the myocardium in addition to producing a generalized decrease in oxygen transport to tissue.

Several methodological problems create difficulties of interpretation for this study. The study design did not use each subject as his own control. Thus, only one exposure was conducted for each subject. Half of the subjects were tobacco smokers who were required to stop smoking, and certainly some of the electrocardiogram changes could have been due to the effects of nicotine withdrawal. Although the subjects were deemed to be normal, no screening stress tests were performed to uncover latent ischaemic heart disease or propensity to arrhythmia. Most importantly, no sustained arrhythmias or measurable effects on the conduction system were noted by the authors. If P-wave changes of clinical significance are representative of a toxic effect of carbon monoxide on the atrium, then an effect on conduction of arrhythmias should be demonstrated.

Knelson (1972) reported that 7 of 26 individuals aged 41–60 years had abnormal electrocardiograms after exposure to 110 mg carbon monoxide/m^3 (100 ppm) for 4 h (carboxyhaemoglobin levels of 5–9%). Two of them developed arrhythmias. No further details were given regarding specifics of these abnormalities. Among 12 younger subjects aged 25–36 years, all electrocardiograms were normal.

Hinderliter et al. (1989) reported on effects of low-level carbon monoxide exposure on resting and exercise-induced ventricular arrhythmias in patients with coronary artery disease and no baseline ectopy. They studied 10 patients with ischaemic heart disease and no ectopy according to baseline monitoring. After an initial training session, patients were exposed to air, 110 mg carbon monoxide/m³ (100 ppm) or 230 mg carbon monoxide/m³ (200 ppm) on successive days in a randomized, double-blinded crossover fashion. Venous carboxyhaemoglobin levels after exposure to 110 and 230 mg carbon monoxide/m³ (100 and 200 ppm) averaged 4% and 6%, respectively. Symptom-limited supine exercise was performed after exposure. Eight of the 10 patients had evidence of exercise-induced ischaemia — angina, ST-segment depression or abnormal left ventricular ejection fraction response — during one or more exposure days. Ambulatory electrocardiograms were obtained for each day and were analysed for arrhythmia frequency and severity. On air and carbon monoxide exposure days, each patient had only 0–1 ventricular premature beats per hour in the 2 h prior to exposure, during the exposure period, during the subsequent exercise test and in the 5 h following exercise. The authors concluded that low-level carbon monoxide exposure is not arrhythmogenic in patients with coronary artery disease and no ventricular ectopy at baseline.

The results of low-level carbon monoxide exposure on patients with higher levels of ectopy were reported by the same investigators (Sheps et al., 1990, 1991). The frequency of a single ventricular premature depolarization per hour was significantly greater after carbon monoxide exposure producing 6% carboxyhaemoglobin (167.72 ± 37.99) compared with exposure to room air (127.32 ± 28.22, $P = 0.03$) and remained significant when adjusted for baseline ventricular premature depolarization levels for all subjects regardless of ventricular premature depolarization frequency category. During exercise, the mean number of multiple ventricular premature depolarizations per hour was greater after carbon monoxide exposure producing 6% carboxyhaemoglobin (9.59 ± 3.70) compared with exposure to room air (3.18 ± 1.67, $P = 0.02$) and remained significant after adjustment for baseline multiple ventricular premature depolarization levels and when all subjects were included regardless of ventricular premature depolarization frequency category. The authors concluded that the number and complexity of ventricular arrhythmias increase significantly during exercise after carbon monoxide exposures producing 6% carboxyhaemoglobin compared with room air exposures, but not after carbon monoxide exposures producing 4% carboxyhaemoglobin. Because statistically significant effects were shown only during the exercise period, however, these reported changes are likely occurring at a lower carboxyhaemoglobin level. In fact, the carboxyhaemoglobin levels during exercise were 1.4% on the air exposure day, 3.7% on the 4% carboxyhaemoglobin target exposure day and 5.3% on the 6% carboxyhaemoglobin target exposure day, reflecting the mean values of the pre- and post-exercise levels. Analysis of dose–response relationships could not be carried out in this study, making it more difficult to determine the strength of the evidence for the effects of carbon monoxide on arrhythmia. In this study, the amount of arrhythmia produced by carbon monoxide exposure was not correlated with measured variables of angina (e.g., time to ST-segment depression and time to angina) or with the clinical descriptors of disease status or medication usage.

Dahms et al. (1993) also studied the effects of low-level carbon monoxide exposure in patients with myocardial ischaemia and a minimum of 30 ventricular ectopic beats over a 20-h period. In total, 28 subjects were exposed in a randomized double-blind fashion to either room air or sufficient carbon monoxide for 1 h to elevate carboxyhaemoglobin levels to 3% and 5%. The frequency of single ventricular ectopic beats at rest was 115 ± 28 in room air, 121 ± 31 at 3% carboxyhaemoglobin and 94 ± 23 at 5% carboxyhaemoglobin. Exposure to carbon monoxide had no additional effect over exercise-induced increases in the frequency of single or multiple ectopic beats. The amount of arrhythmia was not related to the severity of myocardial ischaemia during normal daily activity.

There are important clinical differences in the patients studied by Sheps et al. (1990) and Dahms et al. (1993) that may account for the different results obtained with exercise. In the Dahms et al. (1993) study, the percentage of patients with ≥50 ventricular ectopic beats per hour was significantly greater and ventricular function of the patients was significantly worse than in the Sheps et al. (1990) study; however, the myocardial ischaemia, as indicated by exercise-induced angina or ischaemic ST-segment depression, was greater in the patients studied by Sheps et al. (1990), and a greater percentage of them were taking β-adrenergic blocking drugs.

The effects of low-level carbon monoxide exposure on exercise-induced ventricular arrhythmia in patients with coronary artery disease and baseline ectopy are dependent on their clinical status. In more severely compromised individuals, exposures to carbon monoxide that produce 6% carboxyhaemoglobin (but not lower carboxyhaemoglobin levels) have been shown to significantly increase the number and complexity of arrhythmias.

8.2.5 Effects on coronary blood flow

The effects of breathing carbon monoxide on myocardial function in patients with and without coronary heart disease have been examined by Ayres et al. (1969, 1970, 1979). Acute elevation of car-boxyhaemoglobin from 0.98% to 8.96% by a bolus exposure using either 1100 mg carbon monoxide/m^3 (1000 ppm) for 8–15 min or 57 000 mg/m^3 (50 000 ppm) for 30–45 s caused a 20% average decrease in coronary sinus oxygen tension without a concomitant increase in coronary blood flow in the patients with coronary artery disease. Observations in patients with coronary disease revealed that acute elevation of carboxyhaemoglobin to approximately 9% decreased the extraction of oxygen by the myocardium. However, overall myocardial oxygen consumption did not change significantly, because an increase in coronary blood flow served as a mechanism to compensate for a lower overall myocardial oxygen extraction. In contrast, patients with non-coronary disease increased their coronary blood flow with an insignificant decrease in coronary sinus oxygen tension as a response to increased carboxyhaemoglobin. The coronary patients also switched from lactate extraction to lactate production. Thus, because of their inability to increase coronary blood flow to compensate for the effects of increased carboxyhaemoglobin, a poten-tial threat exists for patients with coronary heart disease who inhale carbon monoxide.

Although the coronary sinus P_{O_2} dropped only slightly in this study (reflecting average coronary venous oxygen tension), it is certainly possible that, in areas beyond a significant coronary arterial stenosis, tissue hypoxia might be precipitated by very low tissue P_{O_2}

values. Tissue hypoxia might be further exacerbated by a coronary-steal phenomenon whereby increased overall coronary flow diverts flow from areas beyond a stenosis to other normal areas. Therefore, the substrate for the worsening of ischaemia and consequent precipitation of arrhythmias is present with carbon monoxide exposure.

2.6 Relationship between carbon monoxide exposure and risk of cardiovascular disease in humans

6.1 Introduction

General population epidemiological studies on the relation between carbon monoxide exposure and ischaemic heart disease are not conclusive. In the USA, early population studies (Goldsmith & Landaw, 1968; Cohen et al., 1969; Hexter & Goldsmith, 1971) suggested an association between atmospheric levels of carbon monoxide and increased mortality from cardiovascular disease in Los Angeles, California, but potential confounders were not effectively controlled. In contrast, a study in Baltimore, Maryland (Kuller et al., 1975), showed no association between ambient carbon monoxide levels and cardiovascular disease or sudden death. A study of emergency room visits for cardiovascular complaints in Denver, Colorado (Kurt et al., 1978), showed a relationship with carbon monoxide exposure levels, but the correlations were relatively weak, and other environmental factors were not evaluated. These early epidemiological data were summarized by Kuller & Radford (1983). They concluded that mortality and morbidity studies have been negative or equivocal in relating carbon monoxide levels to health effects, but studies in human subjects with compromised coronary circulation support an effect of acute exposure to carbon monoxide at blood levels corresponding to a carbon monoxide exposure level of about 20 ppm over several hours. They calculate that, based on health surveys, probably over 10 million subjects in the USA are exposed to potentially deleterious levels of carbon monoxide and that perhaps 1250 excess deaths related to low-dose environmental carbon monoxide exposure occur each year.

6.2 Daily mortality

More recent time-series studies in North and South America and in Europe have also been equivocal in relating day-to-day variations in carbon monoxide levels with daily mortality. No relationship was found between carbon monoxide and daily mortality in Los Angeles, California, or Chicago, Illinois (Ito et al., 1995; Kinney et al., 1995; Ito & Thurston, 1996), after adjusting for particulate matter with mass median aerodynamic diameter less than 10 μm (PM_{10}). Verhoeff et al. (1996) found no relationship between 24-h average carbon monoxide concentrations and daily mortality in Amsterdam, Netherlands, with or without adjustment for PM_{10} and other pollutants. Saldiva et al. (1994, 1995) found no association between carbon monoxide and daily mortality among children or the elderly in San Paulo, Brazil, after adjusting for nitrogen oxides and PM_{10}, respectively. Three other studies (Touloumi et al., 1994; Salinas & Vega, 1995; Wietlisbach et al., 1996) showed small, statistically significant relationships between carbon monoxide and daily mortality; however, other pollutants (e.g., total suspended particulates, sulfur dioxide, nitrogen dioxide, black smoke) and other environmental variables (e.g., temperature and relative humidity) were also significant. Further studies and analyses will be needed to determine if low-level carbon monoxide exposure is actually increasing mortality, particularly in the elderly population, or if carbon monoxide is a surrogate marker for some other mobile-source pollutant.

Touloumi et al. (1994) investigated the association of air pollution with daily all-cause mortality in Athens, Greece, for the years 1984 through 1988. Daily mean pollution indicators for sulfur dioxide, black smoke and carbon monoxide were averaged over all the available monitoring stations. Auto-regressive models with log-transformed daily mortality as the dependent variable were used to adjust for temperature, relative humidity, year, season, day of week and serial correlations in mortality. Separate models for log(sulfur dioxide), log(smoke) and log(carbon monoxide) produced statistically significant ($P < 0.001$) coefficients. Air pollution data lagged by 1 day had the strongest association with daily mortality. Multiple regression modelling showed that sulfur dioxide and smoke were independent predictors of mortality, although to a lesser extent than temperature and relative humidity. The inclusion of carbon monoxide in this model did not further improve the association with daily mortality, suggesting that carbon monoxide may be a surrogate marker for other mobile-source pollutants.

Daily mortality results of the European Community multicentre APHEA (Short-term effects of Air Pollution on Health: a European

Approach using epidemiologic time-series data) study (Touloumi et al., 1996) in Athens for the years 1987 through 1991 show that for 8-h carbon monoxide concentrations in ambient air (median 6.1 mg/m^3 [5.3 ppm], mean 6.6 mg/m^3 [5.8 ppm] and maximum 24.9 mg/m ^{3} [21.7 ppm]) compiled from three fixed monitoring sites, the relative risk (RR) of dying from a 10 mg/m^3 (8.7 ppm) increase in the daily carbon monoxide concentration in ambient air is 1.10 (95% confidence interval [CI] = 1.05–1.15). The strongest effect was observed during the winter, when higher levels of sulfur dioxide were observed. This new result has not yet been confirmed in other epidemiological studies. It may be explained by yet unknown health effects of low levels of carbon monoxide, by the presence of highly compromised susceptible groups in the population or by carbon monoxide being merely a surrogate of other combustion-generated air pollutants.

Salinas & Vega (1995) determined the effect of urban air pollution on daily mortality in metropolitan Santiago, Chile, from 1988 to 1991. Data on maximum 8-h average carbon monoxide, maximum hourly ozone, daily mean sulfur dioxide, PM$_{10}$ and PM$_{2.5}$ (particles with diameter <2.5 μm), and meteorological variables were obtained from five monitoring stations. Total and respiratory disease-specific deaths were compared, calculating the risk of death by municipality and month of the year using age-adjusted standardized mortality ratios and controlling for socioeconomic status. Daily counts of deaths were regressed using a Poisson model on the pollutants, controlling for temperature and relative humidity. A clear pattern in the geographic distribution of risk of death was found, both for total mortality and for disease-specific mortality (e.g., pneumonia, chronic obstructive pulmonary disease, asthma), regardless of socioeconomic and living conditions. The number of deaths was significantly associated directly with humidity, carbon monoxide and suspended particles and indirectly with temperature when the model included all days with available data during the 4-year period. The associations remained significant for those days with fine suspended particle levels below 150 μg/m^3.

Wietlisbach et al. (1996) assessed the association between daily mortality and air pollution in three Swiss metropolitan areas of Zurich, Basle and Geneva for the period 1984 through 1989. Daily counts were obtained for total mortality, mortality for persons 65 years of age or older and respiratory and cardiovascular disease mortality. Daily weather variables and pollution data for total suspended particulates, sulfur dioxide, nitrogen dioxide, carbon monoxide and ozone were obtained from the respective reference stations. Daily counts of death were regressed using a Poisson model on the pollutants, controlling for time trends, seasonal factors and weather variables. A positive, statistically significant association was found between daily mortality and total suspended particulates, sulfur dioxide and nitrogen dioxide. The strongest association was observed with a 3-day moving average. Somewhat smaller associations were observed in each city between mortality in persons 65 years of age or older and measured carbon monoxide concentrations (mean = 1.1–2.3 mg/m^3 [1–2 ppm]; maximum = 5–8 mg/m^3 [4–7 ppm]). Associations with ozone were very weak and inconsistent. When all pollutants were included in the model together, the regression coefficients were unstable and statistically insignificant.

8.2.6.3   Hospital admissions

Two recent studies in the USA (Morris et al., 1995; Schwartz & Morris, 1995), a similar study in Canada (Burnett et al., 1997) and one in Greece (Pantazopoulou et al., 1995) have suggested that day-to-day variations in ambient carbon monoxide concentrations are related to cardiovascular hospital admissions, especially for persons 65 years of age or over.

A time-series analysis of ambient levels of gaseous air pollutants (carbon monoxide, nitrogen dioxide, sulfur dioxide, ozone) and Medicare hospital admissions for congestive heart failure was performed for seven US cities (Chicago, Illinois; Detroit, Michigan; Houston, Texas; Los Angeles, California; Milwaukee, Wisconsin; New York, New York; Philadelphia, Pennsylvania) during the 4-year period from 1986 through 1989 by Morris et al. (1995). Maximum daily carbon monoxide levels (mean ± standard deviation [SD]) ranged from 2.1 ± 1.1 mg/m^3 (1.8 ± 1.0 ppm) in Milwaukee to 6.4 ± 1.9 mg/m^3 (5.6 ± 1.7 ppm) in New York. The relative risk of admissions associated with an 11 mg/m^3 (10 ppm) increase in carbon monoxide ranged from 1.10 in New York to 1.37 in Los Angeles. All seven cities showed similar patterns of increasing admissions with increasing ambient carbon monoxide concentrations. Approximately 3250 hospital admissions for congestive heart failure (5.7% of all such admissions) each year can therefore be attributed to the observed association with carbon monoxide levels. It is possible, however, that the observed association represents the impact of some other, unmeasured pollutant or group of pollutants covarying in time with carbon monoxide.

The association between air pollution and cardiovascular hospital admissions for persons aged 65 years or older was examined in the Detroit, Michigan, metropolitan area during the years 1986 through 1989 by Schwartz & Morris (1995). Air quality data were available for PM_{10} on 82% and for ozone on 85% of possible days. Data were available for sulfur dioxide and carbon monoxide on all days during the study period. The mean PM_{10} was 48.0 $\mu g/m^3$, the mean ozone was 82.0 $\mu g/m^3$ (41.0 ppb), the mean sulfur dioxide was 66.0 $\mu g/m^3$ (25.4 ppb) and the mean carbon monoxide was 2.7 mg/m^3 (2.4 ppm). A Poisson auto-regressive model was used to analyse the data with dummy variables for temperature and dew point, month, and linear and quadratic time trends. Daily admissions for ischaemic heart disease were associated with a 32 $\mu g/m^3$ increase in PM_{10} (RR = 1.018; 95% CI = 1.005–1.032), a 47 $\mu g/m^3$ (18 ppb) increase in sulfur dioxide (RR = 1.014; 95% CI = 1.003–1.026) and a 1.47 mg/m^3 (1.28 ppm) increase in carbon monoxide (RR = 1.010; 95% CI = 1.001–1.018); however, both sulfur dioxide and carbon monoxide became insignificant after controlling for PM_{10}, whereas PM_{10} remained significant after controlling for the other pollutants. Daily admissions for heart failure were independently associated with a 1.28 mg/m^3 increase in PM_{10} (RR = 1.024; 95% CI = 1.004–1.044) and carbon monoxide (RR = 1.022; 95% CI = 1.110–1.034). Ozone was not a significant risk factor for cardiovascular hospital admissions, and no pollutant was a significant risk factor for dysrhythmia admissions.

A number of issues need to be resolved before the results of Morris et al. (1995) and Schwartz & Morris (1995) can be fully understood. Congestive heart failure is the most common indication for hospitalization among adults 65 years of age or over in the USA; however, elderly patients with heart failure have demonstrated high rates of readmission, ranging from 29% to 47%. Behavioural factors (e.g., non-compliance with medication and diet) and social factors (e.g., isolation) frequently contribute to early readmission, suggesting that many hospital admissions for congestive heart failure could be prevented. Hospital admission for congestive heart failure could also be indicated for severe pulmonary oedema, pneumonia or generalized oedema. Additional information on the admission criteria (e.g., clinical evidence of acute myocardial ischaemia, oxygen saturation or symptomatic syncope), on cigarette and tobacco use and on indoor carbon monoxide exposures is needed.

The relative risks reported by Morris et al. (1995) are small, ranging from 1.10 to 1.37; the city with the highest carbon monoxide levels (New York) had the lowest relative risk for congestive heart failure admission. Also, the model without any lag provided the strongest association, suggesting that cumulative exposure was not important. The carbon monoxide concentrations measured by stationary monitors are also very low; any carboxyhaemoglobin levels produced by a 1-h exposure to <11 mg carbon monoxide/m^3 (<10 ppm), for example, would be difficult to measure. If cumulative exposures were important, even 8 h of exposure to 11 mg carbon monoxide/m^3 (10 ppm) with moderate exercise (20 litres/min) would be expected to produce only 1.5% carboxyhaemoglobin. In addition, carbon monoxide data from stationary monitors are not highly correlated with personal exposures, and individuals with coronary heart disease would not be expected to be in locations where carbon monoxide monitors exist. It is possible that carbon monoxide could be a surrogate for automobile pollution, in general. Carbon monoxide is highly correlated with particles during the winter months. Particles (PM$_{10}$) were found to be correlated in the Schwartz & Morris (1995) study, but were not included in the Morris et al. (1995) analysis. Also, particles have previously been shown to be associated with hospital admissions for both heart failure and ischaemic heart disease in Ontario (Burnett et al., 1995).

Burnett et al. (1997) examined temporal relationships between ambient air pollutants and hospitalizations among the elderly (persons 65 years of age or older) in 10 Canadian cities for the 11-year period from 1981 through 1991. A time-series analysis adjusted for long-term time trends, seasonal and subseasonal variations, and day-of-week effects was used to explore the association between cardiopulmonary illness and the ambient air pollutants carbon monoxide, nitrogen dioxide, sulfur dioxide, ozone and coefficient of haze. After stratifying for months of the year and adjusting for temperature, dew point and other pollutants, the log of the daily 1-h maximum carbon monoxide concentration recorded on the day of admission had the strongest and most consistent statistical association with hospitalization for congestive heart failure. The relative risk was 1.065 (95% CI = 1.028–1.104) for an increase from 1.1 to 3.4 mg carbon monoxide/m^3 (1 to 3 ppm, the 25th and 75th percentiles of the exposure distribution). Except for ozone, the other pollutants were clearly confounded in this analysis; however, carbon monoxide alone accounted for 90% of the daily excess hospitalizations attributable to the entire mix. The authors noted that this relationship may not be causal because of possible misclassification of exposure to carbon monoxide and the likelihood that carbon monoxide may be acting as a surrogate for pollution from transportation sources in general.

Pantazopoulou et al. (1995) studied the daily number of emergency outpatient visits and admissions for cardiac and respiratory causes to all major hospitals in the greater Athens area during 1988. Concentrations of air pollutants (smoke, carbon monoxide and nitrogen dioxide) were obtained from the Ministry of the Environment. Mean levels of carbon monoxide for all available monitoring stations were 4.5 ± 1.6 (SD) mg/m³ (3.9 ± 1.4 ppm) in winter and 3.4 ± 1.0 (SD) mg/m³ (3.0 ± 0.9 ppm) in summer. Multiple linear regression modelling was used to look for statistical relationships, controlling for the potential effects of meteorological and chronological variables, separately for winter and summer. A positive association was found between the daily number of emergency admissions for cardiac and respiratory causes and all measured pollutants during the winter, but not during the summer.

6.4 Occupational exposures

Early studies of occupational exposure to carbon monoxide (Jones & Sinclair, 1975; Redmond, 1975; Redmond et al., 1979) failed to identify any increased risk of cardiovascular disease associated with carbon monoxide exposure. In a Finnish study (Hernberg et al., 1976; Koskela et al., 1976), the prevalence of angina among foundry workers showed an exposure–response relationship with regard to carbon monoxide exposure, but no such result was found for ischaemic electrocardiogram findings during exercise.

Stern et al. (1981) reported a study performed by NIOSH. They investigated the health effects of chronic exposure to low concentrations of carbon monoxide by conducting a historical prospective cohort study of mortality patterns among 1558 white male motor vehicle examiners in New Jersey, USA. The examiners were exposed to 11–27 mg carbon monoxide/m³ (10–24 ppm). The carboxyhaemoglobin levels were determined in 27 volunteers. The average carboxyhaemoglobin level before a work shift was 3.3% in the whole group, and the post-shift level was 4.7%; these levels were 2.1% and 3.7%, respectively, in non-smokers only. The death rates were compared with the rates in the US population based on vital statistics. The cohort demonstrated a slight overall increase in cardiovascular deaths, but a more pronounced excess was observed within the first 10 years following employment. The study has several important limitations, however, including the use of historical controls, lack of knowledge about smoking habits and the fact that the individuals' carboxyhaemoglobin levels were not known.

Stern et al. (1988) investigated the effect of occupational exposure to carbon monoxide on mortality from arteriosclerotic heart disease in a retrospective cohort study of 5529 New York City, New York, USA, bridge and tunnel officers. Among former tunnel officers, the standardized mortality ratio (SMR) was 1.35 (90% CI = 1.09–1.68) compared with the New York City population. Using the proportional hazards model, the authors compared the risk of mortality from arteriosclerotic heart disease among tunnel workers with that of the less exposed bridge officers. They found an elevated risk in the tunnel workers that declined within as few as 5 years after cessation of exposure. The 24-h average carbon monoxide level in the tunnel was around 57 mg/m^3 (50 ppm) in 1961 and around 46 mg/m^3 (40 ppm) in 1968. However, higher values were recorded during rush hours. In 1971, the ventilation was further improved and the officers were allowed clean air breaks. Although the authors concluded that carbon monoxide exposure may play an important role in the pathophysiology of cardiovascular mortality, other factors must be taken into consideration. Mortality from arteriosclerotic heart disease has a complex multifactor etiology. The presence of other risk factors, such as cigarette smoke, hypertension, hyperlipidaemia, family history of heart disease, obesity, socioeconomic status and sedentary living, can all increase the risk of developing coronary heart disease. In addition, detailed exposure monitoring was not done in this study. The bridge and tunnel workers were exposed not only to carbon monoxide but also to other compounds emitted from motor vehicle exhaust and to the noise and stress of their environment. These other factors could have contributed to the findings.

Hansen (1989) reported the results of a 10-year follow-up study on mortality among 583 Danish male automobile mechanics between

15 and 74 years of age. The number of deaths expected for the automobile mechanics was compared with those for a similar group of Danish men employed as carpenters, electricians and other skilled workers free from occupational exposure to automobile exhaust, petrochemical products, asbestos and paint pigments. The number of deaths observed among the automobile mechanics exceeded the expected number by 21%. Although the increased mortality was not confined to any single cause of death, the author reported a remarkable excess of deaths attributed to ischaemic heart disease where the standardized mortality ratio was 121 and the 95% confidence interval was 102–145. The only other significant category of death was that due to external causes (SMR = 131; 95% CI = 113–153). No significant differences were found among the automobile mechanics for other diseases except for an increase in pancreatic cancer (SMR = 219; 95% CI = 128–351). Exposure to carbon monoxide and poly-cyclic aromatic hydrocarbons through the inhalation of automobile exhaust and the handling of solvents and oils may have accounted for the difference in ischaemic heart disease deaths between the auto-mobile mechanics and the comparison group; however, other occupational exposures or other lifestyle factors, as indicated above, may also have contributed to the findings.

6.5 Carbon monoxide poisoning

Intoxication with carbon monoxide that induces carboxyhaemo-globin levels around 50–60% is often lethal; however, even levels around 20% carboxyhaemoglobin have been associated with death, mainly coronary events, in patients with severe coronary artery disease. Balraj (1984) reported on 38 cases of individuals dying immediately or within a few days following carbon monoxide exposures producing 10–50% carboxyhaemoglobin, usually non-lethal levels of carbon monoxide. All of the subjects had coronary artery disease, and 29 of them had severe cases. The author concluded that the carbon monoxide exposure, resulting in carboxyhaemoglobin levels between 10% and 30% in 24 cases, triggered the lethal event in subjects with a restricted coronary flow reserve. Similar associations between carbon monoxide exposure and death or myocardial infarction have been reported by several other authors. Atkins & Baker (1985) reported two cases with 23% and 30% carboxyhaemoglobin, McMeekin & Finegan (1987) reported one case with 45% carboxy-haemoglobin, Minor & Seidler (1986) reported one case with 19% carboxyhaemoglobin and Ebisuno et al. (1986) reported one case with

21% carboxyhaemoglobin. For a more complete discussion of carbon monoxide poisoning, see section 8.3.

Forycki et al. (1980) described electrocardiogram changes in 880 patients treated for acute poisoning. Effects were observed in 279 cases, with the most marked changes in cases with carbon monoxide poisoning. In those, the most common change was a T-wave abnormality; in six cases, a pattern of acute myocardial infarction was present. Conduction disturbances were also common in carbon monoxide poisoning, but arrhythmias were less common.

Elsasser et al. (1995) reported that 78 myocardial infarction patients with Q-wave infarction had more arrhythmias and higher creatine kinase levels after acute carbon monoxide exposures that raised carboxyhaemoglobin levels to ≥5%. The carboxyhaemoglobin concentration was measured at admission to the coronary care unit of a university hospital and 4 h later. The authors concluded that carbon monoxide was associated with a more severe course of acute myocardial infarction; however, causation could not be determined.

8.2.6.6 Tobacco smoke

The association between smoking and cardiovascular disease is fully established. Although little is known about the relative importance of carbon monoxide compared with other components of tobacco smoke, such as nicotine and polycyclic aromatic hydrocarbons, most researchers consider all of them to be important. The nicotine component clearly aggravates the decrease in oxygen capacity induced by carbon monoxide through an increase in the oxygen demand of the heart, and polycyclic aromatic hydrocarbons have been implicated in the atherosclerotic process (Glantz & Parmley, 1991).

Passive smoking exposes an individual to all components in the cigarette smoke, but the carbon monoxide component dominates heavily, because only 1% or less of the nicotine is absorbed from sidestream smoke, compared with 100% in an active smoker (Jarvis, 1987; Wall et al., 1988). Therefore, exposure to sidestream smoke will be the closest to pure carbon monoxide exposure, even if the resultant levels of carboxyhaemoglobin are low (about 1–2%) (Jarvis, 1987). The relationship between passive smoking and increased risk of coronary heart disease is controversial. Early studies on this relationship were reviewed in the 1986 report of the Surgeon General

(Surgeon General of the United States, 1986) and by the US National Research Council (NRC, 1986a). Since that time, the epidemiological evidence linking passive smoking exposure to heart disease has rapidly expanded. The available literature, to date, on the relationship between passive exposure to environmental tobacco smoke in the home and the risk of heart disease death in the non-smoking spouse of a smoker consists of 19 published reports (Gillis et al., 1984; Hirayama, 1984; Garland et al., 1985; Lee et al., 1986; Svendsen et al., 1987; Helsing et al., 1988; He, 1989; Hole et al., 1989; Kawachi et al., 1989; Sandler et al., 1989; Humble et al., 1990; Butler, 1991; Dobson et al., 1991; La Vecchia et al., 1993; He et al., 1994; Muscat & Wynder, 1995; Steenland et al., 1996; Kawachi et al., 1997) and 4 review articles (Glantz & Parmley, 1991; Steenland, 1992; Wells, 1994; LeVois & Layard, 1995). All but two of the studies yielded relative risks greater than 1.0; however, six studies in men and eight studies in women had 95% confidence intervals that included 1.0, indicating that the risk of passive smoking for heart disease was not statistically significant. By combining the studies to improve the power to detect an effect, Glantz & Parmley (1991) reported a combined relative risk of 1.3 (95% CI = 1.2–1.4) for 10 studies, Wells (1994) reported a combined relative risk of 1.23 (95% CI = 1.12–1.35) for 12 studies and LeVois & Layard (1995) reported a combined relative risk of 1.29 (95% CI = 1.18–1.41) for 14 studies. Even though it is impossible to rule out an effect of the other components in sidestream smoke, the data suggest an increase in risk of coronary heart disease associated with prolonged exposure to low levels of carbon monoxide.

In a cross-sectional study of 625 smokers aged 30–69, Wald et al. (1973) reported that the incidence of cardiovascular disease was higher in subjects with carboxyhaemoglobin greater than 5% than in subjects with carboxyhaemoglobin below 3%, a relative risk of 21.2 (95% CI = 3.3–734.3). Even if all of the subjects were smokers, the association between carboxyhaemoglobin and cardiovascular disease might be due to the fact that percent carboxyhaemoglobin is a measure of smoke exposure.

Low to intermediate levels of carboxyhaemoglobin might interfere with the early course of an acute myocardial infarction. The increase in carboxyhaemoglobin can be due to recent smoking or environmental exposure. Mall et al. (1985) reported on a prospective study in smoking and non-smoking patients with an acute myocardial infarction who were separated by their baseline carboxyhaemoglobin levels. Sixty-six patients were studied in total. Thirty-one patients were found to have a carboxyhaemoglobin level of 1.5%, and 35 were found to have a level of 4.5%. In the group with elevated carboxy-haemoglobin, more patients developed transmural infarction, but the difference was not significant. Patients with transmural infarction had higher maximum creatine phosphokinase values when carboxyhaemo-globin was over 2%. During the first 6 h after admission to the hospi-tal, these patients needed an antiarrhythmic treatment significantly more frequently. Differences in rhythm disorders were still present at a time when nicotine, owing to its short half-life, was already elimi-nated. The authors concluded that moderately elevated levels of carboxyhaemoglobin may aggravate the course of an acute myocardial infarction.

8.2.7 Effects of exposure during pregnancy and early childhood

8.2.7.1 Pregnancy

It is thought that carbon monoxide crosses the placenta by simple diffusion and that the concentration of carboxyhaemoglobin in the fetus is dependent on maternal carboxyhaemoglobin concentrations, the rate of fetal carbon monoxide production, the diffusion capacity of the placenta for carbon monoxide and the relative affinity of the haemoglobin for carbon monoxide (Longo, 1977). Maternal carboxy-haemoglobin levels in non-smokers range between 0.5% and 1.0% and those of fetal blood between 0.7% and 2.5% (Longo, 1970, 1977). Fetal uptake of carbon monoxide takes place more slowly than mater-nal uptake. Following acute exposure, maternal carboxyhaemoglobin levels reach half the steady-state value in 2 h; in the fetus, on the other hand, steady-state values are not reached until 7 h after the onset of exposure. Final equilibrium is reached 36–48 h after the onset of exposure, and the carboxyhaemoglobin level in fetal blood at this time is 10% higher than maternal levels due to the increased affinity of fetal haemoglobin (Hill et al., 1977; Longo, 1977).

A fairly wide range of neonate and maternal carboxyhaemoglobin levels has been published for humans, probably as a result of wide differences in cigarette smoking patterns prior to and during labour. Maternal smoking during pregnancy exposes the fetus to greater than normal concentrations of carbon monoxide. Mean concentrations of

2.0–8.3% carboxyhaemoglobin and 2.4–7.6% carboxyhaemoglobin for maternal and fetal blood, respectively, have been reported (Longo, 1970).

In a study by Bureau et al. (1982), the measurement of fetal cord blood in the offspring of cigarette smokers who smoked during labour showed that fetal carboxyhaemoglobin levels were 2.55 times higher than in maternal blood. Cord blood averaged 10.1% carboxyhaemoglobin at delivery, whereas maternal blood averaged 5.6% carboxyhaemoglobin on the mother's arrival at the hospital and 4.1% carboxyhaemoglobin at delivery. Most of the observations on carbon monoxide exposure in pregnancy are based upon reports of cases of accidental or deliberate maternal carbon monoxide intoxication. These studies provide valuable information on the consequences of high peak carbon monoxide concentrations on pregnancy outcome but say nothing about the effects of short- or long-term low-level exposure to carbon monoxide.

Desclaux et al. (1951) reported a case of carbon monoxide intoxication in the fifth month of pregnancy where the subject was unconscious for 3 days. The child was born at term with a birth weight of 3500 g. Encephalography revealed gross ventricular dilatation, and the child was mentally retarded and had general hypertonia.

Muller & Graham (1995) reviewed eight cases from the literature in which the mother had been intoxicated with carbon monoxide in pregnancy and delivered living offspring. The common features in the offspring were psychomotor disturbances and mental retardation. The authors presented a case of carbon monoxide intoxication shortly before the expected date of delivery. The child was stillborn with cherry red-coloured skin and organs. Fetal blood was 49% saturated with carboxyhaemoglobin.

Copel et al. (1982) reported a case of carbon monoxide intoxication with a maternal blood carboxyhaemoglobin level of 24.5% in the first trimester of pregnancy. Pregnancy was uncomplicated until the 38th week, when amniocentesis revealed a greater than 3:1 lecithin:sphingomyelin ratio and the presence of phosphatidyl glycerol; the amniotic fluid was also meconium-stained. The child was delivered prematurely by caesarean section with a birth weight of

1950 g. The child developed normally and showed no signs of central nervous system damage at 6 months of age.

The association between carbon monoxide exposure and reduced fetal growth was studied by Astrup et al. (1972). Blood carboxyhaemoglobin was measured in 176 smoking and 177 non-smoking pregnant women. Blood carboxyhaemoglobin levels in 97% of non-smokers and 77% of smokers were below 3%; the mean carboxyhaemoglobin concentration in each group was 0.87% and 1.92%, respectively. The average birth weight of the newborns of smoking women (2990 g) was lower than that of the newborns of non-smoking women (3225 g). No evidence of a dose–effect relationship between maternal blood carboxyhaemoglobin and infant birth weight was presented, but the correlation coefficient between birth weight and the mean carboxyhaemoglobin measurements of each individual was statistically significant ($P < 0.05$). No attempts were made to adjust for other potential growth-retarding factors in cigarette smoke (e.g., nicotine), which may also have effects on birth weight and cannot be disregarded.

There is one case reported in which a 17-year-old healthy female of 37 weeks' gestation was exposed to carbon monoxide while a passenger in a car. The exposure was for two periods of 3 h each and resulted in headache, nausea, vomiting and chest pain; she became unresponsive while being transferred to hospital. On examination, physical signs were normal except that she was orientated to person and time but not to place. Oxygen partial pressure was 30.5 kPa (229 mmHg), and carboxyhaemoglobin was 47.2%. Results of fetal monitoring were consistent with acute fetal hypoxia. The patient was treated with supplemental oxygen immediately and approximately 2 h after admission to the hospital; hyperbaric oxygen treatment was initiated (100% oxygen at 243 kPa [2.4 atm] absolute for 90 min). After treatment, the subject's carboxyhaemoglobin was 2.4% (within the normal range for a pregnant female) and fetal heart rate was 140 beats per minute, with normalized beat-to-beat variability. The patient was delivered at term of a health female child (3600 g) with Apgar scores of 9 at 1 min and 10 at 5 min post-delivery. The newborn's physical condition, including neurological findings, was normal, and mother and child were discharged from hospital 2 days postpartum. Assessment of the child at 2 and 6 months revealed normal growth and development (Van Hoesen et al., 1989).

Koren et al. (1991), in a multicentre prospective study of fetal outcome following accidental carbon monoxide poisoning in pregnancy, reported on a total of 40 pregnancies, which included 3 twin births, 1 termination of pregnancy and 4 ongoing pregnancies. Exposure to carbon monoxide occurred in the first trimester in 12 pregnancies, the second trimester in 14 pregnancies and the third trimester in 14 pregnancies. The analyses were based on 38 babies. The trimester at the time of carbon monoxide exposure did not affect mean birth weight (3.4 ± 0.5 kg, excluding the twin births and offspring of mothers who smoked ≥20 cigarettes a day). The exposure was stratified into grades based on clinical symptoms and signs and carboxyhaemoglobin levels where available. Adverse fetal outcome occurred only after grade 4 or 5 poisoning (carboxyhaemoglobin ≥21%). There were two grade 5 cases; one resulted in stillbirth at 29 weeks of gestation (26% carboxyhaemoglobin, treated with high-flow oxygen), and the second resulted in fetal death at term followed by maternal death. Of the three cases of grade 4 severity (which included the case described by van Hoesen et al., 1989), two were treated with hyperbaric oxygen, and their infants were developing well at 1 year of age. The third case was exposed to carbon monoxide at 23 weeks of gestation and had a carboxyhaemoglobin of 25% 2 h after the exposure. She was treated with high-flow oxygen for 2 h; although she was delivered of a normal infant at term (birth weight 4 kg), the child at 8 months of age had poor head control and developmental delay, which were judged to be compatible with post-anoxic encephalopathy. One patient with grade 2 poisoning at 30 weeks' gestation (carboxyhaemoglobin 13.8%) was treated with high-flow oxygen for 7 h followed by hyperbaric oxygen for 2 h. She delivered a 3.2-kg infant, who was developing normally at 3 weeks of age. The other grade 2 case was exposed at 20 weeks' gestation, and the baby, born at 25 weeks, had respiratory distress syndrome and jaundice; at 3 months of age, the infant was developing appropriately. All of the infants of mothers with grade 1 symptoms (10 treated with high-flow oxygen, 19 untreated) developed normally.

In a prospective study designed to assess hyperbaric oxygen tolerance in pregnancy, 44 pregnant women who had been exposed to carbon monoxide were treated with hyperbaric oxygen (203 kPa [2 atm] absolute) for 2 h followed by 4 h of normobaric oxygen within 5.3 ± 3.7 h of the exposure, irrespective of the clinical severity of the intoxication and the stage of the pregnancy. Six patients were lost to follow-up, two sustained spontaneous abortion and one elected to terminate the pregnancy for reasons unrelated to the intoxication. Thirty-four women gave birth to normal newborns, and one gave birth to a child with Down's syndrome. There was no evidence that the use of hyperbaric oxygen was implicated in either of the spontaneous abortions. It was concluded that hyperbaric oxygen treatment may be used in pregnant women acutely poisoned with carbon monoxide (Elkharrat et al., 1991).

A prospective study was carried out in Lille, France, in which every pregnant woman admitted to the hyperbaric oxygen unit between January 1983 and December 1989 with carbon monoxide poisoning was evaluated. According to the protocol of the unit, every patient was treated with hyperbaric oxygen. Follow-up data were obtained on 86 of 90 women; when compared with a matched population of carbon monoxide-poisoned non-pregnant women, no difference was observed in source of carbon monoxide, clinical severity, carboxyhaemoglobin or plasma bicarbonate concentrations. Short-term complications were more common in the pregnant women compared with the matched group, but long-term outcome did not differ. In five cases, carbon monoxide intoxication led to fetal death (a fourfold increase in relative risk), 77 women (89.5%) had a successful pregnancy, and the prematurity, fetal hypotrophy and malformation rate were the same as those of the general population. It was concluded that although carbon monoxide intoxication induced an increase in the short-term maternal complication rate and the fetal death rate, the long-term outcome for both the mother and infant where hyperbaric oxygen had been used was not different from that of the general population (Mathieu et al., 1996a).

In summary, carbon monoxide is transferred slowly to the fetus, and fetal haemoglobin has a higher affinity for carbon monoxide than does that of the adult. The oxygen dissociation curve is shifted to the left, making the fetal hypoxia more pronounced than the maternal tissue hypoxia. The severity of fetal intoxication cannot be assessed by the maternal state. There is a difference between the rate of formation of carboxyhaemoglobin in the mother's and the fetus's blood. Moreover, in comparison with the mother's dissociation curve, the curve in the fetus is shifted to the right. These two mechanisms result in a significant delay in elimination of carbon monoxide from the fetus, thus prolonging exposure. High doses of carbon monoxide are without doubt able to cause fetal death (Norman & Halton, 1990), developmental disorders and reduced fetal growth. The dose–response functions for these effects are, however, not known, and no safe level of exposure has been defined based upon scientific data. Case reports have generally been based upon high and toxic doses of carbon monoxide. There is a lack of follow-up studies on women with low and chronic carbon monoxide exposure, especially on central nervous system development in their children.

*.7.2   Reduced birth weight

Studies relating human carbon monoxide exposure from ambient sources or cigarette smoking to reduced birth weight have frequently failed to take into account all sources of carbon monoxide exposure. Alderman et al. (1987), for example, studied the relationship between birth weight and maternal carbon monoxide exposure based upon neighbourhood ambient carbon monoxide data obtained from stationary air monitoring sites in Denver, Colorado, USA. They failed to show a relationship between these factors, but they also failed to control for maternal cigarette smoking or possible occupational exposures to carbon monoxide. Carboxyhaemoglobin measurements were not made among either the mothers or their offspring to estimate net exposure levels. A similar design problem is found in the study of Wouters et al. (1987), in which cord blood carboxyhaemoglobin and birth weight were correlated. The authors reported a significant correlation between cigarette smoking and reduced birth weight, but no correlation between cord blood carboxyhaemoglobin and birth weight. Such data might be interpreted to mean that carbon monoxide is not the component in cigarette smoke responsible for reduced birth weight. Such a conclusion appears to be unjustified based upon Wouters et al. (1987), because carboxyhaemoglobin is a good estimate of recent carbon monoxide exposure only. Thus, it may indicate only how recently women in this study smoked their last cigarette before delivery of the child rather than estimating smoking rates or history throughout pregnancy.

Other studies have related indirect exposure to smoke in pregnancy with reduced birth weights. Martin & Bracken (1986) showed an association between passive smoking (exposure to cigarette smoke for at least 2 h per day) and reduced birth weight. Unfortunately, sidestream smoke contains significant nicotine as well as carbon monoxide, so it is not possible to relate this effect to carbon monoxide exposure.

Mochizuki et al. (1984) attempted to evaluate the role of maternal nicotine intake in reduced birth weight and presented evidence of possibly impaired utero-placental circulation among smokers. These changes were not related specifically to the nicotine content of the cigarettes and failed, moreover, to take into account the possible synergistic effects between reduced perfusion that might have resulted from the vasoconstrictive effects of nicotine and the reduced oxygen availability that might have resulted from carbon monoxide exposure. As noted in the section of this report that deals with the effects of high altitude (section 8.2.8), many of the outcomes of maternal carbon monoxide exposure are also observed in offspring of women living at high altitude. These include reduced birth weight, increased risk of perinatal mortality and increased risk of placental abnormalities. Limited data exist on the possibility of increased risk of carbon monoxide exposure to the fetus being carried at high altitude. Such findings are considered in the section on high altitude.

8.2.7.3 Sudden infant death syndrome

There have been a number of studies linking maternal cigarette smoking with sudden infant death syndrome (Schoendorf & Kiely, 1992; Mitchell et al., 1993; Scragg et al., 1993; Klonoff-Cohen et al., 1995; Blair et al., 1996; Hutter & Blair, 1996), but it is uncertain what the role of carbon monoxide might be in such a relationship.

It has been suggested that carbon monoxide may be a causative factor in sudden infant death syndrome. Hoppenbrouwers et al. (1981) reported a statistical association between the frequency of sudden infant death syndrome and levels of several airborne pollutants, including carbon monoxide, sulfur dioxide, nitrogen dioxide and hydrocarbons. Sudden infant death syndrome was reported more commonly in the winter, at a time when the burning of fossil fuels for heating would be greatest. It is interesting to note that there is a phase lag of approximately 7 weeks between the increase in pollutant levels and the increase of sudden infant death syndrome. Further correlations were obtained between sudden infant death syndrome and the predicted level of carbon monoxide and lead for the child's birth month and between sudden infant death syndrome and the level of pollution at the reporting station closest to the infant's home. These correlations are not compelling without more information on the methods by which other possible risk factors were controlled in making the geographical correlations. Although it is technically difficult, it would be very useful to obtain carboxyhaemoglobin levels close to the time of death in sudden infant death syndrome victims, as this would greatly assist in determining the incidence of elevated carbon monoxide exposure in such cases.

The current data from human children suggesting a link between environmental carbon monoxide exposures and sudden infant death syndrome are weak, but further study should be encouraged.

2.7.4   Neurobehavioural effects

Behaviour is an essential function of the nervous system, and abnormalities in this outcome can be diagnostic for particular neuro-logical disorders or for nervous system dysfunction. Because at present no data are available on behavioural changes of children with chronically elevated carboxyhaemoglobin, this section discusses only the findings of acute studies. However, these studies are not adequate for evaluating the dose–effect relationship to be used in ambient air standard setting. Because carboxyhaemoglobin levels are almost always determined at hospital admission time, they will be lower than the carboxyhaemoglobin levels at exposure time. Therefore, except for the confirmation of carbon monoxide inhalation, it is difficult to use these carboxyhaemoglobin levels to estimate severity of poisoning at time of exposure.

Crocker & Walker (1985) reported on the consequences of acute carbon monoxide exposure in 28 children, 16 of which had carboxy-haemoglobin levels over 15% and were considered to have had "potentially toxic" carboxyhaemoglobin levels. These children were between the ages of 8 months and 14 years. The authors reported nausea, vomiting, headache, lethargy and syncope to be the most common signs and symptoms. A very limited follow-up investigation was performed with these children, so no conclusions can be drawn from this work concerning persisting effects. In addition to very large differences in the nature (dose and duration) of exposure, the extreme variability in patient age limits the potential value of the data presented in this work. The absence of any reports from children having carboxyhaemoglobin levels of ≤15% (a very high carboxy-haemoglobin level) is regrettable, because these are the children one

317

must study to develop an understanding of the relationship between dose and effect for the purpose of setting standards for ambient air.

Klees et al. (1985) conducted a more comprehensive study of the consequences of childhood carbon monoxide exposure on subsequent behavioural development. They reported that the age at which exposure occurred, its severity and also the child's intellectual level at the time of exposure also play a role in the outcome.

The authors suggested that the severity of neurobehavioural effects depends on the exposure period when it occurs at a critical time of the development of certain neurobehavioural functions. Subjects who had higher intellectual function prior to accidental exposure also appeared to fare better after carbon monoxide exposure. However, the authors stressed that long-term perceptual and intellectual consequences of carbon monoxide exposure may occur that are not well identified in short-term cursory examinations, because certain invisible visuo-spatial disorders persisting for years may seriously impair adaptive functioning of the child. Because of the difficulty of interpreting carboxyhaemoglobin levels on hospital admission, the authors adopted the following criteria to evaluate intensity of intoxication: (1) light: hypotonia, vertigo, vomiting, etc.; (2) medium: loss of consciousness (short period); (3) severe: any of the previous symptoms followed by coma. Of 14 children followed up for 2–11 years after intoxication, only 1 (medium) showed no sequelae (despite carboxyhaemoglobin levels of 42% on admission to hospital at the age of 9 years 10 months). Seven children (four light, three medium) had impairment of visual memory and concentration, but normal IQ scores. Six children (two light, one medium, three severe) who had serious learning disorders did not have more severe carbon monoxide exposures as judged from their carboxyhaemoglobin levels. They include several cases where exposures did occur at a young age and children who had psychological difficulties prior to carbon monoxide exposure. This study leaves some question concerning the relative vulnerability of children to carbon monoxide as a function of their age, because several of the youngest children did make full recovery, whereas others did not. It seems likely that the child's age may have an influence, as well as the promptness of hospitalization and efficacy of treatment. Further study of the outcomes of childhood carbon monoxide exposures will be useful in determining whether there are differences with respect to vulnerability to carbon monoxide level.

Venning et al. (1982) reported on a case of acute carbon monoxide poisoning in a 13-week-old baby who had profoundly elevated carboxyhaemoglobin levels (60% 2 h after removal from the automobile in which she had been accidentally exposed to carbon monoxide). Her parents had much lower carboxyhaemoglobin values, although this may reflect differences in concentration of carbon monoxide inhaled. The child was reported to be unconscious for 48 h, to go through convulsions over the next 18 days, but to show recovery from "minor neurological abnormalities" by 6 weeks later.

1.2.8 High-altitude effects

.8.1 Introduction

Although there are many studies comparing and contrasting inhaling carbon monoxide with exposure to altitude, there are relatively few reports on the effects of inhaling carbon monoxide at altitude. There are data to support the possibility that the effects of these two hypoxia episodes are at least additive. These data were obtained at carbon monoxide concentrations that are too high to have much significance for regulatory concerns. There are also data that indicate decrements in visual sensitivity and flicker fusion frequency in subjects exposed to carbon monoxide (5–10% carboxyhaemoglobin) at higher altitudes. These data, however, are somewhat controversial.

There are even fewer studies of the long-term effects of carbon monoxide at high altitude. These studies generally indicate few changes at carbon monoxide concentrations below 110 mg/m^3 (100 ppm) and altitudes below 4570 m. A provocative study by McDonagh et al. (1986) suggests that the increase in ventricular capillarity seen with altitude exposure may be blocked by carbon monoxide. The fetus may be particularly sensitive to the effects of carbon monoxide at altitude; this is especially true with the high levels of carbon monoxide associated with maternal smoking.

Precise estimates of the number of people exposed to carbon monoxide at high altitude are not readily available. As of 1980, however, more than 4.2 million people (T.K. Lindsey, letter to Dr. J. McGrath dated 4 July 1989, Forestville, Maryland, USA) were living at altitudes in excess of 1525 m. Moreover, estimates obtained from several US states with mountainous regions (i.e., California, Nevada,

Hawaii and Utah) indicate that more than 35 million tourists may sojourn in high-altitude areas during the summer and winter months.

The potential effects on human health of inhaling carbon monoxide at high altitudes are complex. Whenever carbon monoxide binds to haemoglobin, it reduces the amount of haemoglobin available to carry oxygen. People at high altitudes already live in a state of hypoxaemia, however, because of the reduced P_{O_2} in the air. Carbon monoxide, by binding to haemoglobin, intensifies the hypoxaemia existing at high altitudes by further reducing transport of oxygen to the tissues. Hence, the effects of carbon monoxide and high altitude are usually considered to be additive.

This consideration does not take into account the fact that within hours of arrival at high altitude, certain physiological adjustments begin to take place. Haemoconcentration occurs, and the increased haemoglobin concentration offsets the decreased oxygen saturation and restores oxygen concentration to pre-ascent levels. Consequently, the simple additive model of carboxyhaemoglobin and altitude hypoxaemia may be valid only during early altitude exposure.

The visitor newly arrived to higher altitudes may be at greater risk from carbon monoxide than the adapted resident, however, because of a non-compensated respiratory alkalosis from hyperventilation, lower arterial haemoglobin saturation without a compensatory absolute polycythaemia (therefore greater hypoxaemia) and hypoxia-induced tachycardia.

Several factors tend to exacerbate ambient carbon monoxide levels at high altitude (Kirkpatrick & Reeser, 1976). For example, carbon monoxide emissions from automobiles are likely to be higher in mountain communities. Early studies on automobile emissions showed that automobiles tuned for driving at 1610 m emit almost 1.8 times more carbon monoxide when driven at 2440 m. Automobiles tuned for driving at sea level emit almost 4 times more carbon monoxide when driven at 2440 m. Moreover, automobile emissions are increased by driving at reduced speeds, along steep grades and under poor driving conditions. Therefore, large influxes of tourists driving automobiles tuned for sea level conditions into high-altitude resort areas may drastically increase pollutant levels in general, and carbon monoxide levels in particular (NRC, 1977). Although emission data comparing sea level and high-altitude conditions for the current automobile fleet are not yet available, newer automobile engine technologies should significantly reduce carbon monoxide emissions in general, as well as carbon monoxide emissions at high altitude. Heating devices (space heaters and fireplaces) used for social effect, as well as warmth, are a second factor contributing to carbon monoxide emissions in mountain resort areas. Finally, population growth in mountain areas is concentrated along valley floors; this factor, combined with the reduced volume of air available for pollutant dispersal in valleys, causes pollutants, including carbon monoxide, to accumulate in mountain valleys.

8.2 Carboxyhaemoglobin formation

The effects of high altitude on carboxyhaemoglobin formation have been considered in a theoretical paper by Collier & Goldsmith (1983). Transforming and rearranging the CFK equation (Coburn et al., 1965), these workers derived an equation expressing carboxy-haemoglobin in terms of endogenous and exogenous sources of carbon monoxide. According to this relationship, a given partial pressure of carbon monoxide will result in a higher percent carboxyhaemoglobin at high altitudes (where P_{O_2} is reduced). Thus, Collier & Goldsmith (1983) calculated an incremental increase in carboxyhaemoglobin at altitude even in the absence of inhaled carbon monoxide that is most likely due to endogenous production of carbon monoxide (see Table 17).

8.3 Cardiovascular effects

There are studies comparing the cardiovascular responses to carbon monoxide with those to high altitude, but there are relatively few studies of the cardiovascular responses to carbon monoxide at high altitude. Forbes et al. (1945) reported that carbon monoxide uptake increased during 6 min of exercise of varying intensity on a bicycle ergometer at an equivalent altitude of 4875 m. The increased carbon monoxide uptake was caused by altitude hyperventilation stimulated by decreased arterial oxygen tension and not by diminished barometric pressure.

Pitts & Pace (1947) reported that pulse rate increased in response to the combined stress of high altitude and carbon monoxide. The subjects were 10 healthy men who were exposed to simulated altitudes

Table 17. Calculated equilibrium values of percent carboxyhaemoglobin and percent oxyhaemoglobin in humans exposed to ambient carbon monoxide at various altitudes[a,b]

Ambient CO		Sea level		1530 m		3050 m		3660 m	
mg/m³	ppm	% COHb	% O₂Hb	% COHb	% O₂Hb	% COHb	% O₂Hb	% COHb	% O₂Hb
0	0	0.2	97.3	0.26	93.6	0.35	82.4	0.37	73.3
5	4	0.8	96.8	0.9	93.0	1.1	82.1	1.1	73.1
9	8	1.4	96.2	1.6	92.5	1.8	81.7	1.8	72.9
14	12	2.1	95.6	2.3	91.9	2.5	81.3	2.5	72.7
18	16	2.7	95.1	2.9	91.4	3.2	80.9	3.2	72.5

[a] The table is for unacclimatized, sedentary individuals at one level of activity (oxygen uptake of 500 ml/min).
[b] Adapted from Collier & Goldsmith (1983).

of 2135, 3050 and 4570 m and inhaled 3400 or 6900 mg carbon monoxide/m^3 (3000 or 6000 ppm) to obtain carboxyhaemoglobin levels of 6% or 13%, respectively. The mean pulse rate during exercise and the mean pulse rate during the first 5 min after exercise were correlated with and increased with the carboxyhaemoglobin concentration and simulated altitude. The authors concluded that the response to a 1% increase in carboxyhaemoglobin level was equivalent to that obtained by raising a normal group of men 100 m in altitude. This relationship was stated for a range of altitudes from 2135 to 3050 m and for increases in carboxyhaemoglobin up to 13%.

Weiser et al. (1978) studied the effects of carbon monoxide on aerobic work at 1610 m in young subjects rebreathing from a closed-circuit system containing a bolus of 100% carbon monoxide until carboxyhaemoglobin levels reached 5%. They reported that this level of carboxyhaemoglobin impaired exercise performance at high altitude to the same extent as that reported at sea level (Horvath et al., 1975). Because these subjects were Denver, Colorado, USA, residents and were fully adapted to this altitude, however, they would have had an arterial oxygen concentration the same as at sea level (about 20 ml oxygen/dl). Hence, 5% carboxyhaemoglobin would lower arterial oxygen concentration about the same amount at both altitudes and impair work performance at altitude to the same extent as at sea level. In the Weiser et al. (1978) study, breathing carbon monoxide during submaximal exercise caused small but significant changes in cardio-respiratory function; the working heart rate increased, and post-exercise left ventricular ejection time did not shorten to the same extent as when filtered air was breathed. Carbon monoxide exposure lowered the anaerobic threshold and increased minute ventilation at work rates heavier than the anaerobic threshold due to increased blood lactate levels.

Wagner et al. (1978) studied young smokers and non-smokers who exercised at 53% of $\dot{V}_{O_2}$ max at 101.3 and 69.7 kPa (760 and 523 torr). Carboxyhaemoglobin levels were raised to 4.2%. While at altitude with these elevated carboxyhaemoglobin levels, non-smokers increased their cardiac output and decreased their arterial–mixed venous oxygen differences. Smokers did not respond in a similar manner. Smokers, with their initial higher haemoglobin concentrations, may have developed some degree of adaptation to carbon monoxide and/or high altitude.

In a complex study involving four altitudes ranging from sea level up to 3050 m and four ambient carbon monoxide concentrations (up to 170 mg/m^3 [150 ppm]), Horvath et al. (1988a,b) evaluated carboxyhaemoglobin levels during a maximal aerobic capacity test. They concluded that $\dot{V}_{O_2}$ max values determined in men and women were only slightly diminished as a result of increased ambient carbon monoxide. Carboxyhaemoglobin concentrations attained at maximum were highest at 55 m (4.42%) and lowest at 3050 m (2.56%) while breathing 170 mg carbon monoxide/m^3 (150 ppm). This was attributed to the reduced partial pressure of carbon monoxide at high altitude. No additional effects that could be attributed to the combined exposure to high altitude and carbon monoxide were found. Independence of the altitude and carbon monoxide hypoxia was demonstrated under the condition of performing a maximum aerobic capacity test. The reductions in $\dot{V}_{O_2}$ max due to high altitude and to the combined exposure of ambient carbon monoxide and high altitude were similar.

Horvath & Bedi (1989) studied 17 non-smoking young men to determine the alterations in carboxyhaemoglobin during exposure to ambient carbon monoxide at 0 or 10 mg/m^3 (0 or 9 ppm) for 8 h at sea level or an altitude of 2135 m. Nine subjects rested during the exposures, and eight exercised for the last 10 min of each hour at a mean ventilation of 25 litres (at body temperature, barometric pressure, saturated conditions). All subjects performed a maximal aerobic capacity test at the completion of their respective exposures. At the low carbon monoxide concentrations studied, the CFK equation estimated carboxyhaemoglobin levels to be 1.4% (Peterson & Stewart, 1975). Carboxyhaemoglobin concentrations fell in all subjects during their exposures to 0 mg carbon monoxide/m^3 (0 ppm) at sea level or 2135 m. During the 8-h exposures to 10 mg carbon monoxide/m^3 (9 ppm), carboxyhaemoglobin levels rose linearly from approximately 0.2% to 0.7%. No significant differences in uptake were found whether the subjects were resting or intermittently exercising. Levels of carboxyhaemoglobin were similar at both altitudes. A portion of the larger estimate of carboxyhaemoglobin determined by the CFK equation could be accounted for by the use of an assumed blood volume. Maximal aerobic capacity was reduced approximately 7–10% consequent to altitude exposure during 0 mg carbon monoxide/m^3 (0 ppm). These values were not altered following 8-h exposure to 10 mg carbon monoxide/m^3 (9 ppm) in either resting or exercising individuals.

In a study of patients with coronary artery disease, Leaf & Klein-man (1996) performed exercise stress tests after random exposures to either carbon monoxide or clean air at sea level or a simulated 2.1-km-high altitude. The carbon monoxide and high-altitude conditions were each selected to reduce the percentage of oxygen saturation of arterial blood by 4%. Levels of carboxyhaemoglobin were increased from an average of 0.62% after clean air exposure to 3.91% after carbon monoxide exposure. The average incidence of exercise-induced ventricular ectopy was approximately doubled after exposure to carbon monoxide or high altitude compared with clean air exposures (from 10 to approximately 20 premature ventricular contractions). There was also a significant trend of increased ectopy with decreased oxygen saturation. No effects were observed in subjects who were free from ectopy. These results indicate that individuals having coronary artery disease and baseline ectopy are susceptible to increased levels of hypoxaemia resulting from either hypoxic or carbon monoxide exposures.

Exposure to carbon monoxide from smoking may pose a special risk to the fetus at altitude. Moore et al. (1982) reported that maternal smoking at 3100 m was associated with a two- to threefold greater reduction in infant birth weight than has been reported at sea level. Moreover, carboxyhaemoglobin levels of 1.8–6.2% measured in all pregnant subjects were inversely related to infant birth weight. Earlier, Brewer et al. (1970, 1974) reported that the mean carboxyhaemo-globin level in smokers at altitude is higher than in smokers at sea level, and that subjects who smoked had greater oxygen affinities than non-smokers. Moreover, cessation of smoking by polycythaemic individuals at altitude results in a marked reduction in carboxyhaemo-globin and a decrease in oxygen–haemoglobin affinity to values less than those reported for normal individuals at sea level.

8.4   Neurobehavioural effects

The neurobehavioural effects following carbon monoxide expo-sure are controversial and should, therefore, be interpreted with extreme caution. Those neurobehavioural studies specifically con-cerned with carbon monoxide exposure at altitude are reviewed briefly in this section.

McFarland et al. (1944) reported changes in visual sensitivity occurring at a carboxyhaemoglobin concentration of 5% or at a simulated altitude of approximately 2430 m. Later, McFarland (1970) expanded on the original study and noted that a pilot flying at 1830 m breathing 0.005% (57 mg/m^3 [50 ppm]) carbon monoxide in air is at an altitude physiologically equivalent to approximately 3660 m. McFarland (1970) stated that sensitivity of the visual acuity test was such that even the effects of small quantities of carbon monoxide absorbed from cigarette smoke were clearly demonstrable. In subjects inhaling smoke from three cigarettes at 2285 m, there was a combined loss of visual sensitivity equal to that occurring at 3050–3355 m. Results from the original study were confirmed by Halperin et al. (1959), who also observed that recovery from the detrimental effects of carbon monoxide on visual sensitivity lagged behind elimination of carbon monoxide from the blood.

Lilienthal & Fugitt (1946) reported that combined exposure to altitude and carbon monoxide decreased flicker fusion frequency (i.e., the critical frequency in cycles per second at which a flickering light appears to be steady). Whereas mild hypoxia (that occurring at 2745–3660 m) alone impaired flicker fusion frequency, carboxy-haemoglobin levels of 5–10% decreased the altitude threshold for onset of impairment to 1525–1830 m.

The psychophysiological effects of carbon monoxide at altitude are a particular hazard in high-performance aircraft (Denniston et al., 1978). Acute ascent to altitude increases ventilation via the stimulating effects of a reduced P_{O_2} on the chemoreceptors. The increased ventilation causes a slight increase in blood pH and a slight leftward shift in the oxyhaemoglobin dissociation curve. Although such a small shift would probably have no physiological significance under normal conditions, it may take on physiological importance for aviators required to fly under a variety of operational conditions and to perform tedious tasks involving a multitude of cognitive processes. The leftward shift of the oxyhaemoglobin dissociation curve may be further aggravated by the persisting alkalosis caused by hyper-ventilation resulting from anxiety. The potential for this effect has been reported by Pettyjohn et al. (1977), who found that respiratory minute volume may be increased by 110% during final landing approaches requiring night vision devices. Thus, the hypoxia-inducing effects of carbon monoxide inhalation would accentuate the cellular hypoxia caused by stress- and altitude-induced hyperventilation.

8.5   Compartmental shifts

The shift of carbon monoxide out of the blood has been demonstrated in studies (Horvath et al., 1988a,b) conducted on both men and women undergoing maximal aerobic capacity tests at altitudes of 55, 1525, 2135 and 3060 m and carbon monoxide concentrations of 0, 57, 110 and 170 mg/m^3 (0, 50, 100 and 150 ppm). Carbon monoxide at maximum work shifted into extravascular spaces and returned to the vascular space within 5 min after exercise stopped.

8.3   Carbon monoxide poisoning

Carbon monoxide is responsible for more than half of the fatal poisonings that are reported all over the world each year (National Safety Council, 1982; Faure et al., 1983; Cobb & Etzel, 1991; Mathieu et al., 1996a). At sublethal levels, carbon monoxide poisoning occurs in a small but important fraction of the population. Certain conditions exist in both the indoor and outdoor ambient environments that cause a small percentage of the population to become exposed to dangerous levels of carbon monoxide. Outdoors, concentrations of carbon monoxide are highest near intersections, in congested traffic, near exhaust gases from internal combustion engines and from industrial combustion sources, and in poorly ventilated areas such as parking garages and tunnels. Indoors, carbon monoxide concentrations in the workplace or in homes that have faulty appliances or downdrafts and backdrafts have been measured in excess of 110 mg/m^3 (100 ppm), resulting in carboxyhaemoglobin levels of greater than 10% for 8 h of exposure. In addition, carbon monoxide is found in the smoke produced by all types of fires. Of the 6000 deaths from burns in the USA each year, more than half are related to inhalation injuries where victims die from carbon monoxide poisoning, hypoxia and smoke inhalation (Heimbach & Waeckerle, 1988). Despite efforts in prevention and in public and medical education, this intoxication remains frequent, severe and too often overlooked (Barret et al., 1985).

Carbon monoxide poisoning is not new, although more attention has been focused on this problem recently in the scientific literature as well as in the popular media. The first scientific studies of the hypoxic effects of carbon monoxide were described by Bernard in 1865. The attachment of carbon monoxide to haemoglobin, producing carboxyhaemoglobin, was evaluated by Douglas et al. (1912), providing the necessary tools for studying human response to carbon monoxide. During the next half century, numerous studies were conducted, with the principal emphasis being on high concentrations of carboxyhaemoglobin. Carbon monoxide poisoning as an occupational hazard (Grut, 1949) received the greatest attention owing to the increased use of natural gas and the potential for leakage of exhaust fumes in homes and industry. Other sources of carbon monoxide have become more important and more insidious. The clinical picture of carbon monoxide poisoning, as described by Grut (1949), relates primarily to the alterations in cardiac and central nervous system function as a result of the extreme hypoxia induced.

Mortality from carbon monoxide exposure is high. In 1985, 1365 deaths due to carbon monoxide exposure were reported in England and Wales (Meredith & Vale, 1988). In the USA, more than 3800 people die annually from carbon monoxide (accidental and intentional), and more than 10 000 individuals seek medical attention or miss at least 1 day of work because of a sublethal exposure (US Centers for Disease Control, 1982). The per capita mortality and morbidity statistics for carbon monoxide are surprisingly similar for the Scandinavian countries and for Canada as well. However, not all instances of carbon monoxide poisoning are reported, and complete up-to-date data are difficult to obtain. In some places, continuous surveillance by recording all cases hospitalized for every hospital in the region covered has been set up by a poison centre that gives annual epidemiological reports at the local level (Mathieu et al., 1996a). Often the individuals suffering from carbon monoxide poisoning are unaware of their exposure, because symptoms are similar to those associated with the flu or with clinical depression. This may result in a significant number of misdiagnoses by medical professionals (Grace & Platt, 1981; Fisher & Rubin, 1982; Barret et al., 1985; Dolan et al., 1987; Heckerling et al., 1987, 1988; Kirkpatrick, 1987). Therefore, the precise number of individuals who have suffered from carbon monoxide intoxication is not known, but it is certainly larger than the mortality figures indicate. Nonetheless, the reported literature available for review indicates the seriousness of this problem.

The symptoms, signs and prognosis of acute poisoning correlate poorly with the level of carboxyhaemoglobin measured at the time of arrival at the hospital (Klees et al., 1985; Meredith & Vale, 1988). Carboxyhaemoglobin levels below 10% are usually not associated with symptoms. At the higher carboxyhaemoglobin saturations of 10–30%, neurological symptoms of carbon monoxide poisoning can occur, such as headache, dizziness, weakness, nausea, confusion, disorientation and visual disturbances. Exertional dyspnoea, increases in pulse and respiratory rates, and syncope are observed with continuous exposure producing carboxyhaemoglobin levels in excess of 30–50%. When carboxyhaemoglobin levels are higher than 50%, coma, convulsions and cardiorespiratory arrest may occur. There are numerous tables giving symptom-associated carboxyhaemoglobin levels and the corresponding carbon monoxide concentrations in the atmosphere (Ellenhorn & Barceloux, 1988). Examples are given in Table 18.

Different individuals experience very different clinical manifestations of carbon monoxide poisoning and, therefore, have different outcomes even under similar exposure conditions. Norkool & Kirkpatrick (1985) found that carboxyhaemoglobin levels in individuals who had never lost consciousness ranged from 5% to 47%. In individuals who were found unconscious but regained consciousness at hospital arrival, the range was 10–64%; for those remaining unconscious, carboxyhaemoglobin levels ranged from 1% to 53%. The large differences in carboxyhaemoglobin levels found in these individuals most likely resulted from differences in time elapsing from exposure to carbon monoxide and admission to the hospital. Considerable differences in exposure duration may also be responsible for the lack of correlation between blood carboxyhaemoglobin and the clinical severity of carbon monoxide poisoning (Sokal, 1985; Sokal & Kralkowska, 1985). These data clearly indicate that carboxyhaemoglobin saturations correlate so poorly with clinical status that they have little prognostic significance.

The level of carbon monoxide in the tissues may have an equal or greater impact on the clinical status of the patient compared with the level of carbon monoxide in the blood (Broome et al., 1988). The extent of tissue toxicity, which becomes significant under hypoxic conditions or with very high levels of carbon monoxide, is likely determined by the length of exposure. For example, a short exposure to carbon monoxide at high ambient concentrations may allow insufficient time for significant increases in tissue levels of carbon monoxide to occur. The syncope observed in individuals with carbon monoxide poisoning who were exposed in this manner may be the

Table 18. Symptoms associated with varying levels of carbon monoxide poisoning[a]

CO in the atmosphere			COHb in blood (%)	Physiological and subjective symptoms
%	mg/m³	ppm		
0.007	80	70	10	No appreciable effect, except shortness of breath on vigorous exertion; possible tightness across the forehead; dilation of cutaneous blood vessel
0.01	140	120	20	Shortness of breath on moderate exertion; occasional headache with throbbing in temples
0.02	250	220	30	Decided headache; irritable; easily fatigued; judgement disturbed; possible dizziness; dimness of vision
0.035–0.052	400–600	350–520	40–50	Headache, confusion; collapse; fainting on exertion
0.080–0.122	900–1400	800–1220	60–70	Unconsciousness; intermittent convulsion; respiratory failure, death if exposure is long continued
0.195	2200	1950	80	Rapidly fatal

[a] Adapted from Winter & Miller (1976).

result of simple hypoxia with rapid recovery despite high carboxy-haemoglobin levels. On the other hand, prolonged exposure to carbon monoxide prior to hospital arrival may allow sufficient uptake of carbon monoxide by tissues to inhibit the function of intracellular compounds such as myoglobin. This effect, in combination with the existing reduction in tissue oxygen, may cause irreversible central nervous system or cardiac damage.

Complications occur frequently in carbon monoxide poisoning. Immediate death from carbon monoxide is most likely cardiac in origin, because myocardial tissue is most sensitive to hypoxic effects of carbon monoxide. Severe poisoning results in marked hypotension and lethal arrhythmias, which may be responsible for a large number of pre-hospital deaths. Rhythm disturbances include sinus tachycardia, atrial flutter and fibrillation, premature ventricular contractions, ventricular tachycardia and fibrillation. Other documented electro-cardiographic changes include decrease in the magnitude of the R-wave, ST elevation, T-wave inversion and heart block. Coronary underperfusion can lead to myocardial infarction, especially if coronary arteries are previously narrowed. Pulmonary oedema is a fairly common feature that makes routine chest X-ray mandatory in each carbon monoxide-poisoned patient (Mathieu & Wattel, 1990). Rhabdomyolosis and its consequence, renal failure, as well as pancreatitis, are also complications that may be seen in carbon monoxide poisoning.

Neurological manifestations of acute carbon monoxide poisoning include disorientation, confusion, coma, cogwheel rigidity, opisthotonic posturing, extremity flaccidity or spasticity, and extensor plantar response. Perhaps the most insidious effect of carbon monoxide poisoning is the delayed development of neuropsychiatric impairment. Within 1–3 weeks of poisoning, 15–40% of patients will manifest inappropriate euphoria, impaired judgement, poor concentration and relative indifference to obvious neurological deficits. Computed axial tomography can show low density in the area of globus pallidus in some patients with poor neurological outcome (Sawday et al., 1980), but this feature is not specific for carbon monoxide poisoning.

Observed ocular effects from case reports on acute carbon monoxide poisoning range from retinal haemorrhages (Dempsey et al., 1976; Kelley & Sophocleus, 1978) to blindness (Duncan & Gumpert, 1983; Katafuchi et al., 1985). In addition, Trese et al. (1980) described a case report of a 57-year-old woman having peripheral neuropathy and tortuous retinal vessels after chronic, intermittent exposure to low levels of carbon monoxide over a 16-month period. These authors speculated that increased blood flow from low-level, chronic exposure to carbon monoxide may lead to the development of a compensatory retinal vascular tortuosity. With high-level, acute exposures to carbon monoxide, the compensation will not take place, and localized vascular haemorrhages result.

Management of carbon monoxide-poisoned patients first consists of removing the patient from exposure to the toxic atmosphere and supplying pure oxygen to accelerate the elimination of carbon monoxide and improve tissue oxygenation. Respiratory and circulatory conditions are assessed rapidly, and resuscitative measures are performed if needed. Evaluation includes neurological status of conscious level, motor response and reflectivity, and a complete physical examination looking for complications, associated trauma or intoxication, and previous disease. Laboratory exams should include a blood gas analysis to look for acidosis and a carboxyhaemoglobin measurement. Carboxyhaemoglobin levels over 5% in a non-smoker and over 10% in a smoker confirm the diagnosis but not the severity of intoxication (Ilano & Raffin, 1990).

Patients with carbon monoxide poisoning respond to treatment with 100% oxygen (Pace et al., 1950). If available, treatment with hyperbaric oxygen at 2.5–3 times atmospheric pressure for 90 min is preferable (Myers, 1986), but the precise conditions requiring treatment have been a topic of debate in the literature (Mathieu et al., 1985; Norkool & Kirkpatrick, 1985; Broome et al., 1988; Brown et al., 1989; James, 1989; Raphael et al., 1989; Roy et al., 1989; Thom & Keim, 1989; Van Hoesen et al., 1989). It has been suggested that if carboxyhaemoglobin is above 25%, hyperbaric oxygen treatment should be initiated (Norkool & Kirkpatrick, 1985), although treatment plans based on specific carboxyhaemoglobin saturations are not well founded (Thom & Keim, 1989). Most hyperbaric centres treat patients with carbon monoxide intoxication when they manifest loss of consciousness or other neurological signs and symptoms (excluding headache), regardless of the carboxyhaemoglobin saturation at presentation (Piantadosi, 1990). The first European Consensus Conference on hyperbaric medicine has concluded that hyperbaric oxygen is highly recommended in every comatose patient, every patient who had lost consciousness during exposure, every patient with abnormal neuropsychological manifestation and every pregnant woman (Mathieu et al., 1996b).

The half-time elimination of carbon monoxide while breathing air is approximately 320 min; when breathing 100% oxygen, it is 80 min; and when breathing oxygen at 304 kPa (3 atm), it is 23 min (Myers et al., 1985). In the case of normobaric oxygen treatment, length of oxygen administration is also controversial, but it appears that it must be long enough to ensure total carbon monoxide detoxication. Proposed durations are often between 12 and 48 h.

Successful removal of carbon monoxide from the blood does not ensure an uneventful recovery with no further clinical signs or symptoms. Neurological problems may develop insidiously weeks after recovery from the acute episode of carbon monoxide poisoning

(Meredith & Vale, 1988). These late neurological sequelae include intellectual deterioration, memory impairment and cerebral, cerebellar and midbrain damage. Smith & Brandon (1973) published a study in which they found 10% of cases with immediate gross neurological sequelae, but also 33% with delayed personality deterioration and 43% with memory disturbances. In a literature review, Ginsberg & Romano (1976) found between 15% and 40% of late cases with neurological sequelae. The explanation for neurological problems may lie in misdiagnosis (30% in a French Poison Control Center study; Mathieu et al., 1985), inadequate therapy (40% of the patients of Smith & Brandon, 1973, did not receive any oxygen in the emergency treatment) or delayed therapy. Complete recovery is obtained in more cases if oxygen is applied in less than 6 h (Barois et al., 1979).

Mathieu et al. (1996a) conducted a study on the long-term consequences of carbon monoxide poisoning and treatment by hyperbaric oxygen therapy on 774 patients, who were divided into five groups: group zero, in which patients suffered only from headache or nausea; group I, with abnormality in neurological examination; group II, where patients had lost consciousness regardless of their clinical state of admission; group III, in which patients were comatose (Glasgow coma scale 6); and group IV, where patients were deeply comatose. Group zero received only normobaric pure oxygen, while groups I, II, III and IV received hyperbaric oxygen. At 1 year only, 4.4% of patients suffered from persistent manifestation, and only 1.6% had major functional impairment. However, persistent neurological manifestations occurred only in patients from groups I and IV.

As in the previous study (Mathieu et al., 1996a) and in accordance with the Klees et al. (1985) conclusions, neither the clinical status nor the carboxyhaemoglobin level on hospital admission can predict the development of delayed neuropsychological changes. Thus, the authors advocate the use of hyperbaric oxygen in every carbon monoxide-poisoned patient who has suffered loss of consciousness during carbon monoxide exposure, who has a neurological abnormality upon clinical examination or who remains comatose upon admission. These results are in accordance with those obtained by Myers et al. (1981).

In conclusion, carbon monoxide poisoning remains frequent and severe with a relatively high risk of immediate death, complications and late sequelae; furthermore, it is too often overlooked. The human data from cases of accidental high-dose carbon monoxide exposures are not adequate for evaluating the relationship between dose and effect and identifying a LOEL and a NOEL, or for setting carbon monoxide standards in ambient air, because of the small number of cases reviewed and problems in documenting levels of exposure. Nevertheless, they suggest that in some circumstances and meteorological conditions, acute poisoning can always occur in a chronic situation. This possibility has to be taken into account in defining standards. As well, such data, if systematically gathered and reported, could be useful in identifying possible ages of special sensitivity to carbon monoxide and cofactors or other risk factors that might identify sensitive subpopulations.

9. EVALUATION OF HEALTH RISKS

9.1   Introduction

Most of the critical information on the human health effects of carbon monoxide discussed in this document has been derived by consideration of data from two carefully defined population groups — young, healthy adults and patients with diagnosed coronary artery disease. In addition to laboratory studies, supporting evidence has been obtained from epidemiological and clinical studies as well as reports of accidental exposures. On the basis of the known effects described, patients with reproducible exercise-induced ischaemia appear to be a well-defined sensitive group within the general population that is at increased risk for experiencing health effects (i.e., decreased exercise duration due to exacerbation of cardiovascular symptoms) of concern at ambient or near-ambient carbon monoxide exposure concentrations that result in carboxyhaemoglobin levels down to 3%. A smaller sensitive group of healthy individuals experiences decreased exercise duration at similar levels of carbon monoxide exposure, but only during short-term maximal exercise. Decrements in exercise duration in the healthy population, therefore, would be mainly of concern to competing athletes rather than to ordinary people carrying out the common activities of daily life.

From both clinical and theoretical work and from experimental research on laboratory animals, certain other groups in the population are identified as being at probable risk from exposure to carbon monoxide. These probable risk groups include (1) fetuses and young infants, (2) pregnant women, (3) the elderly, especially those with compromised cardiopulmonary or cerebrovascular functions, (4) individuals with obstructed coronary arteries, but not yet manifesting overt symptomatology of coronary artery disease, (5) individuals with congestive heart failure, (6) individuals with peripheral vascular or cerebrovascular disease, (7) individuals with haematological diseases (e.g., anaemia) that affect oxygen carrying capacity or transport in the blood, (8) individuals with genetically unusual forms of haemoglobin associated with reduced oxygen carrying capacity, (9) individuals with chronic obstructive lung diseases, (10) individuals using medicinal or recreational drugs having effects on the brain or cerebrovasculature, (11) individuals exposed to other pollutants (e.g., methylene chloride) that increase endogenous formation of carbon monoxide and

(12) individuals who have not been adapted to high altitude and are exposed to a combination of high altitude and carbon monoxide.

Little empirical evidence is currently available by which to specify health effects associated with ambient or near-ambient carbon monoxide exposures for most of these probable risk groups. Whereas the previous chapters dealt with documented evidence of carbon monoxide exposure through controlled or natural laboratory investigations, in this chapter an effort will be made to determine the anticipated effects of carbon monoxide in special subpopulations that form a significant proportion of the population at large.

As well, guideline values for ambient air carbon monoxide exposure and recommendations for further research and protection of human health are provided in this chapter.

9.2 Age, gender and pregnancy as risk factors

The fetus and newborn infant are theoretically susceptible to carbon monoxide exposure for several reasons. Fetal circulation is likely to have a higher carboxyhaemoglobin level than the maternal circulation as a result of differences in uptake and elimination of carbon monoxide from fetal haemoglobin. Because the fetus also has a lower oxygen tension in the blood than adults, any further drop in fetal oxygen tension owing to the presence of carboxyhaemoglobin could have a potentially serious effect. The newborn infant with a comparatively high rate of oxygen consumption and lower oxygen transport capacity for haemoglobin than most adults would also be potentially susceptible to the hypoxic effects of increased carboxyhaemoglobin. Newer data from laboratory animal studies on the developmental toxicity of carbon monoxide suggest that prolonged exposure to high levels (>110 mg/m^3 [>100 ppm]) of carbon monoxide during gestation may produce a reduction in birth weight, cardiomegaly and delayed behavioural development. Human data are scant and more difficult to evaluate, but further research is warranted. Additional studies are therefore needed to determine if chronic exposure to carbon monoxide, particularly at low, near-ambient levels, can compromise the already marginal conditions existing in the fetus and newborn infant.

The effects of carbon monoxide on maternal–fetal relationships are not well understood. In addition to fetuses and newborn infants, pregnant women also represent a susceptible group, because pregnancy is associated with increased alveolar ventilation and an increased rate of oxygen consumption that serves to increase the rate of carbon monoxide uptake from inspired air. Perhaps a more important factor is that pregnant women experience haemodilution as a result of the disproportionate increase in plasma volume compared with erythrocyte volume. This group, therefore, should be studied to evaluate the effects of carbon monoxide exposure and elevated carboxyhaemoglobin levels.

Changes in metabolism with age may make the aging population particularly susceptible to the effects of carbon monoxide. Maximal oxygen uptake declines steadily with age. The rate of decline in the population is difficult to determine, however, partly because of the wide range of values reported in the cross-sectional and longitudinal studies published in the literature, and partly because of confounding factors such as heredity, changes in body weight and composition, and level of physical activity.

By the time an average healthy, non-smoking male reaches the age of 65 years, the maximal oxygen uptake will be about 23 ± 5 ml/kg body weight per minute. At 75 years of age, the maximal oxygen uptake will be about 17 ± 5 ml/kg body weight per minute. The decline in maximal oxygen uptake with age seems to be the same in females, as well. However, because females have about 20–25% lower maximal oxygen uptake, the corresponding values will occur about 5–8 years earlier in females. In physically active individuals, the corresponding values will occur about 10–15 years later than in the average sedentary person.

9.3 Risk of carbon monoxide exposure in individuals with pre-existing disease

9.3.1 Subjects with coronary artery disease

Coronary heart disease is one of the major causes of death and disability in the world, especially in industrialized societies. In a 1996 WHO estimate, 7.2 million deaths globally were caused by coronary heart disease, which ranked first in leading causes of death (WHO,

1997b). Although deaths from coronary heart disease constitute only about 14% of the global total number of deaths, they are responsible for about one-third of all deaths in industrialized societies. The coronary heart disease epidemic began in industrialized societies in the early decades of the 20th century. Its death rates peaked in the 1960s and early 1970s and have since declined dramatically. On the other hand, coronary heart disease is now increasing in developing countries as their populations age and adopt unhealthy habits and behaviours. The world's highest coronary heart disease mortality rates are now found in eastern and central Europe.

According to data compiled by the American Heart Association (1989), persons with diagnosed coronary artery disease numbered 5 million in 1987 and 7 million in 1990, the latter being about 3% of the total US population (Collings, 1988; US Department of Health and Human Services, 1990). These persons have myocardial ischaemia, which occurs when the heart muscle receives insufficient oxygen delivered by the blood. For some, exercise-induced angina pectoris can occur. In all patients with diagnosed coronary artery disease, however, the predominant type of ischaemia, as identified by ST-segment depression, is asymptomatic (i.e., silent). In other words, patients who experience angina usually have more ischaemic episodes that are asymptomatic. Unfortunately, some individuals in the population have coronary artery disease but are totally asymptomatic. It has been estimated that 5% of middle-aged men develop a positive exercise test, one of the signs of ischaemia. A significant number of these men will have angiographic evidence of coronary artery disease (Cohn, 1988; Epstein et al., 1988, 1989). For example, in the USA, more than 1 million heart attacks occur each year, half of them being fatal (American Heart Association, 1989). About 10–15% of all myocardial infarctions are silent (Kannel & Abbott, 1984; Epstein et al., 1988). Of the 500 000 survivors of hospitalized myocardial infarction, about 10% are asymptomatic but have signs of ischaemia. Thus, many more persons, as many as 3–4 million Americans (American Heart Association, 1989), are not aware that they have coronary heart disease and may constitute a high-risk group.

Persons with both asymptomatic and symptomatic coronary artery disease have a limited coronary flow reserve and therefore will be sensitive to a decrease in oxygen carrying capacity induced by carbon monoxide exposure. In addition, carbon monoxide might exert a direct effect on vascular smooth muscle, particularly in those individuals with an already damaged vascular endothelium. Naturally occurring vasodilators like acetylcholine cause a release of endothelium-derived relaxing factor that precedes the onset of vascular smooth muscle relaxation. Oxyhaemoglobin and oxymyoglobin will antagonize these smooth muscle relaxant effects. Although no clinical studies have been done, *in vitro* studies suggest that carbon monoxide may inhibit the effects of oxyhaemoglobin on the action of acetylcholine. Carbon monoxide exposure in patients with a diseased endothelium, therefore, could accentuate acetylcholine-induced vasospasm and aggravate silent ischaemia.

.3.2 Subjects with congestive heart failure

Congestive heart failure is a major and growing public health problem. Because the prevalence of heart failure is known to increase with age, improvements in the average life expectancy of the general population would be expected to increase the magnitude of the problem over the next few decades. It was reported that about 75% of patients with heart failure are above the age of 60 years in the USA (Brody et al., 1987).

Worldwide morbidity and mortality data for heart failure are not available. In the USA, about 400 000 new cases of heart failure are diagnosed every year, resulting in about 1.6 million hospitalizations. The mortality rate is high, between 15% and 60% per year. The onset of death is often sudden, and, because about 65% of heart failure patients have serious arrhythmias, this sudden death is thought to be due to arrhythmia (Brody et al., 1987).

Patients with congestive heart failure have a markedly reduced circulatory capacity and therefore may be very sensitive to any limitations in oxygen carrying capacity. Thus, exposure to carbon monoxide will certainly reduce their exercise capacity and will even be dangerous, especially because of its arrhythmogenic activity. The etiology of heart failure is diverse, but the dominating disease is coronary artery disease. The large portion of heart failure patients with coronary artery disease, therefore, might be even more sensitive to carbon monoxide exposure.

9.3.3   Subjects with other vascular diseases

Vascular disease including cerebrovascular disease is present in both the male and female population and is more prevalent above 65 years of age. Both of these conditions are often found in subjects with coronary artery disease. Both conditions are also associated with a limited blood flow capacity and therefore should be sensitive to carbon monoxide exposure. It is not clear, however, how low levels of exposure to carbon monoxide will affect these individuals. Only one study has been reported on patients with peripheral vascular disease.

9.3.4   Subjects with anaemia and other haematological disorders

Clinically diagnosed low levels of haemoglobin, characterized as anaemia, are a relatively prevalent condition throughout the world. If the anaemia is mild to moderate, an inactive person is often asymptomatic. However, owing to the limitation in the oxygen carrying capacity resulting from the low haemoglobin levels, an anaemic person should be more sensitive to low-level carbon monoxide exposure than a person with normal haemoglobin levels. If anaemia is combined with other prevalent diseases, such as coronary artery disease, the individual will also be at an increased risk from carbon monoxide exposure. Anaemia is more prevalent in women and in the elderly, already two potentially high-risk groups. An anaemic person will also be more sensitive to the combination of carbon monoxide exposure and high altitude.

Individuals with haemolytic anaemia often have higher baseline levels of carboxyhaemoglobin, because the rate of endogenous carbon monoxide production from haem catabolism is increased. One of the many causes of anaemia is the presence of abnormal haemoglobin in the blood. For example, in sickle-cell disease, the average life span of red blood cells with abnormal haemoglobin S is 12 days compared with an average of 88 days in healthy individuals with normal haemoglobin (haemoglobin A). As a result, baseline carboxyhaemoglobin levels can be as high as 4%. In subjects with haemoglobin Zurich, where affinity for carbon monoxide is 65 times that of normal haemoglobin, carboxyhaemoglobin levels range from 4% to 7%.

3.5   *Subjects with obstructive lung disease*

Chronic obstructive pulmonary disease is a prevalent disease, especially among smokers, and a large number (>50%) of these individuals have limitations in their exercise performance, demonstrated by a decrease in oxygen saturation during mild to moderate exercise. In spite of their symptoms, many of them (~30%) continue to smoke and already may have carboxyhaemoglobin levels of 4–8%. Subjects with hypoxia are also more likely to have a progression of the disease, resulting in severe pulmonary insufficiency, pulmonary hypertension and right heart failure. Studies suggest that individuals with hypoxia due to chronic lung disease such as bronchitis and emphysema may be susceptible to carbon monoxide during submaximal exercise typically found during normal daily activity.

Hospital admissions for asthma have increased considerably in the past few years, particularly among individuals less than 18 years of age. Because asthmatics can also experience exercise-induced airflow limitation, it is likely that they would also be susceptible to hypoxia. It is not known, however, how exposure to carbon monoxide would affect these individuals.

9.4   Subpopulations at risk from combined exposure to carbon monoxide and other chemical substances

4.1   *Interactions with drugs*

There is almost a complete lack of data on the possible toxic consequences of combined carbon monoxide exposure and drug use. Because of the diverse classes of both cardiovascular and psychoactive drugs, and because many other classes have not been examined at all, it must be concluded that this is an area of concern for which it is difficult at the present time to make recommendations that will have an effect on air quality guidelines.

Because data are generally lacking on carbon monoxide–drug interactions, it should be useful to speculate on some of the mechanisms by which carbon monoxide might be expected to alter drug effects, or vice versa, and discuss possible populations at risk because of these potential interaction effects.

9.4.2 Interactions with other chemical substances in the environment

Besides direct ambient exposure to carbon monoxide, there are other chemical substances in the environment that can lead to increased carboxyhaemoglobin saturation when inhaled. Halogenated hydrocarbons used as organic solvents undergo metabolic breakdown by cytochrome P-450 to form carbon monoxide and inorganic halide. Possibly the greatest concern regarding potential risk in the population comes from exposure to one of these halogenated hydrocarbons, methylene chloride, and some of its derivatives.

9.5 Subpopulations exposed to carbon monoxide at high altitudes

For patients with coronary artery disease, restricted coronary blood flow limits oxygen delivery to the myocardium. Carbon monoxide also has the potential for compromising oxygen transport to the heart. For this reason, such patients have been identified as the subpopulation most sensitive to the effects of carbon monoxide. A reduction in the partial pressure of oxygen in the atmosphere, as at high altitude, also has the potential for compromising oxygen transport. Therefore, patients with coronary artery disease who visit higher elevations might be unusually sensitive to the added effects of atmospheric carbon monoxide.

It is important to distinguish between the long-term resident of high altitude and the newly arrived visitor from low altitude. Specifically, the visitor will be more hypoxaemic than the fully adapted resident.

One would postulate that the combination of high altitude with carbon monoxide would pose the greatest risk to persons newly arrived at high altitude who have underlying cardiopulmonary disease, particularly because they are usually older individuals. Surprisingly, this hypothesis has never been tested adequately.

It is known that low birth weights occur in both infants born at altitudes above 1830 m as well as infants born near sea level whose mothers had elevated carboxyhaemoglobin levels as a result of cigarette smoking. It has also been shown that carboxyhaemoglobin levels in smokers at high altitude are higher than in smokers at sea level. Although it is probable that the combination of hypoxic hypoxia and hypoxia resulting from ambient exposure to carbon monoxide could further reduce birth weight at high altitude and possibly modify future development, no data are currently available to support this hypothesis.

9.6 Carbon monoxide poisoning

The majority of this document deals with the relatively low concentrations of carbon monoxide that induce effects in humans at or near the lower margin of detection by current medical technology. Yet the health effects associated with exposure to this pollutant range from the more subtle cardiovascular and neurobehavioural effects at low ambient concentrations, as identified in the preceding sections, to unconsciousness and death after acute exposure to high concentrations of carbon monoxide. The morbidity and mortality resulting from the latter exposures are described briefly here to complete the picture of carbon monoxide exposure in present-day society.

Carbon monoxide is reported to be the cause of more than half of the fatal poisonings that are reported in many countries. Fatal cases are also grossly under-reported or misdiagnosed by medical professionals. Therefore, the precise number of individuals who have suffered from carbon monoxide intoxication is not known. In the USA, the National Center for Health Statistics (1986) estimates that between 700 and 1000 deaths per year are due to accidental carbon monoxide poisoning.

The symptoms, signs and prognosis of acute poisoning correlate poorly with the level of carboxyhaemoglobin measured at the time of hospital admission; however, because carbon monoxide poisoning is a diagnosis frequently overlooked, the importance of the early symptoms (headache, dizziness, weakness, nausea, confusion, disorientation and visual disturbances) has to be emphasized, especially if they recur with a certain periodicity or in certain circumstances. Complications occur frequently in carbon monoxide poisoning. Immediate death is most likely cardiac in origin, because myocardial tissues are most sensitive to the hypoxic effects of carbon monoxide. Severe poisoning results in marked hypotension and lethal arrhythmias and other electrocardiographic changes. Pulmonary oedema is a fairly common feature. Neurological manifestation of acute carbon monoxide poisoning includes disorientation, confusion and coma. Perhaps the most insidious effect of carbon monoxide poisoning is the delayed development of neuropsychiatric impairment within 1–3 weeks and the neurobehavioural consequences, especially in children. Carbon monoxide poisoning during pregnancy results in high risk for the mother, by increasing the short-term complication rate, and for the fetus, by causing fetal death, developmental disorders and chronic cerebral lesions.

In conclusion, carbon monoxide poisoning occurs frequently, has severe consequences, including immediate death, involves complications and late sequelae and is often overlooked. Efforts in prevention in public and medical education should be encouraged.

9.7 Recommended guideline values

9.7.1 *Environmental sources*

Carbon monoxide comes from both natural and anthropogenic processes. About half of the carbon monoxide is released at the Earth's surface, and the rest is produced in the atmosphere. About 60% of the carbon monoxide is from human activities, whereas natural processes account for the remaining 40%. Recent reports showed that global carbon monoxide concentrations started to decline rapidly from 1988 to 1993. Since 1993, the downward trend in global carbon monoxide has levelled off, and it is not clear if carbon monoxide concentrations will continue to decline.

9.7.2 *Environmental concentrations*

The natural background carbon monoxide concentrations in remote areas of the southern hemisphere are around 0.05 mg/m^3 (0.04 ppm), primarily as a result of natural processes. In the northern hemisphere, background concentrations are 2–3 times higher because of a greater human population density.

Carbon monoxide concentrations in ambient air monitored from fixed-site stations are generally below 10 mg/m^3 (9 ppm, 8-h average). The annual mean carbon monoxide concentrations are less than 10 mg/m^3 (9 ppm). However, short-term peak carbon monoxide concentrations occur in traffic environments; carbon monoxide concentrations up to 57 mg/m^3 (50 ppm) are reported on heavily travelled roads. Air quality data from fixed-site monitoring stations seem to underestimate the short-term peak carbon monoxide level in traffic environments. Personal monitoring data from pedestrians and street workers showed that the carbon monoxide level could reach 10–16 mg/m^3 (9–14 ppm) and 11–57 mg/m^3 (10–50 ppm), respectively.

Indoor and in-transit concentrations of carbon monoxide may be significantly different from ambient carbon monoxide concentrations. The carbon monoxide levels in homes are usually lower than 10 mg/m^3 (9 ppm); however, the peak value in homes with gas stoves could be higher than 10 mg/m^3 (9 ppm) and up to 21 mg/m^3 (18 ppm) in the kitchen, from 1.1 to 34 mg/m^3 (1 to 30 ppm) with wood combustion and from 2.3 to 8 mg/m^3 (2 to 7 ppm) with a kerosene heater. A report from homes with unvented gas-fired water heaters showed that the indoor carbon monoxide levels range from less than 11 mg/m^3 (10 ppm) to more than 110 mg/m^3 (100 ppm).

The carbon monoxide concentrations inside motor vehicles are generally around 10–29 mg/m^3 (9–25 ppm) and occasionally over 40 mg/m^3 (35 ppm). Carbon monoxide levels in parking garages, tunnels and ice skating rinks are higher than in common indoor environments.

The differences between indoor and outdoor air quality and the different amounts of time people spend indoors and outdoors explain why using ambient air quality measurements alone will not provide accurate estimates of population exposure. The measurement of carbon monoxide from personal monitors more accurately reflects the real exposure.

Occupational exposure levels are generally below 34 mg/m^3 (30 ppm). Workers exposed to vehicle exhaust may have peak exposures over 230 mg/m^3 (200 ppm).

9.7.3 *Carboxyhaemoglobin concentrations in the population*

Carbon monoxide diffuses rapidly across the alveolar and capillary membrane and more slowly across the placental membrane. Approximately 85% of the absorbed carbon monoxide binds with haemoglobin to form carboxyhaemoglobin, which is a specific bio-marker of exposure in blood. The remaining 15% is distributed extravascularly. During an exposure to a fixed ambient concentration of carbon monoxide, the carboxyhaemoglobin concentration increases rapidly at the onset of exposure, starts to level off after 3 h and approaches a steady state after 6–8 h of exposure. An 8-h value would be representative of any longer continuous exposure value. In real-life situations, the prediction of individual carboxyhaemoglobin levels is difficult because of large spatial and temporal variations in both indoor and outdoor levels of carbon monoxide. Typical carboxyhaemoglobin concentrations in the non-smoking population range from 0.5% to 2.0%; levels are slightly higher in pregnant women.

A common source of carbon monoxide for the general population is tobacco smoke. Exposure to tobacco smoke not only increases carboxyhaemoglobin concentrations in smokers, but under some circumstances can also affect non-smokers. In many of the studies currently cited to justify the formulation of exposure limits, neither the smoking habits of the subjects nor their exposure to passive smoking has been taken into account. In addition, as the result of higher baseline carboxyhaemoglobin levels, smokers may actually be excreting more carbon monoxide into the air than they are inhaling from the ambient environment. Smokers may even show an adaptive response to the elevated carboxyhaemoglobin levels, as evidenced by increased red blood cell volumes and reduced plasma volumes. As a consequence, it is not clear if incremental increases in carboxyhaemoglobin caused by environmental exposure would actually be additive to the chronically elevated carboxyhaemoglobin levels due to tobacco smoke. Thus, the exposure limits are recommended primarily for the protection of non-smokers.

9.7.4 *General population exposure*

The environmental health criteria used in arriving at a recommendation for an exposure limit for the general population were mainly those data obtained from the exposure of non-smoking subjects with coronary artery disease to carbon monoxide while exercising and potential effects in fetuses of non-smoking pregnant mothers exposed to ambient sources of carbon monoxide. The principal cause of carbon monoxide-induced effects at low levels of ambient exposure is thought to be increased carboxyhaemoglobin formation. Therefore, the primary exposure limits are derived on the basis of carboxyhaemoglobin.

In order to protect non-smoking population groups with documented or latent coronary artery disease from acute ischaemic heart attacks and to protect fetuses of non-smoking pregnant mothers from untoward hypoxic effects, a carboxyhaemoglobin level of 2.5% should not be exceeded.

9.7.5 Working population exposure

Non-smoking people in certain occupations (e.g., car, bus and taxi drivers, policemen, firemen, traffic wardens, street workers, garage and tunnel workers, mechanics) can have long-term carboxyhaemoglobin levels up to 5%. On the basis of present knowledge, the Task Group unanimously agreed on maintaining carboxyhaemoglobin levels not exceeding 5%, because working populations comprise individuals who are assumed to be healthy, physiologically resilient and under regular supervision.

9.7.6 Derived guideline values for carbon monoxide concentrations in ambient air

It is important, wherever possible, to have both biological and environmental assessments of human exposure to pollutants. Although the biological measurements may be more relevant in relation to effects in population groups, they may be more difficult to use in practice. For carbon monoxide, the relationship between concentrations in air and carboxyhaemoglobin levels is affected by several physiological and environmental variables, including exposure time. The appropriate carbon monoxide guidelines are based on the CFK exponential equation, which takes into account all known physiological variables affecting carbon monoxide uptake (see chapter 6).

The following guideline values (ppm values rounded) and periods of time-weighted average exposures have been determined in such a way that the carboxyhaemoglobin level of 2.5% is not exceeded, even when a normal subject engages in light or moderate exercise:

100 mg/m^3 (87 ppm) for 15 min
60 mg/m^3 (52 ppm) for 30 min
30 mg/m^3 (26 ppm) for 1 h
10 mg/m^3 (9 ppm) for 8 h

It should be emphasized that analyses of carbon monoxide in air and of carboxyhaemoglobin in blood are complementary and should be in no way regarded as alternative methods of monitoring. Obviously, air monitoring has its uses in the planning and implementation of control measures and for warning purposes, but such measurements have limited value in estimating the actual human exposure defined by carboxyhaemoglobin levels. It must also be recognized that complete protection of all persons, at all times, cannot reasonably be sought by ambient environmental control.

9.8 Recommendations

9.8.1 Further research

Further research is needed to:

- evaluate the effects of long-term carbon monoxide exposure on pregnant women and newborns;

- study the health effects of chronic exposure to low-level carbon monoxide, with emphasis on the fetus and newborn infant;

- determine the effects of carbon monoxide exposure on patients with asthma, chronic obstructive pulmonary disease, anaemia and haematological disorders;

- study the interactions of carbon monoxide and other outdoor and indoor environmental pollutants;

- study the interactions of carbon monoxide and medication;

- study the influence of environmental factors, including high altitude and heat stress, on health effects from exposure to carbon monoxide, both in the general population and in high-risk groups, including the elderly, children and those suffering from cardiorespiratory diseases;

- identify the biomarkers for delayed neurological sequelae of carbon monoxide exposure and for epidemiological studies;

- study the mechanism of carbon monoxide toxicity at cellular and subcellular levels;

- study factors influencing endogenous carbon monoxide production; and

- improve estimates of population exposure to carbon monoxide, especially by the use of personal exposure monitors.

.8.2   *Protection of human health*

For protection of human health, it is desirable to:

- educate the general population, especially those with cardiovascular and respiratory diseases, about the risk of carbon monoxide exposure;

- improve monitoring of carbon monoxide in the workplace and in public places (streets, shops, restaurants, car parks);

- increase awareness of health risks from carbon monoxide exposure for those potentially exposed to high peak carbon monoxide levels from internal combustion engines used in enclosed environments;

- increase efforts to reduce carbon monoxide emissions from combustion engines;

- encourage the development of better methods for carbon monoxide detection and promotion of the use of carbon monoxide detectors; and

- improve the awareness of medical professionals, as well as the general population, of the dangers of carbon monoxide exposure during pregnancy.

REFERENCES

ACGIH (1987) In: Cook WA ed. Occupational exposure limits worldwide. Cincinnati, Ohio, American Conference of Governmental Industrial Hygienists.

ACGIH (1999) TLVs (Threshold Limit Values) and BEI (Biological Exposure Indices). Cincinnati, Ohio, American Conference of Governmental Industrial Hygienists.

Acton LL, Griggs M, Hall GD, Ludwig CB, Malkmus W, Hesketh WD, & Reichle H (1973) Remote measurement of carbon monoxide by a gas filter correlation instrument. J Am Inst Aeronaut Astronaut, **11**: 899-900.

Adams JD, Erickson HH, & Stone HL (1973) Myocardial metabolism during exposure to carbon monoxide in the conscious dog. J Appl Physiol, **34**: 238-242.

Adams KF, Koch G, Chatterjee B, Goldstein GM, O'Neil JJ, Bromberg PA, Sheps DS, McAllister S, Price CJ, & Bissette J (1988) Acute elevation of blood carboxyhemoglobin to 6% impairs exercise performance and aggravates symptoms in patients with ischemic heart disease. J Am Coll Cardiol, **12**: 900-909.

Agostoni A, Stabilini R, Viggiano G, Luzzana M, & Samaja M (1980) Influence of capillary and tissue P_{O_2} on carbon monoxide binding to myoglobin: a theoretical evaluation. Microvasc Res, **20**: 81-87.

Akland GG, Hartwell TD, Johnson TR, & Whitmore RW (1985) Measuring human exposure to carbon monoxide in Washington, D.C., and Denver, Colorado, during the winter of 1982-1983. Environ Sci Technol, **19**: 911-918.

Alcindor LG, Belegaud J, Aalam H, Piot MC, Heraud C, & Boudene C (1984) Teneur en cholesterol des lipoprotéines légères chez le lapin hypercholestérolémique intoxiqué ou non par l'oxyde de carbone. Ann Nutr Metab, **28**: 117-122.

Alderman BW, Baron AE, & Savitz DA (1987) Maternal exposure to neighborhood carbon monoxide and risk of low infant birth weight. Public Health Rep, **102**: 410-414.

Allred EN, Bleecker ER, Chaitman BR, Dahms TE, Gottlieb SO, Hackney JD, Pagano M, Selvester RH, Walden SM, & Warren J (1989a) Short-term effects of carbon monoxide exposure on the exercise performance of subjects with coronary artery disease. N Engl J Med, **321**: 1426-1432.

Allred EN, Bleecker ER, Chaitman BR, Dahms TE, Gottlieb SO, Hackney JD, Hayes D, Pagano M, Selvester RH, Walden SM, & Warren J (1989b) Acute effects of carbon monoxide exposure on individuals with coronary artery disease. Cambridge, Massachusetts, Health Effects Institute (Research Report No. 25).

Allred EN, Bleecker ER, Chaitman BR, Dahms TE, Gottlieb SO, Hackney JD, Pagano M, Selvester RH, Walden SM, & Warren J (1991) Effects of carbon monoxide on myocardial ischemia. Environ Health Perspect, **91**: 89-132.

Alm S, Reponen A, Mukala K, Pasanen P, Tuomist J, & Jantunen MJ (1994) Personal exposures of preschool children to carbon monoxide: Roles of ambient air quality and gas stoves. Atmos Environ, **28**(22): 3577-3580.

Alm S, Reponen A, Jurvelin J, Salonen RO, Pekkanen J, & Jantunen MJ (1995) Time activity patterns and levels of carbon monoxide in different microenvironments among Finnish primary school children. Epidemiology, 6(suppl 4): S51.

Amendola AA & Hanes NB (1984) Characterization of indoor carbon monoxide levels produced by the automobile. In: Berglund B, Lindvall T, & Sundell J ed. Indoor air. Proceedings of the 3rd International Conference on Indoor Air Quality and Climate — Volume 4: Chemical characterization and personal exposure. Stockholm, Sweden, Swedish Council for Building Research, pp 97-102 (Publication No. PB85-104214, available from NTIS, Springfield, Virginia, USA).

American Heart Association (1989) 1990 Heart and stroke facts. Dallas, Texas, American Heart Association.

American Industrial Hygiene Association (1972) Intersociety committee methods for ambient air sampling and analysis: Report II. Am Ind Hyg Assoc J, **33**: 353-359.

Anders MW & Sunram JM (1982) Transplacental passage of dichloromethane and carbon monoxide. Toxicol Lett, **12**: 231-234.

Anderson EW, Andelman RJ, Strauch JM, Fortuin NJ, & Knelson JH (1973) Effect of low-level carbon monoxide exposure on onset and duration of angina pectoris: a study in ten patients with ischemic heart disease. Ann Intern Med, **79**: 46-50.

Andrecs L, Stenzler A, Steinberg H, & Johnson T (1979) Carboxyhemoglobin levels in garage workers. Am Rev Respir Dis, **119**: 199.

Annau Z & Dyer RS (1977) Effects of environmental temperature upon body temperature in the hypoxic rat. Fed Proc, **36**: 579.

Antonini E & Brunori M (1971) The partition constant between two ligands. In: Hemoglobin and myoglobin in their reactions with ligands. Amsterdam, North-Holland Publishing Company, pp 174-175.

Armitage AK, Davies RF, & Turner DM (1976) The effects of carbon monoxide on the development of atherosclerosis in the White Carneau pigeon. Atherosclerosis, **23**: 333-344.

Aronow WS & Cassidy J (1975) Effect of carbon monoxide on maximal treadmill exercise: a study in normal persons. Ann Intern Med, **83**: 496-499.

Aronow WS & Isbell MW (1973) Carbon monoxide effect on exercise-induced angina pectoris. Ann Intern Med, **79**: 392-395.

Aronow WS, Harris CN, Isbell MW, Rokaw SN, & Imparato B (1972) Effect of freeway travel on angina pectoris. Ann Intern Med, **77**: 669-676.

Aronow WS, Ferlinz J, & Glauser F (1977) Effect of carbon monoxide on exercise performance in chronic obstructive pulmonary disease. Am J Med, **63**: 904-908.

Aronow WS, Stemmer EA, Wood B, Zweig S, Tsao K-P, & Raggio L (1978) Carbon monoxide and ventricular fibrillation threshold in dogs with acute myocardial injury. Am Heart J, **95**: 754-756.

Aronow WS, Stemmer EA, & Zweig S (1979) Carbon monoxide and ventricular fibrillation threshold in normal dogs. Arch Environ Health, **34**: 184-186.

Aronow WS, Schlueter WJ, Williams MA, Petratis M, & Sketch MH (1984) Aggravation of exercise performance in patients with anemia by 3% carboxyhemoglobin. Environ Res, **35**: 394-398.

ASTM (1995) Annual book of ASTM standards, Section 11: Water and environmental technology, Volume 11.03: Atmospheric analysis — D3162-94: Standard test method for carbon monoxide in the atmosphere (continuous measurement by nondispersive infra-red spectrometry). Philadelphia, Pennsylvania, American Society for Testing and Materials.

Astrup P, Kjeldsen K, & Wanstrup J (1967) Enhancing influence of carbon monoxide on the development of atheromatosis in cholesterol-fed rabbits. J Ateroscler Res, **7**: 343-354.

Astrup P, Olsen HM, Trolle D, & Kjeldsen K (1972) Effect of moderate carbon-monoxide exposure on fetal development. Lancet, **1972**(7789): 1220-1222.

Atkins EH & Baker EL (1985) Exacerbation of coronary artery disease by occupational carbon monoxide exposure: a report of two fatalities and a review of the literature. Am J Ind Med, **7**: 73-79.

Ayres SM, Criscitiello A, & Giannelli S Jr (1966) Determination of blood carbon monoxide content by gas chromatography. J Appl Physiol, **21**: 1368-1370.

Ayres SM, Mueller HS, Gregory JJ, Giannelli S Jr, & Penny JL (1969) Systemic and myocardial hemodynamic responses to relatively small concentrations of carboxyhemoglobin (COHb). Arch Environ Health, **18**: 699-709.

Ayres SM, Giannelli S Jr, & Mueller H (1970) Myocardial and systemic responses to carboxyhemoglobin. In: Coburn RF ed. Biological effects of carbon monoxide. Ann NY Acad Sci, **174**: 68-293.

Ayres SM, Evans RG, & Buehler ME (1979) The prevalence of carboxyhemoglobinemia in New Yorkers and its effects on the coronary and systemic circulation. Prev Med, **8**: 323-332.

Balraj EK (1984) Atherosclerotic coronary artery disease and "low" levels of carboxyhemoglobin: Report of fatalities and discussion of pathophysiologic mechanisms of death. J Forensic Sci, **29**: 1150-1159.

Bambach GA, Penney DG, & Negendank NG (1991) *In situ* assessment of the rat heart during chronic carbon monoxide exposure using magnetic resonance imaging. J Appl Toxicol, **11**: 43-49.

Barois A, Brosbuis S, & Goulon M (1979) Les intoxications aiguës par l'oxyde de carbon et les gaz de chauffage. Rev Prat, **29**: 1211-1231.

Barret L, Danel V, & Faure J (1985) Carbon monoxide poisoning, a diagnosis frequently overlooked. J Toxicol Clin Toxicol, **23**: 309-313.

Bartholomew GW & Alexander M (1981) Soil as a sink for atmospheric carbon monoxide. Science, **212**: 1389-1391.

Bartle ER & Hall G (1977) Airborne HCl–CO sensing system: Final report. La Jolla, California, Science Applications, Inc. (Report No. SAI-76-717-LJ, available from NTIS, Springfield, Virginia, USA).

Bassan MM (1990) Sudden cardiac death [letter to the editor]. N Engl J Med, **322**: 272.

Bauer K, Conrad R, & Seiler W (1980) Photooxidative production of carbon monoxide by phototrophic microorganisms. Biochim Biophys Acta, **589**: 46-55.

Bay HW, Blurton KF, Lieb HC, & Oswin HG (1972) Electrochemical measurement of carbon monoxide. Am Lab, **4**: 57-58, 60-61.

Bay HW, Blurton KF, Sedlak JM, & Valentine AM (1974) Electrochemical technique for the measurement of carbon monoxide. Anal Chem, **46**: 1837-1839.

Beard RR & Grandstaff N (1970) Carbon monoxide exposure and cerebral function. Ann NY Acad Sci, **174**: 385-395.

Becker LC & Haak ED Jr (1979) Augmentation of myocardial ischemia by low level carbon monoxide exposure in dogs. Arch Environ Health, **34**: 274-279.

Bender W, Goethert M, & Malorny G (1972) Effect of low carbon monoxide concentrations on psychological functions. Staub-Reinhalt Luft, **32**: 54-60.

Benesch W, Migeotte M, & Neven L (1953) Investigations of atmospheric CO at the Jungfraujoch. J Opt Soc Am, **43**: 1119-1123.

Benignus VA (1984) EEG as a cross species indicator of neurotoxicity. Neurobehav Toxicol Teratol, **6**: 473-483.

Benignus VA (1994) Behavioural effects of carbon monoxide: meta analyses and extrapolations. Appl Physiol, **76**: 1310-1316.

Benignus VA, Otto DA, Prah JD, & Benignus G (1977) Lack of effects of carbon monoxide on human vigilance. Percept Mot Skills, **45**: 1007-1014.

Benignus VA, Muller KE, Barton CN, & Prah JD (1987) Effect of low level carbon monoxide on compensatory tracking and event monitoring. Neurotoxicol Teratol, **9**: 227-234.

Benignus VA, Muller KE, Smith MV, Pieper KS, & Prah JD (1990a) Compensatory tracking in humans with elevated carboxyhemoglobin. Neurotoxicol Teratol, **12**: 105-110.

Benignus VA, Muller KE, & Malott CM (1990b) Dose-effects functions for carboxyhemoglobin and behavior. Neurotoxicol Teratol, **12**: 111-118.

Benignus VA, Petrovick MK, Newlin Clapp L, & Prah JD (1992) Carboxyhemoglobin and brain blood flow in humans. Neurotoxicol Teratol, **14**: 285-290.

Benignus VA, Hazucha MJ, Smith MV, & Bromberg PA (1994) Prediction of carboxyhemoglobin formation due to transient exposure to carbon monoxide. J Appl Physiol, **76**(4): 1739-1745.

Benzie TP, Bossart CJ, & Poli AA (1977) US Patent 4,030,887: Carbon monoxide detection apparatus and method — catalysts. Washington, DC, US Patent Office (Mine Safety Appliances Co., assignee).

Bergman I, Coleman JE, & Evans D (1975) A simple gas chromatograph with an electrochemical detector for the measurement of hydrogen and carbon monoxide in the parts per million range, applied to exhaled air. Chromatographia, **8**: 581-583.

Berk PD, Rodkey FL, Blaschke TF, Collison HA, & Waggoner JG (1974) Comparison of plasma bilirubin turnover and carbon monoxide production in man. J Lab Clin Med, **83**: 29-37.

Berk PD, Blaschke TF, Scharschmidt BF, Waggoner JG, & Berlin NI (1976) A new approach to quantitation of the various sources of bilirubin in man. J Lab Clin Med, **87**: 767-780.

Bidwell RGS & Fraser DE (1972) Carbon monoxide uptake and metabolism by leaves. Can J Bot, **50**: 1435-1439.

Bing RJ, Sarma JSM, Weishaar R, Rackl A, & Pawlik G (1980) Biochemical and histological effects of intermittent carbon monoxide exposure in cynomolgus monkeys (*Macaca fascicularis*) in relation to atherosclerosis. J Clin Pharmacol, **20**: 487-499.

Bissette J, Carr G, Koch GG, Adams KF, & Sheps DS (1986) Analysis of (events/time at risk) ratios from two-period crossover studies. In: American Statistical Association 1986 Proceedings of the Biopharmaceutical Section. Washington, DC, American Statistical Association, pp 104-108.

Bissonnette JM & Wickham WK (1977) Placental diffusing capacity for carbon monoxide in unanesthetized guinea pigs. Respir Physiol, **31**: 161-168.

Blackmore DJ (1974) Aircraft accident toxicology: U.K. experience 1967-1972. Aerosp Med, **45**: 987-994.

Blair PS, Fleming PJ, Bensley D, Smith I, Bacon C, Taylor E, Berry J, Golding J, & Tripp J (1996) Smoking and the sudden infant death syndrome: Results from 1993-5 case control study for confidential inquiry into stillbirths and deaths in infancy. Br Med J, **313**(7051): 195-198.

Bondi KR, Very KR, & Schaefer KE (1978) Carboxyhemoglobin levels during a submarine patrol. Undersea Biomed Res, **5**: 17-18.

Brewer GJ, Eaton JW, Weil JV, & Grover RF (1970) Studies of red cell glycolysis and interactions with carbon monoxide, smoking, and altitude. In: Brewer GJ ed. Red cell metabolism and function: Proceedings of the First International Conference on Red Cell Metabolism and Function, Ann Arbor, Michigan, October 1969. New York, Plenum Press, pp 95-114 (Advances in Experimental Medicine and Biology series, Volume 6).

Brewer GJ, Sing CF, Eaton JW, Weil JV, Brewer LF, & Grover RF (1974) Effects on hemoglobin oxygen affinity of smoking in residents of intermediate altitude. J Lab Clin Med, **84**: 191-205.

Breysse PA & Bovee HH (1969) Use of expired air-carbon monoxide for carboxyhemoglobin determinations in evaluating carbon monoxide exposures resulting from the operation of gasoline fork lift trucks in holds of ships. Am Ind Hyg Assoc J, **30**: 477-483.

Brian JE Jr, Heistad DD, & Faraci FM (1994) Effect of carbon monoxide on rabbit cerebral arteries. Stroke, **25**: 639-643.

Bridge DP & Corn M (1972) Contribution to the assessment of exposure of nonsmokers to air pollution from cigarette and cigar smoke in occupied spaces. Environ Res, **5**: 192-209.

Britten JS & Myers RAM (1985) Effects of hyperbaric treatment on carbon monoxide elimination in humans. Undersea Biomed Res, **12**: 431-438.

Brody JA, Brock DB, & Williams TF (1987) Trends in the health of the elderly population. Annu Rev Public Health, **8**: 211-234.

Broome JR, Pearson RR, & Skrine H (1988) Carbon monoxide poisoning: forgotten not gone! Br J Hosp Med, **39**: 298, 300, 302, 304-305.

Brown LJ (1980) A new instrument for the simultaneous measurement of total hemoglobin, % oxyhemoglobin, % carboxyhemoglobin, % methemoglobin, and oxygen content in whole blood. IEEE Trans Biomed Engl, 27: 132-138.

Brown SD & Piantadosi CA (1990) *In vivo* binding of carbon monoxide to cytochrome *c* oxidase in rat brain. J Appl Physiol, 68: 604-610.

Brown S & Piantadosi C (1992) Recovery of energy metabolism in rat brain after carbon monoxide hypoxia. J Clin Invest, 89: 666-672.

Brown SD, Piantadosi CA, Gorman DF, Gilligan JEF, Clayton DG, Neubauer RA, Gottlieb SF, Raphael J-C, Elkharrat D, Jars-Guincestre M-C, Chastang C, Chasles V, Vercken J-B, & Gajdos P (1989) Hyperbaric oxygen for carbon monoxide poisoning [letter to the editor]. Lancet, Oct 28(8670): 1032-1033.

Bruene B & Ullrich V (1987) Inhibition of platelet aggregation by carbon monoxide is mediated by activation of guanylate cyclase. Mol Pharmacol, 32: 497-504.

Brunekreef B, Smit HA, Biersteker K, Boleij JSM, & Lebret E (1982) Indoor carbon monoxide pollution in The Netherlands. Environ Int, 8: 193-196.

Bruner F, Ciccioli P, & Rastelli R (1973) The determination of carbon monoxide in air in the parts per billion range by means of a helium detector. J Chromatogr, 77: 125-129.

Bunnell DE & Horvath SM (1988) Interactive effects of physical work and carbon monoxide on cognitive task performance. Aviat Space Environ Med, 59: 1133-1138.

Bunnell DE & Horvath SM (1989) Interactive effects of heat, physical work, and CO exposure on metabolism and cognitive task performance. Aviat Space Environ Med, 60: 428-432.

Burch DE & Gryvnak DA (1974) Cross-stack measurement of pollutant concentrations using gas-cell correlation spectroscopy. In: Stevens RK & Herget WF ed. Analytical methods applied to air pollution measurements. Ann Arbor, Michigan, Ann Arbor Science Publishers, Inc., pp 193-231.

Burch DE, Gates FJ, & Pembrook JD (1976) Ambient carbon monoxide monitor. Research Triangle Park, North Carolina, US Environmental Protection Agency, Environmental Sciences Research Laboratory (EPA-600/2-76-210).

Bureau MA, Monette J, Shapcott D, Pare C, Mathieu J-L, Lippe J, Blovin D, Berthiaume Y, & Begin R (1982) Carboxyhemoglobin concentration in fetal cord blood and in blood of mothers who smoked during labor. Pediatrics, 69: 371-373.

Burnett RT, Dales RE, Brook JR, Raizenne ME, & Krewski D (1997) Association between ambient carbon monoxide levels and hospitalizations for congestive heart failure in the elderly in 10 Canadian cities. Epidemiology, 8(2): 162-167.

Burns TR, Greenberg SD, Cartwright J, & Jachimczyk JA (1986) Smoke inhalation: an ultrastructural study of reaction to injury in the human alveolar wall. Environ Res, 41: 447-457.

Busey WM (1972) Chronic exposure of albino rats to certain airborne pollutants. Vienna, Virginia, Hazleton Laboratories, Inc. (Unpublished document).

Butler WJ (1991) Commentary on "Links between passive smoking and disease: A best evidence synthesis — A report of the Working Group on Passive Smoking." Clin Invest Med, **14**(5): 484-489.

Butt J, Davies GM, Jones JG, & Sinclair A (1974) Carboxyhaemoglobin levels in blast furnace workers. Ann Occup Hyg, **17**: 57-63.

Cagliostro DE & Islas A (1982) The effects of reduced oxygen and of carbon monoxide on performance in a mouse pole-jump apparatus. J Combust Toxicol, **9**: 187-193.

Calverley PMA, Leggett RJE, & Flenley DC (1981) Carbon monoxide and exercise tolerance in chronic bronchitis and emphysema. Br Med J, **283**: 878-880.

Camici P, Araujo LI, Spinks T, Lammertsma AA, Kaski JC, Shea MJ, Selwyn AP, Jones T, & Maseri A (1986) Increased uptake of ^{18}F-fluorodeoxyglucose in postischemic myocardium of patients with exercise-induced angina. Circulation, **74**: 81-88.

Campbell JA (1934) Growth, fertility, etc. in animals during attempted acclimatization to carbon monoxide. Exp Physiol, **24**: 271-281.

Carratú MR, Renna G, Giustino A, De Salvia MA, & Cuomo V (1993) Changes in peripheral nervous system activity produced in rats by prenatal exposure to carbon monoxide. Arch Toxicol, **67**: 297-301.

Carratú MR, Cagiano R, De Salvia MA, Trabace L, & Cuomo V (1995) Developmental neurotoxicity of carbon monoxide. Arch Toxicol, **17**(suppl): 295-301.

Castelli WP, Garrison RJ, McNamara PM, Feinleib M, & Faden E (1982) The relationship of carbon monoxide in expired air of cigarette smokers to the tar nicotine content of the cigarette they smoke: The Framingham study. In: 55th Scientific Sessions of the American Heart Association, Dallas, Texas, November. Circulation, **66**: 11-315 (American Heart Association Monograph No. 91).

Caughey WS (1970) Carbon monoxide bonding in hemeproteins. In: Coburn RF ed. Biological effects of carbon monoxide. Ann NY Acad Sci, **174**: 148-153.

CEC (1993) Carbon monoxide. Luxembourg, Commission of the European Communities, Scientific Expert Group on Occupational Exposure Limits (SEG/CDO/44A).

Celsing F, Svedenhag J, Pihlstedt P, & Ekblom B (1987) Effects of anemia and stepwise-induced polycythaemia on maximal aerobic power in individuals with high and low haemoglobin concentrations. Acta Physiol Scand, **129**: 47-54.

Chace DH, Goldbaum LR, & Lappas NT (1986) Factors affecting the loss of carbon monoxide from stored blood samples. J Anal Toxicol, **10**: 181-189.

Chan LY & Wu HWY (1993) A study of bus commuter and pedestrian exposure to traffic air pollution in Hong Kong. Environ Int, **19**: 121-132.

Chan WH, Uselman WM, Calvert JG, & Shaw JH (1977) The pressure dependence of the rate constant for the reaction: $OH + CO \rightarrow H + CO_2$. Chem Phys Lett, **45**: 240-244.

Chan C-C, Ozkaynak H, Spengler JD, Sheldon L, Nelson W, & Wallace L (1989) Commuter's exposure to volatile organic compounds, ozone, carbon monoxide, and nitrogen dioxide. Presented at the 82nd Annual Meeting of the Air and Waste Management Association, Anaheim,

California, June 1989. Pittsburgh, Pennsylvania, Air and Waste Management Association (Paper No. 9-34A.4).

Chan C-C, Ozkaynak H, Spengler JD, & Sheldon L (1991) Driver exposure to volatile organic compounds, CO, ozone, and NO_2 under different driving conditions. Environ Sci Technol, 25(5): 964-972.

Chance B, Erecinska M, & Wagner M (1970) Mitochondrial responses to carbon monoxide toxicity. In: Coburn RF ed. Biological effects of carbon monoxide. Ann NY Acad Sci, 174: 193-204.

Chaney LW (1978) Carbon monoxide automobile emissions measured from the interior of a traveling automobile. Science, 199: 1203-1204.

Chaney LW & McClenny WA (1977) Unique ambient carbon monoxide monitor based on gas filter correlation: Performance and application. Environ Sci Technol, 11: 1186-1190.

Chapin FS Jr (1974) Human activity patterns in the city. New York, Wiley-Interscience Publishers.

Chapman RW, Santiago TV, & Edelman NH (1980) Brain hypoxia and control of breathing: neuromechanical control. J Appl Physiol Respir Environ Exercise Physiol, 49: 497-505.

Chen S, Weller MA, & Penney DG (1982) A study of free lung cells from young rats chronically exposed to carbon monoxide from birth. Scanning Electron Microsc, 2: 859-867.

Chevalier RB, Krumholz RA, & Ross JC (1966) Reaction of nonsmokers to carbon monoxide inhalation: Cardiopulmonary responses at rest and during exercise. J Am Med Assoc, 198: 1061-1064.

Choi KD & Oh YK (1975) [A teratological study on the effects of carbon monoxide exposure upon the fetal development of albino rats.] Chungang Uihak, 29: 209-213 (in Korean with summary in English).

Chovin P (1967) Carbon monoxide: analysis of exhaust gas investigations in Paris. Environ Res, 1: 198-216.

Christensen CL, Gliner JA, Horvath SM, & Wagner JA (1977) Effects of three kinds of hypoxias on vigilance performance. Aviat Space Environ Med, 48: 491-496.

Christensen P, Gronlund J, & Carter AM (1986) Placental gas exchange in the guinea-pig: Fetal blood gas tensions following the reduction of maternal oxygen capacity with carbon monoxide. J Dev Physiol, 8: 1-9.

Clark BJ & Coburn RF (1975) Mean myoglobin oxygen tension during exercise at maximal oxygen uptake. J Appl Physiol, 39: 135-144.

Clark RT Jr & Otis AB (1952) Comparative studies on acclimatization of mice to carbon monoxide and to low oxygen. Am J Physiol, 169: 285-294.

Clausen GH, Fanger PO, Cain WS, & Leaderer BP (1985) The influence of aging, particle filtration and humidity on tobacco smoke odor. In: Fanger PO ed. CLIMA 2000: Proceedings of the World Congress on Heating, Ventilating and Air-Conditioning, Copenhagen, August 1985 — Volume 4: Indoor climate. Copenhagen, Denmark, VVS Kongres, pp 345-350.

Clerbaux T, Willems E, & Brasseur L (1984) Evaluation d'une méthode chromatographique de référence pour la détermination du taux sanguin de carboxyhémoglobine. Pathol Biol, **32**: 813-816.

Clubb FJ Jr, Penney DG, Baylerian MS, & Bishop SP (1986) Cardiomegaly due to myocyte hyperplasia in perinatal rats exposed to 200 ppm carbon monoxide. J Mol Cell Cardiol, **18**: 477-486.

Cobb N & Etzel RA (1991) Unintentional carbon monoxide-related deaths in the United States, 1979 through 1988. J Am Med Assoc, **266**: 659-663.

Coburn RF (1970) Enhancement by phenobarbital and diphenylhydantoin of carbon monoxide production in normal man. N Engl J Med, **283**: 512-515.

Coburn RF & Forman HJ (1987) Carbon monoxide toxicity. In: Fishman AP, Farhi LE, Tenney SM, & Geiger SR ed. Handbook of physiology: a critical, comprehensive presentation of physiological knowledge and concepts, Volume IV: Gas Exchange, Section 3: The respiratory system. Bethesda, Maryland, American Physiological Society, pp 439-456.

Coburn RF & Mayers LB (1971) Myoglobin O_2 tension determined from measurements of carboxymyoglobin in skeletal muscle. Am J Physiol, **220**: 66-74.

Coburn RF, Williams WJ, & Forster RE (1964) Effect of erythrocyte destruction on carbon monoxide production in man. J Clin Invest, **43**: 1098-1103.

Coburn RF, Forster RE, & Kane PB (1965) Considerations of the physiological variables that determine the blood carboxyhemoglobin concentration in man. J Clin Invest, **44**: 1899-1910.

Coburn RF, Williams WJ, & Kahn SB (1966) Endogenous carbon monoxide production in patients with hemolytic anemia. J Clin Invest, **45**: 460-468.

Coburn RF, Wallace HW, & Abboud R (1971) Redistribution of body carbon monoxide after hemorrhage. Am J Physiol, **220**: 868-873.

Coburn RF, Ploegmakers F, Gondrie P, & Abboud R (1973) Myocardial myoglobin oxygen tension. Am J Physiol, **224**: 870-876.

Coburn RF, Grubb B, & Aronson RD (1979) Effect of cyanide on oxygen tension-dependent mechanical tension in rabbit aorta. Circ Res, **44**: 368-378.

Cohen SI, Deane M, & Goldsmith JR (1969) Carbon monoxide and survival from myocardial infarction. Arch Environ Health, **19**: 510-517.

Cohn PF (1988) Detection and prognosis of the asymptomatic patient with silent myocardial ischemia. Am J Cardiol, **61**: 4B-6B.

Cole RP (1982) Myoglobin function in exercising skeletal muscle. Science, **216**: 523-525.

Collier CR & Goldsmith JR (1983) Interactions of carbon monoxide and hemoglobin at high altitude. Atmos Environ, **17**: 723-728.

Collins JG (1988) Prevalence of selected chronic conditions, United States, 1983-85. Hyattsville, Maryland, US Department of Health and Human Services, National Center for Health Statistics (Advance data from vital and health statistics No.1/55) (DHHS Publication No. (PHS) 88-1250).

Collison HA, Rodkey FL, & O'Neal JD (1968) Determination of carbon monoxide in blood by gas chromatography. Clin Chem, **14**: 162-171.

Colwill DM & Hickman AJ (1980) Exposure of drivers to carbon monoxide. J Air Pollut Control Assoc, **30**: 1316-1319.

Comroe JH Jr (1974) Physiology of respiration: an introductory text, 2nd ed. Chicago, Illinois, Year Book Medical Publishers, Inc.

Conrad R & Seiler W (1982) Arid soils as a source of atmospheric carbon monoxide. Geophys Res Lett, **9**: 1353-1356.

Cooke RA (1986) Blood lead and carboxyhaemoglobin levels in roadside workers. J Soc Occup Med, **36**: 102-103.

Cooper KR & Alberti RR (1984) Effect of kerosene heater emissions on indoor air quality and pulmonary function. Am Rev Respir Dis, **129**: 629-631.

Cooper RL, Dooley BS, McGrath JJ, McFaul SJ, & Kopetzky MT (1985) Heart weights and electrocardiograms in rats breathing carbon monoxide at altitude. Fed Proc, **44**: 1048.

Copel JA, Bowen F, & Bolognese RJ (1982) Carbon monoxide intoxication in early pregnancy. Obstet Gynecol, **59**(suppl): 26S-28S.

Cortese AD & Spengler JD (1976) Ability of fixed monitoring stations to represent personal carbon monoxide exposure. J Air Pollut Control Assoc, **26**: 1144-1150.

Costantino AG, Park J, & Caplan YH (1986) Carbon monoxide analysis: a comparison of two CO-oximeters and headspace gas chromatography. J Anal Toxicol, **10**: 190-193.

Council on Environmental Quality (1980) Environmental quality: the eleventh annual report of the Council on Environmental Quality. Washington, DC, Council on Environmental Quality.

Cox BD & Whichelow MJ (1985) Carbon monoxide levels in the breath of smokers and nonsmokers: effect of domestic heating systems. J Epidemiol Commun Health, **39**: 75-78.

Crane CR (1985) Are the combined toxicities of CO and CO_2 synergistic? J Fire Sci, **3**: 143-144.

Crocker PJ & Walker JS (1985) Pediatric carbon monoxide toxicity. J Emerg Med, **3**: 443-448.

Czeplak G & Junge C (1974) Studies of interhemispheric exchange in the troposphere by a diffusion model. In: Frenkiel FN & Munn RE ed. Turbulent diffusion in environmental pollution: Proceedings of a symposium, Charlottesville, Virginia, April 1973. New York, London, Academic Press, pp 57-72.

Dagnall RM, Johnson DJ, & West TS (1973) A method for the determination of carbon monoxide, carbon dioxide, nitrous oxide and sulphur dioxide in air by gas chromatography using an emissive helium plasma detector. Spectrosc Lett, **6**: 87-95.

Dahms TE & Horvath SM (1974) Rapid, accurate technique for determination of carbon monoxide in blood. Clin Chem, **20**: 533-537.

Dahms TE, Younis LT, Wiens RD, Zarnegar S, Byers SL, & Chaitman BR (1993) Effects of carbon monoxide exposure in patients with documented cardiac arrhythmias. J Am Coll Cardiol, **21**(2): 442-450.

Daughtrey WC & Norton S (1982) Morphological damage to the premature fetal rat brain after acute carbon monoxide exposure. Exp Neurol, **78**: 26-37.

Davidson SB & Penney DG (1988) Time course of blood volume change with carbon monoxide inhalation and its contribution to the overall cardiovascular response. Arch Toxicol, **61**: 306-313.

Davidson CI, Borrazzo JE, & Hendrickson CT (1987) Pollutant emission factors for gas stoves: a literature survey. Research Triangle Park, North Carolina, US Environmental Protection Agency, Air and Energy Engineering Research Laboratory (EPA-600/9-87-005).

Davies DM & Smith DJ (1980) Electrocardiographic changes in healthy men during continuous low-level carbon monoxide exposure. Environ Res, **21**: 197-206.

Davies RF, Topping DL, & Turner DM (1976) The effect of intermittent carbon monoxide exposure on experimental atherosclerosis in the rabbit. Atherosclerosis, **24**: 527-536.

Davis GL & Gantner GE Jr (1974) Carboxyhemoglobin in volunteer blood donors. J Am Med Assoc, **230**: 996-997.

DeBias DA, Banerjee CM, Birkhead NC, Harrer WV, & Kazal LA (1973) Carbon monoxide inhalation effects following myocardial infarction in monkeys. Arch Environ Health, **27**: 161-167.

DeBias DA, Banerjee CM, Birkhead NC, Greene CH, Scott SD, & Harrer WV (1976) Effects of carbon monoxide inhalation on ventricular fibrillation. Arch Environ Health, **31**: 42-46.

Delivoria-Papadopoulos M, Coburn RF, & Forster RE (1970) Cyclical variation of rate of heme destruction and carbon monoxide production ($\dot{V}_{co}$) in normal women. Physiologist, **13**: 178.

De Luca A, Pierno S, Tricarico D, Carratú MR, Cagiano R, Cuomo V, & Conte Camerino D (1996) Developmental changes of membrane electrical properties of rat skeletal muscle fibers produced by prenatal exposure to carbon monoxide. Environ Toxicol Pharmacol, **2**: 213-221.

DeLucia AJ, Whitaker JH, & Bryant LR (1983) Effects of combined exposure to ozone and carbon monoxide (CO) in humans. In: Lee SD, Mustafa MG, & Mehlman MA ed. International Symposium on the Biomedical Effects of Ozone and Related Photochemical Oxidants, Pinehurst, NC, March 1983. Princeton, New Jersey, Princeton Scientific Publishers, Inc., pp 145-159 (Advances in Modern Environmental Toxicology series, Volume 5).

DeMore WB, Margitan JJ, Molina MJ, Watson RT, Golden DM, Hampson RF, Kurylo MJ, Howard CJ, & Ravishankara AR (1985) Chemical kinetics and photochemical data for use in stratospheric modeling: evaluation number 7. Washington, DC, National Aeronautics and Space Administration (JPL Publication No. 85-37, available from NTIS, Springfield, Virginia, USA).

Dempsey RM, LaConti AB, Nolan ME, Torkildsen RA, Schnakenberg G, & Chilton E (1975) Development of fuel cell CO detection instruments for use in a mine atmosphere. Washington, DC, US Department of the Interior, Bureau of Mines (Report No. 77-76, available from NTIS, Springfield, Virginia, USA).

Dempsey LC, O'Donnell JJ, & Hoff JT (1976) Carbon monoxide retinopathy. Am J Ophthalmol, **82**: 692-693.

Dennis RC & Valeri CR (1980) Measuring percent oxygen saturation of hemoglobin, percent carboxyhemoglobin and methemoglobin, and concentrations of total hemoglobin and oxygen in blood of man, dog, and baboon. Clin Chem, **26**: 1304-1308.

Denniston JC, Pettyjohn FS, Boyter JK, Kelliher JC, Hiott BF, & Piper CF (1978) The interaction of carbon monoxide and altitude on aviator performance: pathophysiology of exposure to carbon monoxide. Fort Rucker, Alabama, US Army Aeromedical Research Laboratory (Report No. 78-7, available from NTIS, Springfield, Virginia, USA).

De Salvia MA, Cagiano R, Carratù MR, Di Giovanni V, Trabace L, & Cuomo V (1995) Irreversible impairment of active avoidance behaviour in rats prenatally exposed to mild concentrations of carbon monoxide. Psychopharmacology, 122: 66-71.

Desclaux P, Soulairac A, & Morlon C (1951) Carbon monoxide intoxication during the course of a pregnancy (fifth month). Arch Fr Pediatr, 8: 316-318.

DFG (German Research Association) (1996) [Maximum allowable workplace concentrations and biological tolerance values for working materials.] Weinheim, Germany, VCH-Verlag, 188 pp (in German).

Dianov-Klokov VI & Yurganov LN (1981) A spectroscopic study of the global space-time distribution of atmospheric CO. Tellus, 33: 262-273.

Dianov-Klokov VI, Fokeyeva YeV, & Yurganov LN (1978) A study of the carbon monoxide content of the atmosphere. Izv Acad Sci USSR Atmos Oceanic Phys, 14: 263-270.

Di Giovanni V, Cagiano R, De Salvia MA, Giustino A, Lacomba C, Renna G, & Cuomo V (1993) Neurobehavioural changes produced in rats by prenatal exposure to carbon monoxide. Brain Res, 616: 126-131.

Doblar DD, Santiago TV, & Edelman NH (1977) Correlation between ventilatory and cerebrovascular responses to inhalation of CO. J Appl Physiol Respir Environ Exercise Physiol, 43: 455-462.

Dobson AJ, Alexander HM, Heller RF, & Lloys DM (1991) Passive smoking and the risk of heart attack or coronary death. Med J Aust, 154(12): 793-797.

Dockery DW & Spengler JD (1981) Personal exposure to respirable particulates and sulfates. J Air Pollut Control Assoc, 31: 153-159.

Dodds RG, Penney DG, & Sutariya BB (1992) Cardiovascular, metabolic and neurologic effects of carbon monoxide and cyanide in the rat. Toxicol Lett, 61: 243-254.

Dolan MC, Haltom TL, Barrows GH, Short CS, & Ferriell KM (1987) Carboxyhemoglobin levels in patients with flu-like symptoms. Ann Emerg Med, 16: 782-786.

Dominick MA & Carson TL (1983) Effects of carbon monoxide exposure on pregnant sows and their fetuses. Am J Vet Res, 44: 35-40.

Donchin E, McCarthy G, & Kutas M (1977) Electro-encephalographic investigations of hemispheric specialization. In: Desmedt JE ed. Progress in clinical neurophysiology, Volume 3: Language and hemispheric specialization in man — cerebral event related potentials. Basel, S. Karger, pp 212-242.

Dor F, LeMoullec Y, & Festy B (1995) Exposure of city residents to carbon monoxide and monocyclic aromatic hydrocarbons during commuting trips in the Paris Metropolitan Area. J Air Waste Manage Assoc, 45: 103-110.

Douglas CG, Haldane JS, & Haldane JBS (1912) The laws of combination of haemoglobin with carbon monoxide and oxygen. J Physiol, **44**: 275-304.

Drägerwerk AG (1994) Dräger tube handbook, 9th ed. Lübeck, Drägerwerk.

Drinkwater BL, Raven PB, Horvath SM, Gliner JA, Ruhling RO, Bolduan NW, & Taguchi S (1974) Air pollution, exercise, and heat stress. Arch Environ Health, **28**: 177-181.

Driscoll JN & Berger AW (1971) Improved chemical methods for sampling and analysis of gaseous pollutants from the combustion of fossil fuels — Volume III: Carbon monoxide. Cincinnati, Ohio, US Environmental Protection Agency, Office of Air Programs (Report No. APTD-1109, available from NTIS, Springfield, Virginia, USA).

Duan N (1982) Models for human exposure to air pollution. Environ Int, **8**: 305-309.

Duncan JS & Gumpert J (1983) A case of blindness following carbon monoxide poisoning, treated with dopamine. J Neurol Neurosurg Psychiatry, **46**: 459.

Dvoryashina EV, Dianov-Klokov VI, & Yurganov YL (1982) Results of carbon monoxide abundance measurements at Zvenigorod, 1970-1982. Moscow, USSR Academy of Sciences.

Dvoryashina EV, Dianov-Klokov VI, & Yurganov YL (1984) Variations of the carbon monoxide content in the atmosphere for 1970-1982. Atmos Ocean Phys, **20**(1): 27-33.

Dyer RS & Annau Z (1978) Carbon monoxide and superior colliculus evoked potentials. In: Otto DA ed. Multidisciplinary perspectives in event-related brain potential research: Proceedings of the Fourth International Congress on Event-Related Slow Potentials of the Brain (EPIC IV), Hendersonville, NC, April 1976. Washington, DC, US Environmental Protection Agency, Office of Research and Development, pp 417-419 (EPA-600/9-77-043).

Ebisuno S, Yasuno M, Yamada Y, Nishino Y, Hori M, Inoue M, & Kamada T (1986) Myocardial infarction after acute carbon monoxide poisoning: case report. Angiology, **37**: 621-624.

Eckardt RE, MacFarland HN, Alarie YCE, & Busey WM (1972) The biologic effect from long-term exposure of primates to carbon monoxide. Arch Environ Health, **25**: 381-387.

Eerens HC, van Velze K, Bezuglaya E, Burn J, & Grønskei K (1995) Air quality in major European cities. Bilthoven, The Netherlands, National Institute of Public Health and Environmental Protection (RIVM Report No. 722401009).

Effeney DJ (1987) Prostacyclin production by the heart: effect of nicotine and carbon monoxide. J Vasc Surg, **5**: 237-247.

Ehhalt DH & Schmidt U (1978) Sources and sinks of atmospheric methane. Pure Appl Geophys, **116**: 452-464.

Einzig S, Nicoloff DM, & Lucus RV Jr (1980) Myocardial perfusion abnormalities in carbon monoxide poisoned dogs. Can J Physiol Pharmacol, **58**: 396-405.

Ekblom B & Huot R (1972) Response to submaximal and maximal exercise at different levels of carboxyhemoglobin. Acta Physiol Scand, **86**: 474-482.

Elkharrat D, Raphael JC, Korach JM, Jars-Guincestre MC, Chastang C, Harboun C, & Gajdos P (1991) Acute carbon monoxide intoxication and hyperbaric oxygen in pregnancy. Intensive Care Med, **17**(5): 289-292.

Ellenhorn MJ & Barceloux DG ed (1988) Medical toxicology diagnosis and treatment of human poisoning. New York, Elsenin, pp 820-829.

Elsasser S, Mall T, Grossenbacher M, Zuber M, Perruchoud AP, & Ritz R (1995) Influence of carbon monoxide (CO) on the early course of acute myocardial infarction. Intensive Care Med, **21**(9): 716-722.

Engen T (1986) The combined effect of carbon monoxide and alcohol on odor sensitivity. Environ Int, **12**: 207-210.

Epstein SE, Quyyumi AA, & Bonow RO (1988) Myocardial ischemia — silent or symptomatic. N Engl J Med, **318**: 1038-1043.

Epstein SE, Quyyumi AA, & Bonow RO (1989) Sudden cardiac death without warning: possible mechanisms and implications for screening asymptomatic populations. N Engl J Med, **321**: 320-324.

Erecinska M & Wilson DF (1982) Regulation of cellular energy metabolism. J Membr Biol, **70**: 1-14.

Estabrook RW, Franklin MR, & Hildebrandt AG (1970) Factors influencing the inhibitory effect of carbon monoxide on cytochrome P-450-catalyzed mixed function oxidation reactions. In: Coburn RF ed. Biological effects of carbon monoxide. Ann NY Acad Sci, **174**: 218-232.

European Environmental Agency (1995) Core inventories air — CORINAIR 90: Summary. Copenhagen, Denmark, European Environmental Agency.

Evans RG, Webb K, Homan S, & Ayres SM (1988) Cross-sectional and longitudinal changes in pulmonary function associated with automobile pollution among bridge and tunnel officers. Am J Ind Med, **14**: 25-36.

Fabian P, Borchers R, Flentje G, Matthews WA, Seiler W, Giehl H, Bunse K, Mueller F, Schmidt U, Volz A, Khedim A, & Johnen FJ (1981) The vertical distribution of stable trace gases at mid-latitudes. J Geophys Res Oceans Atmos, **C86**: 5179-5184.

Faith WL, Renzetti NA, & Rogers LH (1959) Fifth technical progress report. San Marino, California, Air Pollution Foundation.

Farber JP, Schwartz PJ, Vanoli E, Stramba-Badiale M, & De Ferrari GM (1990) Carbon monoxide and lethal arrhythmias. Cambridge, Massachusetts, Health Effects Institute (Research Report No. 36).

Faure J, Arsac P, & Chalandre P (1983) Prospective study in carbon monoxide intoxication by gas water heaters. Hum Toxicol, **2**: 422-425.

Fechter LD (1988) Interactions between noise exposure and chemical asphyxiants: evidence for potentiation of noise induced hearing loss. In: Berglund B, Berglund U, Karlsson J, & Lindvall T ed. Noise as a public health problem, Volume 3: Performance, behaviour, animal, combined agents, and community responses — Proceedings of the 5th International Congress on Noise as a Public Health Problem, Stockholm, August 1988. Stockholm, Sweden, Swedish Council for Building Research, pp 117-122.

Fechter LD & Annau Z (1977) Toxicity of mild prenatal carbon monoxide exposure. Science, **197**: 680-682.

Fechter LD & Annau Z (1980a) Persistent neurotoxic consequences of mild prenatal carbon monoxide exposure. In: Multidisciplinary approach to brain development. Amsterdam, Elsevier/North-Holland Biomedical Press, pp 111-112.

Fechter LD & Annau Z (1980b) Prenatal carbon monoxide exposure alters behavioral development. Neurobehav Toxicol, **2**: 7-11.

Fechter LD, Thakur M, Miller B, Annau Z, & Srivastava U (1980) Effects of prenatal carbon monoxide exposure on cardiac development. Toxicol Appl Pharmacol, **56**: 370-375.

Fechter LD, Mactutus CF, & Storm JE (1986) Carbon monoxide and brain development. Neurotoxicology, **7**: 463-473.

Fechter LD, Karpa MD, Proctor B, Lee AG, & Storm JE (1987) Disruption of neostriatal development in rats following perinatal exposure to mild, but chronic carbon monoxide. Neurotoxicol Teratol, **9**: 277-281.

Fechter LD, Young JS, & Carlisle L (1988) Potentiation of noise induced threshold shifts and hair cell loss by carbon monoxide. Hear Res, **34**: 39-47.

Federal Aviation Administration (1988) FAA air traffic activity. Washington, DC, US Department of Transportation, Office of Management Systems.

Federal Environmental Agency (1997) [Data about the environment, 1994/1995.] Berlin, Erich Schmidt Verlag (in German).

Federal Register (1985) Review of the national ambient air quality standards for carbon monoxide: Final rule. Fed Reg, **50**: 37484-37501.

Fein A, Grossman RF, Jones JG, Hoeffel J, & McKay D (1980) Carbon monoxide effect on alveolar epithelial permeability. Chest, **78**: 726-731.

Feldstein M (1967) Methods for the determination of carbon monoxide. Prog Chem Toxicol, **3**: 99-119.

Fenn WO (1970) The burning of CO in tissues. In: Coburn RF ed. Biological effects of carbon monoxide. Ann NY Acad Sci, **174**: 64-71.

Fenn WO & Cobb DM (1932) The burning of carbon monoxide by heart and skeletal muscle. Am J Physiol, **102**: 393-401.

Fernadenz-Bremauntz AA & Ashmore MR (1995a) Exposure of commuters to carbon monoxide in Mexico City: I. Measurement of in-vehicle concentrations. Atmos Environ, **29**: 525-532.

Fernadenz-Bremauntz AA & Ashmore MR (1995b) Exposure of commuters to carbon monoxide in Mexico City: II. Comparison of in-vehicle and fixed site concentrations. J Expo Anal Environ Epidemiol, **5**: 497-510.

Fink HJ & Klais O (1978) Global distribution of fluorocarbons. Ber Bunsenges Phys Chem, **82**: 1147-1150.

Fisher AB & Dodia C (1981) Lung as a model for evaluation of critical intracellular P_{O_2} and P_{CO}. Am J Physiol, **241**: E47-E50.

Fisher J & Rubin KP (1982) Occult carbon monoxide poisoning. Arch Intern Med, **142**: 1270-1271.

Fisher AB, Hyde RW, Baue AE, Reif JS, & Kelly DF (1969) Effect of carbon monoxide on function and structure of the lung. J Appl Physiol, **26**: 4-12.

Fisher GL, Chrisp CE, & Raabe OG (1979) Physical factors affecting the mutagenicity of fly ash from a coal-fired power plant. Science, **204**: 879-881.

Fitzgerald RS & Traystman RJ (1980) Peripheral chemoreceptors and the cerebral vascular response to hypoxemia. Fed Proc, **39**: 2674-2677.

Flachsbart PG (1989) Effectiveness of priority lanes in reducing travel time and carbon monoxide exposure. Inst Transport Eng J, **59**: 41-45.

Flachsbart PG & Brown DE (1989) Employee exposure to motor vehicle exhaust at a Honolulu shopping center. J Archit Plan Res, **6**: 19-33.

Flachsbart PG & Ott WR (1986) A rapid method for surveying CO concentrations in high-rise buildings. In: Berglund B, Berglund U, Lindvall T, Spengler J, & Sundell J ed. Indoor air quality: Papers from the 3rd International Conference on Indoor Air Quality and Climate, Stockholm, August 1984. Environ Int, **12**: 255-264.

Flachsbart PG, Mack GA, Howes JE, & Rodes CE (1987) Carbon monoxide exposures of Washington commuters. J Air Pollut Control Assoc, **37**: 135-142.

Fodor GG & Winneke G (1972) Effect of low CO concentrations on resistance to monotony and on psychomotor capacity. Staub-Reinhalt Luft, **32**: 46-54.

Forbes WH, Dill DB, De Silva H, & Van deVenter FM (1937) The influence of moderate carbon monoxide poisoning upon the ability to drive automobiles. J Ind Hyg Toxicol, **19**: 598-603.

Forbes WH, Sargent F, & Roughton FJW (1945) The rate of carbon monoxide uptake by normal men. Am J Physiol, **143**: 594-608.

Forster RE (1964) Diffusion of gases. In: Fenn WO & Rahn H ed. Handbook of physiology: a critical, comprehensive presentation of physiological knowledge and concepts, Section 3: Respiration. Washington, DC, American Physiological Society, vol 1, pp 839-872.

Forster RE (1987) Diffusion of gases across the alveolar membrane. In: Fishman AP, Farhi LE, Tenney SM, & Geiger SR ed. Handbook of physiology: a critical, comprehensive presentation of physiological knowledge and concepts, Volume IV: Gas exchange, Section 3: The respiratory system. Bethesda, Maryland, American Physiological Society, pp 71-88.

Forycki Z, Swica P, Krasnowiecki A, Dubicki J, & Panow A (1980) [Electrocardiographic changes during acute poisonings.] Pol Tyg Lek, **35**: 1941-1944 (in Polish).

Foster JR (1981) Arrhythmogenic effects of carbon monoxide in experimental acute myocardial ischemia: lack of slowed conduction and ventricular tachycardia. Am Heart J, **102**: 876-882.

Fozo MS & Penney DG (1993) Dibromomethane and carbon monoxide in the rat: Comparison of the cardiovascular and metabolic effects. J Appl Toxicol, **13**: 147-151.

Fraser PJ, Hyson P, Rasmussen RA, Crawford AJ, & Khalil MAK (1986) Methane, carbon monoxide and methylchloroform in the southern hemisphere. J Atmos Chem, **4**: 3-42.

Frey TM, Crapo RO, Jensen RL, & Elliott CG (1987) Diurnal variation of the diffusing capacity of the lung: is it real? Am Rev Respir Dis, **136**: 1381-1384.

Fristedt B & Akesson B (1971) [Health hazards from automobile exhausts at service facilities of multistory garages.] Hyg Revy, **60**: 112-118 (in Swedish).

Gannon BJ, Fleming BP, Heesch CM, Barron KW, & Diana JN (1988) Effects of acute carbon monoxide exposure on precapillary vessels in the rat cremaster muscle. FASEB J, **2**: A743.

Garland C, Barrett-Connor E, Suarez L, Criqui MH, & Wingard DL (1985) Effects of passive smoking on ischemic heart disease mortality of nonsmokers. Am J Epidemiol, **121**: 645-650.

Garvey DJ & Longo LD (1978) Chronic low level maternal carbon monoxide exposure and fetal growth and development. Biol Reprod, **19**: 8-14.

Gautier H & Bonora M (1983) Ventilatory response of intact cats to carbon monoxide hypoxia. J Appl Physiol Respir Environ Exercise Physiol, **55**: 1064-1071.

Gilli G, Scursatore E, Vanini G, & Vercellotti E (1979) Serum thiocyanate values as a discriminating element in assessment of exposure to carbon monoxide. Minerva Med, **70**: 2803-2810.

Gillis CR, Hole DJ, Hawthorne VM, & Boyle P (1984) The effect of environmental tobacco smoke in two urban communities in the west of Scotland. In: Rylander R, Peterson Y, & Snella M-C ed. ETS — Environmental tobacco smoke: Report from a Workshop on Effects and Exposure Levels, Geneva, Switzerland, March 1983. Eur J Respir Dis, **65**(suppl 133): 121-126.

Ginsberg R & Romano J (1976) Carbon monoxide encephalopathy: Need for appropriate treatment. Am J Psychol, **133**: 317-320.

Girman JR, Apte MG, Traynor GW, Allen JR, & Hollowell CD (1982) Pollutant emission rates from indoor combustion appliances and sidestream cigarette smoke. Environ Int, **8**: 213-221.

Giustino A, Cagiano R, Carratú MR, De Salvia MA, Panaro MA, Jirillo E, & Cuomo V (1993) Immunological changes produced in rats by prenatal exposure to carbon monoxide. Pharmacol Toxicol, **73**: 274-278.

Giustino A, Carratú MR, Siro-Brigiani G, De Salvia MA, Pellegrino NM, Steardo L, Jirillo E, & Cuomo V (1994) Changes in the frequency of splenic immunocompetent cells in rats exposed to carbon monoxide during gestation. Immunopharmacol Immunotoxicol, **16**: 281-292.

Glantz SA & Parmley WW (1991) Passive smoking and heart disease: epidemiology, physiology, and biochemistry. Circulation, **83**: 1-12.

Gliner JA, Raven PB, Horvath SM, Drinkwater BL, & Sutton JC (1975) Man's physiologic response to long-term work during thermal and pollutant stress. J Appl Physiol, **39**: 628-632.

Gliner JA, Horvath SM, & Mihevic PM (1983) Carbon monoxide and human performance in a single and dual task methodology. Aviat Space Environ Med, **54**: 714-717.

Godin G & Shephard RJ (1972) On the course of carbon monoxide uptake and release. Respiration, **29**: 317-329.

Godin G, Wright G, & Shephard RJ (1972) Urban exposure to carbon monoxide. Arch Environ Health, **25**: 305-313.

Goethert M (1972) Factors influencing the CO content of tissues. Staub-Reinhalt Luft, **32**: 15-20.

Goethert M, Lutz F, & Malorny G (1970) Carbon monoxide partial pressure in tissue of different animals. Environ Res, **3**: 303-309.

Goldbaum LR, Chace DH, & Lappas NT (1986) Determination of carbon monoxide in blood by gas chromatography using a thermal conductivity detector. J Forensic Sci, **31**: 133-142.

Goldsmith JR (1970) Contribution of motor vehicle exhaust, industry, and cigarette smoking to community carbon monoxide exposures. In: Coburn RF ed. Biological effects of carbon monoxide. Ann NY Acad Sci, **174**: 122-134.

Goldsmith JR & Landaw SA (1968) Carbon monoxide and human health. Science, **162**: 1352-1358.

Goldstein HW, Bortner MH, Grenda RN, Dick R, & Barringer AR (1976) Correlation interferometric measurement of carbon monoxide and methane from the Canada Centre for Remote Sensing Falcon Fan-Jet Aircraft. Can J Remote Sens, **2**: 30-41.

Gong CH & Wang LJZ (1995) The effects of CO at low concentration on lipid peroxidation and glutathione peroxidation in rats. Environ Health, **12**: 111-114.

Grace TW & Platt FW (1981) Subacute carbon monoxide poisoning: another great imitator. J Am Med Assoc, **246**: 1698-1700.

Greenberg JP, Zimmerman PR, & Chatfield RB (1985) Hydrocarbons and carbon monoxide in African savannah air. Geophys Res Lett, **12**: 113-116.

Groll-Knapp E, Wagner H, Hauck H, & Haider M (1972) Effects of low carbon monoxide concentrations on vigilance and computer-analyzed brain potentials. Staub-Reinhalt Luft, **32**: 64-68.

Groll-Knapp E, Haider M, Hoeller H, Jenkner H, & Stidl HG (1978) Neuro- and psycho-physiological effects of moderate carbon monoxide exposure. In: Otto DA ed. Multidisciplinary perspectives in event-related brain potential research: Proceedings of the 4th International Congress on Event-Related Slow Potentials of the Brain (EPIC IV), Hendersonville, NC, April 1976. Washington, DC, US Environmental Protection Agency, Office of Research and Development, pp 424-430 (EPA-600/9-77-043).

Groll-Knapp E, Haider M, Jenkner H, Liebich H, Neuberger M, & Trimmel M (1982) Moderate carbon monoxide exposure during sleep: neuro- and psychophysiological effects in young and elderly people. Neurobehav Toxicol Teratol, **4**: 709-716.

Groll-Knapp E, Haider M, Kienzl K, Handler A, & Trimmel M (1988) Changes in discrimination learning and brain activity (ERP's) due to combined exposure to NO and CO in rats. Toxicology, **49**: 441-447.

Grösslinger E & Radunsky K ed. (1995) Corinair 90 summary: Summary report No. 1. European Topic Centre on Air Emissions (Report to the European Environmental Agency).

Grut A (1949) Chronic carbon monoxide poisoning: a study in occupational medicine. Copenhagen, Ejnar Munksgaard.

Gryvnak DA & Burch DE (1976a) Monitoring of pollutant gases in aircraft exhausts by gas-filter correlation methods. Presented at the AIAA 14th Aerospace Sciences Meeting, Washington, DC, January 1976. New York, American Institute of Aeronautics and Astronautics (AIAA Paper No. 76-110).

Gryvnak DA & Burch DE (1976b) Monitoring NO and CO in aircraft jet exhaust by a gas-filter correlation technique. Dayton, Ohio, Wright-Patterson Air Force Base, US Air Force, Aeropropulsion Laboratory (Report No. AFAPL-TR-75-101, available from NTIS, Springfield, Virginia, USA).

Guest ADL, Duncan C, & Lawther PJ (1970) Carbon monoxide and phenobarbitone: a comparison of effects on auditory flutter fusion threshold and critical flicker fusion threshold. Ergonomics, **13**: 587-594.

Guillot JG, Weber JP, & Savoie JY (1981) Quantitative determination of carbon monoxide in blood by head-space gas chromatography. J Anal Toxicol, **5**: 264-266.

Gurtner GH, Traystman RJ, & Burns B (1982) Interactions between placental O_2 and CO transfer. J Appl Physiol Respir Environ Exercise Physiol, **52**: 479-487.

Guyatt AR, Holmes MA, & Cumming G (1981) Can carbon monoxide be absorbed from the upper respiratory tract in man? Eur J Respir Dis, **62**: 383-390.

Guyatt AR, Kirkham AJT, Mariner DC, & Cumming G (1988) Is alveolar carbon monoxide an unreliable index of carboxyhaemoglobin changes during smoking in man? Clin Sci, **74**: 29-36.

Guyton AC & Richardson TQ (1961) Effect of hematocrit on venous return. Circ Res, **9**: 157-164.

Gvozdjakova A, Bada V, Sany L, Kucharska J, Kruty F, Bozek P, Trstansky L, & Gvozdjak J (1984) Smoke cardiomyopathy: disturbance of oxidative processes in myocardial mitochondria. Cardiovasc Res, **18**: 229-232.

Hackney JD, Kaufman GA, Lashier H, & Lynn K (1962) Rebreathing estimate of carbon monoxide hemoglobin. Arch Environ Health, **5**: 300-307.

Hackney JD, Linn WS, Mohler JG, Pedersen EE, Breisacher P, & Russo A (1975a) Experimental studies on human health effects of air pollutants: II. Four-hour exposure to ozone alone and in combination with other pollutant gases. Arch Environ Health, **30**: 379-384.

Hackney JD, Linn WS, Law DC, Karuza SK, Greenberg H, Buckley RD, & Pedersen EE (1975b) Experimental studies on human health effects of air pollutants: III. Two-hour exposure to ozone alone and in combination with other pollutant gases. Arch Environ Health, **30**: 385-390.

Hagberg M, Kolmodin-Hedman B, Lindahl R, Nilsson C-A, & Norstrom A (1985) Irritative complaints, carboxyhemoglobin increase and minor ventilatory function changes due to exposure to chain-saw exhaust. Eur J Respir Dis, **66**: 240-247.

Hagen DF & Holiday GW (1964) The effects of engine operating and design variables on exhaust emissions. In: Vehicle emissions (selected SAE papers). New York, Society of Automotive Engineers, Inc., pp 206-223 (Technical Progress Series, Volume 6).

Haggendal E & Norback B (1966) Effect of viscosity on cerebral blood flow. Acta Chir Scand, **364**(suppl): 13-22.

Haider M, Groll-Knapp E, Hoeller H, Neuberger M, & Stidl H (1976) Effects of moderate CO dose on the central nervous system — electrophysiological and behaviour data and clinical relevance. In: Finkel AJ & Duel WC ed. Clinical implications of air pollution research: Air Pollution Medical Research Conference, San Francisco, California, December 1974. Acton, Massachusetts, Publishing Sciences Group, Inc., pp 217-232.

Haldane J (1898) Some improved methods of gas analysis. J Physiol (Lond), **22**: 465-480.

Haldane JS & Priestley JG (1935) Respiration. New Haven, Connecticut, Yale University Press.

Halebian P, Barie P, Robinson N, & Shires GT (1984a) Effects of carbon monoxide on pulmonary fluid accumulation. Curr Surg, **41**: 369-371.

Halebian P, Sicilia C, Hariri R, Inamdar R, & Shires GT (1984b) A safe and reproducible model of carbon monoxide poisoning. Ann NY Acad Sci, **435**: 425-428.

Halperin MH, McFarland RA, Niven JI, & Roughton FJW (1959) The time course of the effects of carbon monoxide on visual thresholds. J Physiol (Lond), **146**: 583-593.

Hameed S & Stewart RW (1979) Latitudinal distribution of the sources of carbon monoxide in the troposphere. Geophys Res Lett, **6**: 841-844.

Hanley DF, Wilson DA, & Traystman RJ (1986) Effect of hypoxia and hypercapnia on neurohypophyseal blood flow. Am J Physiol, **250**: H7-H15.

Hansen ES (1989) Mortality of auto mechanics: a ten-year follow-up. Scand J Work Environ Health, **15**: 3-46.

Hanst PL, Spence JW, & Edney EO (1980) Carbon monoxide production in photooxidation of organic molecules in the air. Atmos Environ, **14**: 1077-1088.

Hare CT & Springer KJ (1973) Exhaust emissions from uncontrolled vehicles and related equipment using internal combustion engines, Part 4: Small air-cooled spark ignition utility engines. Ann Arbor, Michigan, US Environmental Protection Agency, Office of Air and Water Programs (Publication No. APTD-1493, available from NTIS, Springfield, Virginia, USA).

Harrison N (1975) A review of techniques for the measurement of carbon monoxide in the atmosphere. Ann Occup Hyg, **18**: 37-44.

Hartwell TD, Clayton CA, Ritchie RM, Whitmore RW, Zelon HS, Jones SM, & Whitehurst DA (1984) Study of carbon monoxide exposure of residents of Washington, DC and Denver, Colorado. Research Triangle Park, North Carolina, US Environmental Protection Agency, Environmental Monitoring Systems Laboratory (EPA-600/4-84-031).

He Y (1989) [Women's passive smoking and coronary heart disease.] Zhonghua Yufang Yixue Zazhi, **23**: 19-22 (in Chinese).

Heckerling PS, Leikin JB, Maturen A, & Perkins JT (1987) Predictors of occult carbon monoxide poisoning in patients with headache and dizziness. Ann Intern Med, **107**: 174-176.

Heckerling PS, Leikin JB, & Maturen A (1988) Occult carbon monoxide poisoning: validation of a prediction model. Am J Med, **84**: 251-256.

Heidt LE, Krasnec JP, Lueb RA, Pollock WH, Henry BE, & Crutzen PJ (1980) Latitudinal distributions of CO and CH_4 over the Pacific. J Geophys Res Oceans Atmos, **C85**: 7329-7336.

Heimbach DM & Waeckerle JF (1988) Inhalation injuries. Ann Emerg Med, **17**: 1316-1320.

Heinold DW, Sacco AM, & Insley EM (1987) Tollbooth operator exposure on the New Jersey Turnpike. Presented at the 80th Annual Meeting of the Air Pollution Control Association, New

York, June 1987. Pittsburgh, Pennsylvania, Air Pollution Control Association (Paper No. 87-84A.12).

Hellums JD (1977) The resistance to oxygen transport in the capillaries relative to that in the surrounding tissue. Microvasc Res, **13**: 131-136.

Helsing KJ, Sandler DP, Comstock GW, & Chee E (1988) Heart disease mortality in nonsmokers living with smokers. Am J Epidemiol, **127**: 915-922.

Henningfield JE, Stitzer ML, & Griffiths RR (1980) Expired air carbon monoxide accumulation and elimination as a function of number of cigarettes smoked. Addict Behav, **5**: 265-272.

Herget WF, Jahnke JA, Burch DE, & Gryvnak DA (1976) Infrared gas-filter correlation instrument for *in situ* measurement of gaseous pollutant concentrations. Appl Opt, **15**: 1222-1228.

Hernberg S, Karava R, Koskela R-S, & Luoma K (1976) Angina pectoris, ECG findings and blood pressure of foundry workers in relation to carbon monoxide exposure. Scand J Work Environ Health, 2(suppl 1): 54-63.

Hersch P (1964) Galvanic analysis. In: Reilley CN ed. Advances in analytical chemistry and instrumentation. New York, Interscience Publishers, vol 3, pp 183-249.

Hersch PA (1966) US Patent 3,258,411: Method and apparatus for measuring the carbon monoxide content of a gas stream. Washington, DC, US Patent Office (Beckman Instruments, Inc., assignee).

Hexter AC & Goldsmith JR (1971) Carbon monoxide: association of community air pollution with mortality. Science, **172**: 265-266.

Hill EP, Hill JR, Power GG, & Longo LD (1977) Carbon monoxide exchanges between the human fetus and mother: a mathematical model. Am J Physiol, **232**: H311-H323.

Hinderliter AL, Adams KF Jr, Price CJ, Herbst MC, Koch G, & Sheps DS (1989) Effects of low-level carbon monoxide exposure on resting and exercise-induced ventricular arrhythmias in patients with coronary artery disease and no baseline ectopy. Arch Environ Health, **44**: 89-93.

Hirayama T (1984) Lung cancer in Japan: effects of nutrition and passive smoking. In: Mizell M & Correa P ed. Lung cancer — causes and prevention: Proceedings of the International Lung Cancer Update Conference, New Orleans, LA, March 1983. New York, Verlag Chemie International, pp 175-195 (Biomedical Advances in Carcinogenesis, Volume 1).

Hirsch GL, Sue DY, Wasserman K, Robinson TE, & Hansen JE (1985) Immediate effects of cigarette smoking on cardiorespiratory responses to exercise. J Appl Physiol, **58**: 1975-1981.

Hoegg UR (1972) Cigarette smoke in closed spaces. Environ Health Perspect, **2**: 117-128.

Hoell JM, Gregory GL, Carroll MA, McFarland M, Ridley BA, Davis DD, Bradshaw J, Rodgers MO, Torres AL, Sachse GW, Hill GF, Condon EP, Rasmussen RA, Campbell MC, Farmer JC, Sheppard JC, Wang CC, & Davis LI (1984) An intercomparison of carbon monoxide, nitric oxide, and hydroxyl measurement techniques: Overview of results. J Geophys Res, **89**: 11819-11825.

Hole DJ, Gillis CR, Chopra C, & Hawthorne VM (1989) Passive smoking and cardiorespiratory health in a general population in the west of Scotland. Br Med J, **299**: 423-427.

Honigman B, Cromer R, & Kurt TL (1982) Carbon monoxide levels in athletes during exercise in an urban environment. J Air Pollut Control Assoc, **32**: 77-79.

Hoofd L & Kreuzer F (1978) Calculation of the facilitation of O_2 or CO transport by Hb or Mb by means of a new method for solving the carrier-diffusion problem. In: Silver IA, Erecinska M, & Bicher HI ed. Oxygen transport to tissue — III. New York, London, Plenum Press, pp 163-168 (Advances in Experimental Medicine and Biology, Volume 94).

Hoppenbrouwers T, Calub M, Arakawa K, & Hodgman JE (1981) Seasonal relationship of sudden infant death syndrome and environmental pollutants. Am J Epidemiol, **113**: 623-635.

Horvath SM (1975) Influence of carbon monoxide on cardiac dynamics in normal and cardiovascular stressed animals (Final report on grant ARB-2096). Santa Barbara, California, University of California, Institute on Environmental Stress (Available from NTIS, Springfield, Virginia, USA).

Horvath SM (1981) Impact of air quality in exercise performance. Exercise Sport Sci Rev, **9**: 265-296.

Horvath SM & Bedi JF (1989) Alteration in carboxyhemoglobin concentrations during exposure to 9 ppm carbon monoxide for 8 hours at sea level and 2134 m altitude in a hypobaric chamber. J Air Pollut Control Assoc, **39**: 1323-1327.

Horvath SM & Roughton FJW (1942) Improvements in the gasometric estimation of carbon monoxide in blood. J Biol Chem, **144**: 747-755.

Horvath SM, Dahms TE, & O'Hanlon JF (1971) Carbon monoxide and human vigilance: a deleterious effect of present urban concentrations. Arch Environ Health, **23**: 343-347.

Horvath SM, Raven PB, Dahms TE, & Gray DJ (1975) Maximal aerobic capacity at different levels of carboxyhemoglobin. J Appl Physiol, **38**: 300-303.

Horvath SM, Bedi JF, Wagner JA, & Agnew J (1988a) Maximal aerobic capacity at several ambient concentrations of CO at several altitudes. J Appl Physiol, **65**: 2696-2708.

Horvath SM, Agnew JW, Wagner JA, & Bedi JF (1988b) Maximal aerobic capacity at several ambient concentrations of carbon monoxide at several altitudes. Cambridge, Massachusetts, Health Effects Institute (Research Report No. 21).

Hosey AD (1970) Priorities in developing criteria for "breathing air" standards. J Occup Med, **12**: 43-46.

Hosko MJ (1970) The effect of carbon monoxide on the visual evoked response in man and the spontaneous electroencephalogram. Arch Environ Health, **21**: 174-180.

Huch R, Huch A, Tuchschmid P, Zijlstra WG, & Zwart A (1983) Carboxyhemoglobin concentration in fetal cord blood. Pediatrics, **71**: 461-462.

Hudnell HK & Benignus VA (1989) Carbon monoxide exposure and human visual detection thresholds. Neurotoxicol Teratol, **11**: 363-371.

Hugod C (1980) The effect of carbon monoxide exposure on morphology of lungs and pulmonary arteries in rabbits: A light- and electron-microscopic study. Arch Toxicol, **43**: 273-281.

Hugod C (1981) Myocardial morphology in rabbits exposed to various gas-phase constituents of tobacco smoke: An ultrastructural study. Atherosclerosis, **40**: 181-190.

Hugod C, Hawkins LH, Kjeldsen K, Thomsen HK, & Astrup P (1978) Effect of carbon monoxide exposure on aortic and coronary intimal morphology in the rabbit. Atherosclerosis, **30**: 333-342.

Humble C, Croft J, Gerber A, Casper M, Hames CG, & Tyroler HA (1990) Passive smoking and 20-year cardiovascular disease mortality among nonsmoking wives, Evans County, Georgia. Am J Public Health, **80**: 599-601.

Humphreys MP, Knight CV, & Pinnix JC (1986) Residential wood combustion impacts on indoor carbon monoxide and suspended particulates. In: Proceedings of the 1986 EPA/APCA Symposium on Measurement of Toxic Air Pollutants, Raleigh, NC, April 1986. Research Triangle Park, North Carolina, US Environmental Protection Agency, Environmental Monitoring Systems Laboratory, pp 736-747 (EPA-600/9-86-013).

Hutcheon DE, Doorley BM, & Oldewurtel HA (1983) Carbon monoxide inhalation and the vulnerability of the ventricles to electrically induced arrhythmias. Fed Proc, **42**: 1114.

Hutter CD & Blair ME (1996) Carbon monoxide — does fetal exposure cause sudden infant death syndrome? Med Hypotheses, **46**(1): 1-4.

Hwang SJ, Cho SH, & Yun DR (1984) [Postexposure relationship between carboxyhemoglobin in blood and carbon monoxide in expired air.] Seoul J Med, 25: 511-516 (in Korean).

Iglewicz R, Rosenman KD, Iglewicz B, O'Leary K, & Hockemeier R (1984) Elevated levels of carbon monoxide in the patient compartment of ambulances. Am J Public Health, **74**: 511-512.

Ilano A & Raffin T (1990) Management of carbon monoxide poisoning. Chest, **97**: 165-169.

Inman RE, Ingersoll RB, & Levy EA (1971) Soil: a natural sink for carbon monoxide. Science, **172**: 1229-1231.

Insogna S & Warren CA (1984) The effect of carbon monoxide on psychomotor function. In: Mital A ed. Trends in ergonomics/human factors: I. Amsterdam, Oxford, New York, Elsevier Science Publishers, pp 331-337.

International Committee for Standardization in Haematology (1978) Recommendations for reference method for haemoglobinometry in human blood (ICSH standard EP 6/2: 1977) and specifications for international haemoglobincyanide reference preparation (ICSH standard EP 6/3: 1977). J Clin Pathol, **31**: 139-143.

IPCS (1979) Environmental health criteria 13: Carbon monoxide. Geneva, World Health Organization, International Programme on Chemical Safety.

Ischiropoulos H, Beers MF, Ohnishi ST, Fisher D, Garner SE, & Thom SR (1996) Nitric oxide production and perivascular nitration in brain after carbon monoxide poisoning in the rat. J Clin Invest, **97**: 2260-2267.

ISO (1989) International standard ISO 8186: Ambient air — Determination of the mass concentration of carbon monoxide: Gas chromatographic method. Geneva, International Organization for Standardization.

ISO (1996) International standard ISO/DIS 4224: Ambient air — Determination of the mass concentration of carbon monoxide: Nondispersive infrared spectrometric method. Geneva, International Organization for Standardization.

Ito K & Thurston GD (1996) Daily PM10/mortality associations: an investigation of at-risk populations. J Expo Anal Environ Epidemiol, 6: 79-96.

Ito K, Kinney P, & Thurston GD (1995) Variations in PM10 concentrations within two metropolitan areas and their implication for health effects analysis. J Inhal Toxicol, 7(5): 735-745.

Iyanagi T, Suzaki T, & Kobayashi S (1981) Oxidation-reduction states of pyridine nucleotide and cytochrome P-450 during mixed-function oxidation in perfused rat liver. J Biol Chem, 256: 12933-12939.

Jaffe LS (1968) Ambient carbon monoxide and its fate in the atmosphere. J Air Pollut Control Assoc, 18: 534-540.

Jaffe LS (1973) Carbon monoxide in the biosphere: sources, distribution, and concentrations. J Geophys Res, 78: 5293-5305.

James PB (1989) Hyperbaric and normobaric oxygen in acute carbon monoxide poisoning. Lancet, 1989(8666): 799-780.

James WE, Tucker CE, & Grover RF (1979) Cardiac function in goats exposed to carbon monoxide. J Appl Physiol Respir Environ Exercise Physiol, 47: 29-434.

Jantunen MJ, Alm S, Mukala K, Reponen A, & Tuomisto J (1995) Personal CO exposures of preschool children in Helsinki, Finland: The influence of transportation, epidemiology. Abstracts presented at the 1995 ISSE/ISEA Annual Conference, Noordvijkerhout, August-September 1995, p S/70.

Jarvis MJ (1987) Uptake of environmental tobacco smoke. In: O'Neill IK, Brunnemann KD, Dodet B, & Hoffmann D ed. Environmental carcinogens: Methods of analysis and exposure measurement, Volume 9 — Passive smoking. Lyon, International Agency for Research on Cancer, pp 43-58 (IARC Scientific Publications No. 81).

Jobsis FF, Keizer JH, LaManna JC, & Rosenthal M (1977) Reflectance spectrophotometry of cytochrome *aa*$_3$ *in vivo*. J Appl Physiol Respir Environ Exercise Physiol, 43: 858-872.

Johnson T (1984) A study of personal exposure to carbon monoxide in Denver, Colorado. Research Triangle Park, North Carolina, US Environmental Protection Agency, Environmental Monitoring Systems Laboratory (EPA-600/4-84-014).

Johnson T (1987) A study of human activity patterns in Cincinnati, Ohio (Contract No. RP940-06). Palo Alto, California, Electric Power Research Institute.

Johnson T & Paul RA (1983) The NAAQS exposure model (NEM) applied to carbon monoxide. Research Triangle Park, North Carolina, US Environmental Protection Agency, Office of Air Quality Planning and Standards (EPA-450/5-83-003).

Johnson CJ, Moran J, & Pekich R (1975a) Carbon monoxide in school buses. Am J Public Health, 65: 1327-1329.

Johnson CJ, Moran JC, Paine SC, Anderson HW, & Breysse PA (1975b) Abatement of toxic levels of carbon monoxide in Seattle ice-skating rinks. Am J Public Health, 65: 1087-1090.

Johnson T, Capel J, & Wijnberg L (1986) Selected data analyses relating to studies of personal carbon monoxide exposure in Denver and Washington, DC (Contract No. 68-02-3496). Research

Triangle Park, North Carolina, US Environmental Protection Agency, Environmental Monitoring Systems Laboratory.

Johnson T, Capel J, & Wijnberg L (1990) The incorporation of serial correlation into a version of NEM applicable to carbon monoxide (Contract No. 68-02-44069). Research Triangle Park, North Carolina, US Environmental Protection Agency, Atmospheric Research and Exposure Assessment Laboratory.

Jones JG & Sinclair A (1975) Arterial disease amongst blast furnace workers. Ann Occup Hyg, **18**: 15-20.

Jones JG & Walters DH (1962) A study of carboxyhaemoglobin levels in employees at an integrated steelworks. Ann Occup Hyg, **5**: 221-230.

Jones RH, Ellicott MF, Cadigan JB, & Gaensler EA (1958) The relationship between alveolar and blood carbon monoxide concentrations during breathholding. J Lab Clin Med, **51**: 553-564.

Jones RA, Strickland JA, Stunkard JA, & Siegel J (1971) Effects on experimental animals of long-term inhalation exposure to carbon monoxide. Toxicol Appl Pharmacol, **19**: 46-53.

Junge CE (1963) In: von Mieghem J & Hales AL ed. Air chemistry and radioactivity — Volume 4. New York, London, Academic Press (International Geophysics series).

Kahn A, Rutledge RB, Davis GL, Altes JA, Gantner GE, Thornton CA, & Wallace ND (1974) Carboxyhemoglobin sources in the metropolitan St. Louis population. Arch Environ Health, **29**: 127-135.

Kane DM (1985) Investigation of the method to determine carboxyhaemoglobin in blood. Downsview, Ontario, Canada, Department of National Defence, Defence and Civil Institute of Environmental Medicine (Report No. DCIEM-85-R-32, available from NTIS, Springfield, Virginia, USA).

Kannel WB & Abbott RD (1984) Incidence and prognosis of unrecognized myocardial infarction: an update on the Framingham study. N Engl J Med, **311**: 1144-1147.

Kanten WE, Penney DG, Francisco K, & Thill JE (1983) Hemodynamic responses to acute carboxyhemoglobinemia in the rat. Am J Physiol, **244**: H320-H327.

Katafuchi Y, Nishimi T, Yamaguchi Y, Matsuishi T, Kimura Y, Otaki E, & Yamashita Y (1985) Cortical blindness in acute carbon monoxide poisoning. Brain Dev, **7**: 516-519.

Katsumata Y, Aoki M, Oya M, Yada S, & Suzuki O (1980) Liver damage in rats during acute carbon monoxide poisoning. Forensic Sci Int, **16**: 119-123.

Katsumata Y, Sato K, & Yada S (1985) A simple and high-sensitive method for determination of carbon monoxide in blood by gas chromatography. Hanzaigaku Zasshi, **51**: 139-144.

Katzman GM & Penney DG (1993) Electrocardiographic responses to carbon monoxide and cyanide in the conscious rat. Toxicol Lett, **69**: 139-153.

Kaufman S (1966) D. Coenzymes and hydroxylases: ascorbate and dopamine-ß-hydroxylase; tetrahydropteridines and phenylalanine and tyrosine hydroxylases. Pharmacol Rev, **18**: 61-69.

Kaul B, Calabro J, & Hutcheon DE (1974) Effects of carbon monoxide on the vulnerability of the ventricles to drug-induced arrhythmias. J Clin Pharmacol, **14**: 25-31.

Kawachi I, Pearce NE, & Jackson RT (1989) Deaths from lung cancer and ischaemic heart disease due to passive smoking in New Zealand. NZ Med J, **102**(871): 337-340.

Kawachi I, Colditz GA, Speizer FE, Manson JE, Stampfer MJ, Willett WE, & Hennekens CH (1997) A prospective study of passive smoking and coronary heart disease. Circulation, **95**(10): 2374-2379.

Keilin D & Hartree EF (1939) Cytochrome and cytochrome oxidase. Proc R Soc Lond, **B127**: 167-191.

Kelley JS & Sophocleus GJ (1978) Retinal hemorrhages in subacute carbon monoxide poisoning: exposures in homes with blocked furnace flues. J Am Med Assoc, **239**: 1515-1517.

Kety SS & Schmidt CF (1948) The effects of altered arterial tensions of carbon dioxide and oxygen on cerebral blood flow and cerebral oxygen consumption of normal young men. J Clin Invest, **27**: 484-492.

Khalil MAK & Rasmussen RA (1984a) Carbon monoxide in the earth's atmosphere: increasing trend. Science, **224**: 54-56.

Khalil MAK & Rasmussen RA (1984b) The global increase of carbon monoxide. In: Aneja VP ed. Environmental impact of natural emissions: Proceedings of an Air Pollution Control Association Specialty Conference, March 1984. Pittsburgh, Pennsylvania, Air Pollution Control Association, pp 403-414.

Khalil MAK & Rasmussen RA (1984c) The atmospheric lifetime of methylchloroform (CH_3CCl_3). Tellus, **B36**: 317-332.

Khalil MAK & Rasmussen RA (1988) Carbon monoxide in the earth's atmosphere: indications of a global increase. Nature (Lond), **332**: 242-245.

Khalil MAK & Rasmussen RA (1990a) The global cycle of carbon monoxide: trends and mass balance. Chemosphere, **20**: 227-242.

Khalil MAK & Rasmussen RA (1990b) Atmospheric carbon monoxide: latitudinal distribution of sources. Geophys Res Lett, **17**: 1913-1916.

Khalil MAK & Rasmussen RA (1994) Global decrease in atmospheric carbon monoxide concentration. Nature (Lond), **370**: 639-641.

Kidder GW III (1980) Carbon monoxide insensitivity of gastric acid secretion. Am J Physiol, **238**: G197-G202.

Killick EM (1937) The acclimatization of mice to atmospheres containing low concentrations of carbon monoxide. J Physiol (Lond), **91**: 279-292.

Killick EM (1940) Carbon monoxide anoxemia. Physiol Rev, **20**: 313-344.

Killick EM (1948) The nature of the acclimatization occurring during repeated exposure of the human subject to atmospheres containing low concentrations of carbon monoxide. J Physiol (Lond), **107**: 27-44.

Kim YC & Carlson GP (1983) Effect of carbon monoxide inhalation exposure in mice on drug metabolism *in vivo*. Toxicol Lett, **19**: 7-13.

King LA (1983) Effect of ethanol in fatal carbon monoxide poisonings. Hum Toxicol, **2**: 155-157.

King CE, Cain SM, & Chapler CK (1984) Whole body and hindlimb cardiovascular responses of the anesthetized dog during CO hypoxia. Can J Physiol Pharmacol, **62**: 769-774.

King CE, Cain SM, & Chapler CK (1985) The role of aortic chemoreceptors during severe CO hypoxia. Can J Physiol Pharmacol, **63**: 509-514.

King CE, Dodd SL, & Cain SM (1987) O_2 delivery to contracting muscle during hypoxic or CO hypoxia. J Appl Physiol, **63**: 726-732.

Kinker JR, Haffor AS, Stephan M, & Clanton TL (1992) Kinetics of CO uptake and diffusing capacity in transition from rest to steady-state exercise. J Appl Physiol, **72**: 1764-1772.

Kinney PL, Ito K, & Thurston GD (1995) A sensitive analysis of mortality/PM-10 associations in Los Angeles: Proceedings of the 1995 Colloquium on Particulate Air Pollution and Human Mortality and Morbidity. Inhal Toxicol, **7**(1): 59-69.

Kirkham AJT, Guyatt AR, & Cumming G (1988) Alveolar carbon monoxide: a comparison of methods of measurement and a study of the effect of change in body posture. Clin Sci, **74**: 23-28.

Kirkpatrick JN (1987) Occult carbon monoxide poisoning. West J Med, **146**: 52-56.

Kirkpatrick LW & Reeser WK Jr (1976) The air pollution carrying capacities of selected Colorado mountain valley ski communities. J Air Pollut Control Assoc, **26**: 992-994.

Kjeldsen K, Astrup P, & Wanstrup J (1972) Ultrastructural intimal changes in the rabbit aorta after a moderate carbon monoxide exposure. Atherosclerosis, **16**: 67-82.

Klausen K, Andersen C, & Nandrup S (1983) Acute effects of cigarette smoking and inhalation of carbon monoxide during maximal exercise. Eur J Appl Physiol Occup Physiol, **51**: 371-379.

Klees M, Heremans M, & Dougan S (1985) Psychological sequelae to carbon monoxide intoxication in the child. Sci Total Environ, **44**: 165-176.

Klein JP, Forster HV, Stewart RD, & Wu A (1980) Hemoglobin affinity for oxygen during short-term exhaustive exercise. J Appl Physiol Respir Environ Exercise Physiol, **48**: 236-242.

Kleinert HD, Scales JL, & Weiss HR (1980) Effects of carbon monoxide or low oxygen gas mixture inhalation on regional oxygenation, blood flow, and small vessel blood content of the rabbit heart. Pfluegers Arch, **383**: 105-111.

Kleinman MT & Whittenberger JL (1985) Effects of short-term exposure to carbon monoxide in subjects with coronary artery disease. Sacramento, California, California State Air Resources Board (Report No. ARB-R-86/276, available from NTIS, Springfield, Virginia, USA).

Kleinman MT, Davidson DM, Vandagriff RB, Caiozzo VJ, & Whittenberger JL (1989) Effects of short-term exposure to carbon monoxide in subjects with coronary artery disease. Arch Environ Health, **44**: 361-369.

Kloner RA, Allen J, Cox TA, Zheng Y, & Ruiz CE (1991) Stunned left ventricular myocardium after exercise treadmill testing in coronary artery disease. Am J Cardiol, **68**: 329-334.

Klonoff-Cohen HS, Edelstein SL, Lefkowitz ES, Srinivasan IP, Jae-Clun-Chang, & Wiley KJ (1995) The effect of passive smoking and tobacco exposure through breast milk on sudden infant death syndrome. J Am Med Assoc, **273**(10): 795-798.

Knelson JH (1972) United States air quality criteria and ambient standards for carbon monoxide. Staub-Reinhalt Luft, **32**: 183-185.

Knisely JS, Rees DC, Salay JM, Balster RL, & Breen TJ (1987) Effects of intraperitoneal carbon monoxide on fixed-ratio and screen-test performance in the mouse. Neurotoxicol Teratol, **9**: 221-225.

Knisely JS, Rees DC, & Balster RL (1989) Effects of carbon monoxide in combination with behaviorally active drugs on fixed-ratio performance in the mouse. Neurotoxicol Teratol, **11**: 447-452.

Ko BH & Eisenberg RS (1987) Prolonged carboxyhemoglobin clearance in a patient with Waldenstrom's macroglobulinemia. Am J Emerg Med, **5**: 503-508.

Koehler RC, Jones MD Jr, & Traystman RJ (1982) Cerebral circulatory response to carbon monoxide and hypoxic hypoxia in the lamb. Am J Physiol, **243**: H27-H32.

Koehler RC, Traystman RJ, Rosenberg AA, Hudak ML, & Jones MD Jr (1983) Role of O_2-hemoglobin affinity on cerebrovascular response to carbon monoxide hypoxia. Am J Physiol, **245**: H1019-H1023.

Koehler RC, Traystman RJ, Zeger S, Rogers MC, & Jones MD Jr (1984) Comparison of cerebrovascular response to hypoxic and carbon monoxide hypoxia in newborn and adult sheep. J Cereb Blood Flow Metab, **4**: 115-122.

Koehler RC, Traystman RJ, & Jones MD Jr (1985) Regional blood flow and O_2 transport during hypoxic and CO hypoxia in neonatal and adult sheep. Am J Physiol, **248**: H118-H124.

Koike A, Wasserman K, Armon Y, & Weiler Ravell D (1991) The work rate-dependent effect of carbon monoxide on ventilatory control during exercise. Respir Physiol, **85**: 169-183.

Koob GF, Annau Z, Rubin RJ, & Montgomery MR (1974) Effect of hypoxic hypoxia and carbon monoxide on food intake, water intake, and body weight in two strains of rats. Life Sci, **14**: 1511-1520.

Koontz MD & Nagda NL (1987) Survey of factors affecting NO_2 concentrations. In: Seifert B, Esdorn H, Fischer M, Rueden H, & Wegner J ed. Indoor air '87: Proceedings of the 4th International Conference on Indoor Air Quality and Climate, Berlin, August 1987 — Volume 1: Volatile organic compounds, combustion gases, particles and fibres, microbiological agents. Berlin, Federal Republic of Germany, Institute for Water, Soil, and Air Hygiene, pp 430-434.

Koren G, Sharav R, Pastuszak A, Garrettson LK, Hill K, Samson I, Rorem M, King A, & Dolgin JE (1991) A multicenter, prospective study of fetal outcome following accidental carbon monoxide poisoning in pregnancy. Reprod Toxicol, **5**: 397-403.

Korner PI (1965) The role of the arterial chemoreceptors and baroreceptors in the circulatory response to hypoxia of the rabbit. J Physiol (Lond), **180**: 279-303.

Koskela R-S, Hernberg S, Karava R, Jarvinen E, & Nurminen M (1976) A mortality study of foundry workers. Scand J Work Environ Health, **2**(suppl 1): 73-89.

Krall AR & Tolbert NE (1957) A comparison of the light dependent metabolism of carbon monoxide by barley leaves with that of formaldehyde, formate and carbon dioxide. Plant Physiol, **32**: 321-326.

Kreisman NR, Sick TJ, LaManna JC, & Rosenthal M (1981) Local tissue oxygen tension — cytochrome a,a_3 redox relationships in rat cerebral cortex *in vivo*. Brain Res, **218**: 161-174.

Kuller LH & Radford EP (1983) Epidemiological bases for the current ambient carbon monoxide standards. Environ Health Perspect, **52**: 131-139.

Kuller LH, Radford EP, Swift D, Perper JA, & Fisher R (1975) Carbon monoxide and heart attacks. Arch Environ Health, **30**: 477-482.

Kurppa K (1984) Carbon monoxide. In: Aitio A, Riihimaki V, & Vainio H ed. Biological monitoring and surveillance of workers exposed to chemicals. New York, Hemisphere Publishing Corporation, pp 159-164.

Kurt TL, Mogielnicki RP, & Chandler JE (1978) Association of the frequency of acute cardiorespiratory complaints with ambient levels of carbon monoxide. Chest, **74**: 10-14.

Kuska J, Kokot F, & Wnuk R (1980) Acute renal failure after exposure to carbon monoxide. Mater Med Pol (Engl. ed.), **12**: 236-238.

Kustov VV, Belkin VI, Abidin BI, Ostapenko OF, Malkuta AN, & Poddubnaja LT (1972) Aspects of chronic carbon monoxide poisoning in young animals. Gig Tr Prof Zabol, **5**: 50-52.

Kwak HM, Yang YH, & Lee MS (1986) Cytogenetic effects on mouse fetus of acute and chronic transplacental *in vivo* exposure to carbon monoxide: induction of micronuclei and sister chromatid exchanges. Yonsei Med J, **27**: 205-212.

Lagemann RT, Nielsen AH, & Dickey FP (1947) The infra-red spectrum and molecular constants of $C^{12}O^{16}$ and $C^{13}O^{16}$. Phys Rev, **72**: 284-289.

Lahiri S & Delaney RG (1976) Effect of carbon monoxide on carotid chemoreceptor activity and ventilation. In: Paintal AS ed. Morphology and mechanisms of chemoreceptors: Proceedings of an International Satellite Symposium, Srinagar, India, October 1974. Delhi, India, University of Delhi, Vallabhbhai Patel Chest Institute, pp 340-344.

Lambert WE, Colome SD, & Wojciechowski SL (1988) Application of end-expired breath sampling to estimate carboxyhemoglobin levels in community air pollution exposure assessments. Atmos Environ, **22**: 2171-2181.

Lamontagne RA, Swinnerton JW, & Linnenbom VJ (1971) Nonequilibrium of carbon monoxide and methane at the air-sea interface. J Geophys Res, **76**: 5117-5121.

Landaw SA (1973) The effects of cigarette smoking on total body burden and excretion rates of carbon monoxide. J Occup Med, **15**: 231-235.

Landaw SA, Callahan EW Jr, & Schmid R (1970) Catabolism of heme *in vivo*: comparison of the simultaneous production of bilirubin and carbon monoxide. J Clin Invest, **49**: 914-925.

Lanston P, Gorman D, Runciman W, & Upton R (1996) The effect of carbon monoxide on oxygen metabolism in the brains of awake sheep. Toxicology, **114**: 223-232.

Laties VG & Merigan WH (1979) Behavioral effects of carbon monoxide on animals and man. Annu Rev Pharmacol Toxicol, **19**: 357-392.

La Vecchia C, D'Avanzo B, Grazia-Franzosi M, & Tognoni G (1993) Passive smoking and the risk of acute myocardial infarction. Lancet, **341**(8843): 505-506.

Leaderer BP, Cain WS, Isseroff R, & Berglund LG (1984) Ventilation requirements in buildings — II. Particulate matter and carbon monoxide from cigarette smoking. Atmos Environ, **18**: 99-106.

Leaf DA & Kleinman MT (1996) Urban ectopy in the mountains: carbon monoxide exposure at high altitude. Arch Environ Health, **51**: 283-290.

Lebowitz MD (1984) The effects of environmental tobacco smoke exposure and gas stoves on daily peak flow rates in asthmatic and non-asthmatic families. In: Rylander R, Peterson Y, & Snella M-C ed. ETS — Environmental tobacco smoke: Report from a Workshop on Effects and Exposure Levels, Geneva, Switzerland, March 1983. Eur J Respir Dis, **65**(suppl 133): 90-97.

Lebowitz MD, Holberg CJ, O'Rourke MK, Corman G, & Dodge R (1983a) Gas stove usage, CO and TSP, and respiratory effects. Research Triangle Park, North Carolina, US Environmental Protection Agency, Health Effects Research Laboratory (EPA-600/D-83-107).

Lebowitz MD, Holberg CJ, & Dodge RR (1983b) Respiratory effects on populations from low-level exposures to ozone. Presented at the 76th Annual Meeting of the Air Pollution Control Association, Atlanta, GA, June 1983. Pittsburgh, Pennsylvania, Air Pollution Control Association (Paper No. 83-12.5).

Lebowitz MD, Corman G, O'Rourke MK, & Holberg CJ (1984) Indoor-outdoor air pollution, allergen and meteorological monitoring in an arid southwest area. J Air Pollut Control Assoc, **34**: 1035-1038.

Lebowitz MD, Holberg CJ, Boyer B, & Hayes C (1985) Respiratory symptoms and peak flow associated with indoor and outdoor air pollutants in the southwest. J Air Pollut Control Assoc, **35**: 1154-1158.

Lebowitz MD, Collins L, & Holberg CJ (1987) Time series analyses of respiratory responses to indoor and outdoor environmental phenomena. Environ Res, **43**: 332-341.

Lebret E (1985) Air pollution in Dutch homes: an exploratory study in environmental epidemiology. Wageningen, The Netherlands, Department of Air Pollution & Department of Environmental and Tropical Health (Reports Nos R-138 and 1985-221).

Lee PN, Chamberlain J, & Alderson MR (1986) Relationship of passive smoking to risk of lung cancer and other smoking-associated diseases. Br J Cancer, **54**: 97-105.

Lee K, Yanagisawa Y, & Spengler JD (1994) Carbon monoxide and nitrogen dioxide exposure in indoor ice skating rinks. J Sports Sci, **12**: 279-283.

Lehnebach A, Kuhn C, & Pankow D (1995) Dichloromethane as an inhibitor of cytochrome *c* oxidase in different tissues of rats. Arch Toxicol, **69**(3): 180-184.

Leichnitz K (1993) Determination of the time-weighted average concentration of carbon monoxide in air using a long term detector tube. In: Seifert B, van de Wiel HJ, Dodet B, & O'Neil IK ed. Environmental carcinogens: Methods of analysis and exposure measurement — Volume 12: Indoor air. Lyon, International Agency for Research on Cancer, pp 346-352 (IARC Scientific Publications No. 109).

Leichter J (1993) Fetal growth retardation due to exposure of pregnant rats to carbon monoxide. Biochem Arch, **8**: 267-272.

Leithe W (1971) The analysis of air pollutants. Ann Arbor, Michigan, Ann Arbor Science Publishers, Inc.

Leniger-Follert E, Luebbers DW, & Wrabetz W (1975) Regulation of local tissue P_{O_2} of the brain cortex at different arterial O_2 pressures. Pfluegers Arch, **359**: 81-95.

Lévesque B, Dewailly E, Lavoie R, Prud'Homme D, & Allaire S (1990) Carbon monoxide in indoor ice skating rinks: evaluation of absorption by adult hockey players. Am J Public Health, **80**: 594-598.

Levin BC, Paabo M, Gurman JL, Harris SE, & Braun E (1987a) Toxicological interactions between carbon monoxide and carbon dioxide. Toxicology, **47**: 135-164.

Levin BC, Paabo M, Gurman JL, & Harris SE (1987b) Effects of exposure to single or multiple combinations of the predominant toxic gases and low oxygen atmospheres produced in fires. Fundam Appl Toxicol, **9**: 236-250.

Levin BC, Gurman JL, Paabo M, Baier L, & Holt T (1988a) Toxicological effects of different time exposures to the fire gases: Carbon monoxide or hydrogen cyanide or to carbon monoxide combined with hydrogen cyanide or carbon dioxide. In: Jason NH & Houston BA ed. Ninth Joint Panel Meeting of the UJNR Panel on Fire Research and Safety, Norwood, MA, May 1987. Gaithersburg, Maryland, US Department of Commerce, National Bureau of Standards, pp 368-383 (Report No. NBSIR 88-3753).

Levin BC, Paabo M, Gurman JL, Clark HM, & Yoklavich MF (1988b) Further studies of the toxicological effects of different time exposures to the individual and combined fire gases — carbon monoxide, hydrogen cyanide, carbon dioxide and reduced oxygen. In: Polyurethanes 88: Proceedings of the 31st Annual Technical/Marketing Conference of the Society of Plastics Industry, Philadelphia, October 1988. Lancaster, Pennsylvania, Technomic Publishing Co., pp 249-252.

Le Vois ME & Layard MW (1995) Publication bias in the environmental tobacco smoke/coronary heart disease epidemiologic literature. Regul Toxicol Pharmacol, **21**(1): 184-191.

Lilienthal JL Jr & Fugitt CH (1946) The effect of low concentrations of carboxyhemoglobin on the "altitude tolerance" of man. Am J Physiol, **145**: 359-364.

Linnenbom VJ, Swinnerton JW, & Lamontagne RA (1973) The ocean as a source for atmospheric carbon monoxide. J Geophys Res, **78**: 5333-5340.

Liss PS & Slater PG (1974) Flux of gases across the air-sea interface. Nature (Lond), **247**: 181-184.

Locke JL & Herzberg L (1953) The absorption due to carbon monoxide in the infrared solar spectrum. Can J Phys, **31**: 504-516.

Lodge JP ed. (1989) Methods of air sampling and analysis, 3rd ed. Chelsea, Michigan, Lewis Publishers.

Logan JA, Prather MJ, Wofsy SC, & McElroy MB (1981) Tropospheric chemistry: a global perspective. J Geophys Res Oceans Atmos, **C86**: 7210-7254.

Longo LD (1970) Carbon monoxide in the pregnant mother and fetus and its exchange across the placenta. In: Coburn RF ed. Biological effects of carbon monoxide. Ann NY Acad Sci, **174**: 313-341.

Longo LD (1976) Carbon monoxide: Effects on oxygenation of the fetus *in utero*. Science, **194**: 523-525.

Longo LD (1977) The biological effects of carbon monoxide on the pregnant woman, fetus, and newborn infant. Am J Obstet Gynecol, **129**: 69-103.

Longo LD & Ching KS (1977) Placental diffusing capacity for carbon monoxide and oxygen in unanesthetized sheep. J Appl Physiol Respir Environ Exercise Physiol, **43**: 885-893.

Longo LD & Hill EP (1977) Carbon monoxide uptake and elimination in fetal and maternal sheep. Am J Physiol, **232**: H324-H330.

Ludbrook GL, Helps SC, Gorman DF, Reilly PL, & North JB (1992) Electrocardiographic responses to carbon monoxide and cyanide in the conscious rat. Toxicology, **75**: 71-80.

Luft KF (1962) [The "UNOR," a new gas analytical device for mining.] Glückauf, **98**: 493-495 (in German).

Lumio JS (1948) Hearing deficiencies caused by carbon monoxide (generator gas). Helsinki, Finland, Oto-Laryngological Clinic of the University.

Luomanmaki K & Coburn RF (1969) Effects of metabolism and distribution of carbon monoxide on blood and body stores. Am J Physiol, **217**: 354-363.

Luria SM & McKay CL (1979) Effects of low levels of carbon monoxide on visions of smokers and nonsmokers. Arch Environ Health, **34**: 38-44.

Lutz LJ (1983) Health effects of air pollution measured by outpatient visits. J Fam Pract, **16**: 307-313.

Lynch AM & Bruce NW (1989) Placental growth in rats exposed to carbon monoxide at selected stages of pregnancy. Biol Neonate, **56**: 151-157.

Lynch SR & Moede AL (1972) Variation in the rate of endogenous carbon monoxide production in normal human beings. J Lab Clin Med, **79**: 85-95.

Maas AHJ, Hamelink ML, & de Leeuw RJM (1970) An evaluation of the spectrophotometric determination of HbO_2, HbCO and Hb in blood with the CO-Oximeter IL 182. Clin Chim Acta, **29**: 303-309.

McCarthy SM, Yarmac RF, & Yocom JE (1987) Indoor nitrogen dioxide exposure: the contribution from unvented gas space heaters. In: Seifert B, Esdorn H, Fischer M, Rueden H, & Wegner J ed. Indoor air '87: Proceedings of the 4th International Conference on Indoor Air Quality and Climate, Berlin, August 1987 — Volume 1: Volatile organic compounds, combustion gases, particles and fibres, microbiological agents. Berlin, Federal Republic of Germany, Institute for Water, Soil, and Air Hygiene, pp 478-482.

McCartney ML (1990) Sensitivity analysis applied to Coburn-Forster-Kane models of carboxy-hemoglobin formation. Am Ind Hyg Assoc J, **51**: 169-177.

McClean PA, Duguid NJ, Griffin PM, Newth CJL, & Zamel N (1981) Changes in exhaled pulmonary diffusing capacity at rest and exercise in individuals with impaired positional diffusion. Clin Respir Physiol, 17: 179-186.

McCredie RM & Jose AD (1967) Analysis of blood carbon monoxide and oxygen by gas chromatography. J Appl Physiol, 22: 863-866.

McDonagh PF, Reynolds JM, & McGrath JJ (1986) Chronic altitude plus carbon monoxide exposure causes left ventricular hypertrophy but an attenuation of coronary capillarity. Fed Proc, 45: 883.

McFarland RA (1970) The effects of exposure to small quantities of carbon monoxide on vision. In: Coburn RF ed. Biological effects of carbon monoxide. Ann NY Acad Sci, 174: 301-312.

McFarland RA (1973) Low level exposure to carbon monoxide and driving performance. Arch Environ Health, 27: 355-359.

McFarland RA, Roughton FJW, Halperin MH, & Niven JI (1944) The effects of carbon monoxide and altitude on visual thresholds. J Aviat Med, 15: 381-394.

McFaul SJ & McGrath JJ (1987) Studies on the mechanism of carbon monoxide-induced vasodilation in the isolated perfused rat heart. Toxicol Appl Pharmacol, 87: 464-473.

McGrath JJ (1982) Physiological effects of carbon monoxide. In: McGrath JJ & Barnes CD ed. Air pollution: Physiological effects. New York, London, Academic Press, pp 147-181.

McGrath JJ (1988) Body and organ weights of rats exposed to carbon monoxide at high altitude. J Toxicol Environ Health, 23: 303-310.

McGrath JJ (1989) Cardiovascular effects of chronic carbon monoxide and high-altitude exposure. Cambridge, Massachusetts, Health Effects Institute (Research Report No. 27).

McMeekin JD & Finegan BA (1987) Reversible myocardial dysfunction following carbon monoxide poisoning. Can J Cardiol, 3: 118-121.

MacMillan V (1975) Regional cerebral blood flow of the rat in acute carbon monoxide intoxication. Can J Physiol Pharmacol, 53: 644-650.

McMillan DE & Miller AT Jr (1974) Interactions between carbon monoxide and d-amphetamine or pentobarbital on schedule-controlled behavior. Environ Res, 8: 53-63.

McNay LM (1971) Coal refuse fires, an environmental hazard. Washington, DC, US Department of the Interior, Bureau of Mines (Information Circular No. 8515).

Mactutus CF & Fechter LD (1984) Prenatal exposure to carbon monoxide: learning and memory deficits. Science, 223: 409-411.

Mactutus CF & Fechter LD (1985) Moderate prenatal carbon monoxide exposure produces persistent, and apparently permanent, memory deficits in rats. Teratology, 31: 1-12.

Madany IM (1992) Carboxyhemoglobin levels in blood donors in Bahrain. Sci Total Environ, 116: 53-58.

Madsen H & Dyerberg J (1984) Cigarette smoking and its effects on the platelet-vessel wall interaction. Scand J Clin Lab Invest, 44: 203-206.

Maehara K, Riley M, Galassetti P, Barstow TJ, & Wasserman K (1997) Effect of hypoxia and carbon monoxide on muscle oxygenation during exercise. Am J Respir Crit Care Med, 155: 229-235.

Mage D (1991) A comparison of the direct and indirect methods of human exposure. New Horizons in Biological Dosimetry, 443-454.

Mage D, Ozolins G, Peterson P, Webster A, Orthofer R, Vandeweerd V, & Gwynne M (1996) Urban air pollution in megacities of the world. Atmos Environ, 30(5): 681-686.

Malinow MR, McLaughlin P, Dhindsa DS, Metcalfe J, Ochsner AJ III, Hill J, & McNulty WP (1976) Failure of carbon monoxide to induce myocardial infarction in cholesterol-fed cynomolgus monkeys (*Macaca fascicularis*). Cardiovasc Res, 10: 101-108.

Mall Th, Grossenbacher M, Perruchoud AP, & Ritz R (1985) Influence of moderately elevated levels of carboxyhemoglobin on the course of acute ischemic heart disease. Respiration, 48: 237-244.

Mansouri A & Perry CA (1982) Alteration of platelet aggregation by cigarette smoke and carbon monoxide. Thromb Haemost, 48: 286-288.

Marshall M & Hess H (1981) [Acute effects of low carbon monoxide concentrations on blood rheology, platelet function, and the arterial wall in the minipig.] Res Exp Med, 178: 201-210 (in German).

Martin TR & Bracken MB (1986) Association of low birth weight with passive smoke exposure in pregnancy. Am J Epidemiol, 124: 633-642.

Martynjuk VC & Dacenko II (1973) [Aminotransferase activity in chronic carbon monoxide poisoning.] Gig Naselennykh Mest, 12: 53-56 (in Russian).

Mason GR, Uszler JM, Effros RM, & Reid E (1983) Rapid reversible alterations of pulmonary epithelial permeability induced by smoking. Chest, 83: 6-11.

Mathieu D & Wattel F ed. (1990) Oxygenotherapie hyperbare et réanimation. Paris, Editions Masson, pp 129-143.

Mathieu D, Nolf M, Durocher A, Saulnier F, Frimat P, Furon D, & Wattel F (1985) Acute carbon monoxide poisoning risk of late sequelae and treatment by hyperbaric oxygen. J Toxicol Clin Toxicol, 23: 315-324.

Mathieu D, Mathieu-Nolf M, & Wattel F (1996a) Intoxication par le monoxyde de carbone: Aspects actuels. Bull Acad Natl Med, 180: 965-973.

Mathieu D, Wattel F, Neviere R, & Mathieu-Nolf M (1996b) In: Oriani G, Marroni A, & Wattel F ed. Handbook on hyperbaric medicine. Berlin, Heidelberg, New York, Springer Verlag, pp 281-296.

Medical College of Wisconsin (1974) Exposure of humans to carbon monoxide combined with ingestion of ethyl alcohol and the comparison of human performance when exposed for varying periods of time to carbon monoxide. Milwaukee, Wisconsin, Medical College of Wisconsin (Report No. MCOW-ENVM-CO-74-2, available from NTIS, Springfield, Virginia, USA).

Melinyshyn MJ, Cain SM, Villeneuve SM, & Chapler CK (1988) Circulatory and metabolic responses to carbon monoxide hypoxia during ß-adrenergic blockade. Am J Physiol, 255: H77-H84.

Mellor AM (1972) Current kinetic modeling techniques for continuous flow combustors. In: Cornelius W & Agnew WG ed. Emissions from continuous combustion systems: Proceedings of a symposium, Warren, Michigan, September 1971. New York, Plenum Press, pp 23-53.

Meredith T & Vale A (1988) Carbon monoxide poisoning. Br Med J, **296**: 77-79.

Meyer B (1983) Indoor air quality. Reading, Massachusetts, Addison-Wesley Publishing Company, Inc.

Michelson W & Reed P (1975) The time budget. In: Michelson W ed. Behavioral research methods in environmental design. Stroudsburg, Pennsylvania, Dowden, Hutchinson, & Ross, Inc., pp 180-234 (Community Development series).

Migeotte MV (1949) The fundamental band of carbon monoxide at 4.7 $\mu m/m^3$ in the solar spectrum. Phys Rev, **75**: 1108-1109.

Migeotte M & Neven L (1952) Récents progrès dans l'observation du spectre infra-rouge du soleil à la station scientifique du Jungfraujoch (Suisse). Mem Soc R Sci Liege, **12**: 165-178.

Mihevic PM, Gliner JA, & Horvath SM (1983) Carbon monoxide exposure and information processing during perceptual-motor performance. Int Arch Occup Environ Health, **51**: 355-363.

Miller AT & Wood JJ (1974) Effects of acute carbon monoxide exposure on the energy metabolism of rat brain and liver. Environ Res, **8**: 107-111.

Mills AK, Skornik WA, Valles LM, O'Rourke JJ, Hennessey RM, & Verrier RL (1987) Effects of carbon monoxide on cardiac electrical properties during acute coronary occlusion. Fed Proc, **46**: 336.

Minor M & Seidler D (1986) Myocardial infarction following carbon monoxide poisoning. West Va Med J, **82**: 25-28.

Minty BD & Royston D (1985) Cigarette smoke induced changes in rat pulmonary clearance of [99m]TcDTPA: a comparison of particulate and gas phases. Am Rev Respir Dis, **132**: 1170-1173.

Minty BD, Jordan C, & Jones JG (1981) Rapid improvement in abnormal pulmonary epithelial permeability after stopping cigarettes. Br Med J, **282**: 1183-1186.

Miranda JM, Konopinski VJ, & Larsen RI (1967) Carbon monoxide control in a high highway tunnel. Arch Environ Health, **15**(1): 16-25.

Mitchell DS, Packham SC, & Fitzgerald WE (1978) Effects of ethanol and carbon monoxide on two measures of behavioral incapacitation of rats. Proc West Pharmacol Soc, **21**: 427-431.

Mitchell EA, Ford RP, Stewart AW, Taylor BJ, Becroft DM, Thompson JM, Scragg R, Hassall IB, Barry DM, & Allen EM (1993) Smoking and the sudden infant death syndrome. Pediatrics, **91**(5): 893-896.

Miyagawa M, Honma T, Sato M, & Hasegawa H (1995) Acute effects of inhalation exposure to carbon monoxide on schedule-controlled operant behavior and blood carboxyhemoglobin levels in rats. Ind Health, **33**: 119-129.

Miyahara S & Takahashi H (1971) Biological CO evolution: carbon monoxide evolution during auto- and enzymatic oxidation of phenols. J Biochem (Tokyo), **69**: 231-233.

Mochizuki M, Maruo T, Masuko K, & Ohtsu T (1984) Effects of smoking on fetoplacental-maternal system during pregnancy. Am J Obstet Gynecol, **149**: 413-420.

Moffatt S (1986) Backdrafting woes. Prog Builder, **December**: 25-36.

Montgomery MR & Rubin RJ (1971) The effect of carbon monoxide inhalation on *in vivo* drug metabolism in the rat. J Pharmacol Exp Ther, **179**: 465-473.

Montgomery MR & Rubin RJ (1973) Oxygenation during inhibition of drug metabolism by carbon monoxide or hypoxic hypoxia. J Appl Physiol, **35**: 505-509.

Moore LG, Rounds SS, Jahnigen D, Grover RF, & Reeves JT (1982) Infant birth weight is related to maternal arterial oxygenation at high altitude. J Appl Physiol Respir Environ Exercise Physiol, **52**: 695-699.

Mordelet-Dambrine M & Stupfel M (1979) Comparison in guinea-pigs and in rats of the effects of vagotomy and of atropine on respiratory resistance modifications induced by an acute carbon monoxide or nitrogen hypoxia. Comp Biochem Physiol Comp Physiol, **A63**: 555-559.

Mordelet-Dambrine M, Stupfel M, & Duriez M (1978) Comparison of tracheal pressure and circulatory modifications induced in guinea pigs and in rats by carbon monoxide inhalation. Comp Biochem Physiol Comp Physiol, **A59**: 65-68.

Moriske HJ, Drews M, Ebert G, Schneller G, Schöndube M, & Konieczny L (1996) Indoor air pollution by different heating system: coal burning, open fireplace and central heating. Toxicol Lett, **88**: 349-354.

Morris GL, Curtis SE, & Simon J (1985a) Perinatal piglets under sublethal concentrations of atmospheric carbon monoxide. J Anim Sci, **61**: 1070-1079.

Morris GL, Curtis SE, & Widowski TM (1985b) Weanling pigs under sublethal concentrations of atmospheric carbon monoxide. J Anim Sci, **61**: 1080-1087.

Morris RD, Naumova EN, & Munasinghe RL (1995) Ambient air pollution and hospitalization for congestive heart failure among elderly people in seven large US cities. Am J Public Health, **85**(10): 1361-1365.

Moschandreas DJ & Zabransky J Jr (1982) Spatial variation of carbon monoxide and oxides of nitrogen concentrations inside residences. Environ Int, **8**: 177-183.

Moschandreas DJ, Relwani SM, O'Neill HJ, Cole JT, Elkins RH, & Macriss RA (1985) Characterization of emission rates from indoor combustion sources. Chicago, Illinois, Gas Research Institute (Report No. GRI 85/0075, available from NTIS, Springfield, Virginia, USA).

Muller GL & Graham S (1995) Intrauterine death of the fetus due to accidental carbon monoxide poisoning. N Engl J Med, **252**: 1075-1078.

Mullin LS & Krivanek ND (1982) Comparison of unconditioned reflex and conditioned avoidance tests in rats exposed by inhalation to carbon monoxide, 1,1,1-trichloroethane, toluene or ethanol. Neurotoxicology, **3**: 126-137.

Murphy SD (1964) A review of effects on animals of exposure to auto exhaust and some of its components. J Air Pollut Control Assoc, **14**: 303-308.

Murray FJ, Schwetz BA, Crawford AA, Henck JW, & Staples RE (1978) Teratogenic potential of sulfur dioxide and carbon monoxide in mice and rabbits. In: Mahlum DD, Sikov MR, Hackett PL, & Andrew FD ed. Developmental toxicology of energy-related pollutants: Proceedings of the Seventeenth Annual Hanford Biology Symposium, Richland, WA, October 1977. Oak Ridge, Tennessee, US Department of Energy, Technical Information Center, pp 469-478 (Available from NTIS, Springfield, Virginia, USA).

Muscat JE & Wynder EL (1995) Exposure to environmental tobacco smoke and the risk of heart attack. Int J Epidemiol, **24**(4): 715-719.

Musselman NP, Groff WA, Yevich PP, Wilinski FT, Weeks MH, & Oberst FW (1959) Continuous exposure of laboratory animals to low concentration of carbon monoxide. Aerosp Med, **30**: 524-529.

Myers RAM (1986) Hyperbaric oxygen therapy: a committee report. Bethesda, Maryland, Undersea and Hyperbaric Medical Society, pp 33-36.

Myers R, Snyder S, Linderberg S, & Cowley A (1981) Value of hyperbaric oxygen in suspected carbon monoxide poisoning. J Am Med Assoc, **246**: 2478-2480.

Myers RAM, Snyder SK, & Emhoff TA (1985) Subacute sequelae of carbon monoxide poisoning. Ann Emerg Med, **14**: 1163-1167.

Nagda NL & Koontz MD (1985) Microenvironmental and total exposures to carbon monoxide for three population subgroups. J Air Pollut Control Assoc, **35**: 134-137.

NAS (1969) Effects of chronic exposure to low levels of carbon monoxide on human health, behavior, and performance. Washington, DC, National Academy of Sciences and National Academy of Engineering.

NAS (1973) Automotive spark ignition engine emission control systems to meet the requirements of the 1970 Clean Air Amendments. Washington, DC, National Academy of Sciences (Available from NTIS, Springfield, Virginia, USA).

National Air Pollution Control Administration (1970) Air quality criteria for carbon monoxide. Washington, DC, US Department of Health, Education, and Welfare, Public Health Service (Report No. NAPCA-PUB-AP-62, available from NTIS, Springfield, Virginia, USA).

National Center for Health Statistics (1986) Vital statistics of the United States 1982 — Volume II: Mortality, Part A. Hyattsville, Maryland, US Department of Health and Human Services, Public Health Service.

National Safety Council (1982) How people died in home accidents, 1981. In: Accident facts. Chicago, Illinois, National Safety Council, pp. 80-84.

Naumann RJ (1975) Patent No. 3,895,912: Carbon monoxide monitor. Washington, DC, US Patent Office.

Neubauer JA, Santiago TV, & Edelman NH (1981) Hypoxic arousal in intact and carotid chemodenervated sleeping cats. J Appl Physiol Respir Environ Exercise Physiol, **51**: 1294-1299.

Newell RE, Boer GJ Jr, & Kidson JW (1974) An estimate of the interhemispheric transfer of carbon monoxide from tropical general circulation data. Tellus, **26**: 103-107.

Niden AH (1971) The effects of low levels of carbon monoxide on the fine structure of the terminal airways. Am Rev Respir Dis, **103**: 898.

Niden AH & Schulz H (1965) The ultrastructural effects of carbon monoxide inhalation on the rat lung. Virchows Arch Pathol Anat Physiol, **339**: 283-292.

Nilsson C-A, Lindahl R, & Norstrom A (1987) Occupational exposure to chain saw exhausts in logging operations. Am Ind Hyg Assoc J, **48**: 99-105.

NIOSH (1972) Criteria for a recommended standard occupational exposure to carbon monoxide. Cincinnati, Ohio, National Institute for Occupational Safety and Health (Report No. NIOSH-TR-007-72).

Norkool DM & Kirkpatrick JN (1985) Treatment of acute carbon monoxide poisoning with hyperbaric oxygen: a review of 115 cases. Ann Emerg Med, **14**: 1168-1171.

Norman CA & Halton DM (1990) Is carbon monoxide a workplace teratogen? A review and evaluation of the literature. Ann Occup Hyg, 34(4): 335-347.

Norris JC, Moore SJ, & Hume AS (1986) Synergistic lethality induced by the combination of carbon monoxide and cyanide. Toxicology, **40**: 121-129.

Novelli PC, Massarie KA, Tans PT, & Lang PA (1994) Recent changes in atmospheric carbon monoxide. Science, **263**: 1587-1590.

NRC (1977) Carbon monoxide. Washington, DC, National Academy of Sciences, National Research Council.

NRC (1986a) Environmental tobacco smoke: measuring exposures and assessing health effects. Washington, DC, National Research Council, National Academy Press.

NRC (1986b) The airliner cabin environment: air quality and safety. Washington, DC, National Research Council, National Academy Press.

Ocak A, Valentour JC, & Blanke RV (1985) The effects of storage conditions on the stability of carbon monoxide in postmortem blood. J Anal Toxicol, **9**: 202-206.

O'Donnell RD, Mikulka P, Heinig P, & Theodore J (1971a) Low level carbon monoxide exposure and human psychomotor performance. Toxicol Appl Pharmacol, **18**: 593-602.

O'Donnell RD, Chikos P, & Theodore J (1971b) Effect of carbon monoxide exposure on human sleep and psychomotor performance. J Appl Physiol, **31**: 513-518.

OECD (1997) Advanced air quality indicators project — final report. Paris, Organisation for Economic Co-operation and Development.

Okeda R, Funata N, Takano T, Miyazaki Y, Yokoyama K, & Manabe M (1981) The pathogenesis of carbon monoxide encephalopathy in the acute phase physiological morphological correlation. Acta Neuropathol, **54**: 1-10.

Okeda R, Funata N, Song SJ, Higashino F, Takano T, & Yokoyama K (1982) Comparative study on pathogenesis of selective cerebral lesions in carbon monoxide poisoning and nitrogen hypoxia in cats. Acta Neuropathol, **56**: 265-272.

Okeda R, Matsuo T, Kuroiwa T, Tajima T, & Takahashi H (1986) Experimental study on pathogenesis of the fetal brain damage by acute carbon monoxide intoxication of the pregnant mother. Acta Neuropathol, **69**: 244-252.

Okeda R, Matsuo T, Kuroiwa T, Nakai M, Tajima T, & Takahashi H (1987) Regional cerebral blood flow of acute carbon monoxide poisoning in cats. Acta Neuropathol, **72**: 389-393.

Omura T & Sato R (1964) The carbon monoxide-binding pigment of liver microsomes: I. Evidence for its hemoprotein nature. J Biol Chem, **239**: 2370-2378.

Oremus RA, Barron KW, Gannon BJ, Fleming BP, Heesch CM, & Diana JN (1988) Effects of acute carbon monoxide (CO) exposure on cardiovascular (CV) hemodynamics in the anesthetized rat. FASEB J, **2**: A1312.

OSHA (Occupational Safety and Health Administration) (1991a) Ambient air monitoring reference and equivalent methods. Code Fed Regul, **40**: 53.

OSHA (Occupational Safety and Health Administration) (1991b) National primary and secondary ambient air quality standards. Code Fed Regul, **40**: 50.

Oshino N, Sugano T, Oshino R, & Chance B (1974) Mitochondrial function under hypoxic conditions: the steady states of cytochrome $a+a_3$ and their relation to mitochondrial energy states. Biochim Biophys Acta, **368**: 298-310.

Ott WR (1971) An urban survey technique for measuring the spatial variation of carbon monoxide concentrations in cities. Stanford, California, Stanford University, Department of Civil Engineering (Ph.D. dissertation). Ann Arbor, Michigan, University Microfilms (Publication No. 72-16,764).

Ott WR (1982) Concepts of human exposure to air pollution. Environ Int, **7**: 179-196.

Ott WR (1984) Exposure estimates based on computer generated activity patterns. J Toxicol Clin Toxicol, **21**: 97-128.

Ott W & Eliassen R (1973) A survey technique for determining the representativeness of urban air monitoring stations with respect to carbon monoxide. J Air Pollut Control Assoc, **23**: 685-690.

Ott W & Flachsbart P (1982) Measurement of carbon monoxide concentrations in indoor and outdoor locations using personal exposure monitors. Environ Int, **8**: 295-304.

Ott W & Mage DT (1975) A method for simulating the true human exposure of critical population groups to air pollutants. In: Proceedings of the International Symposium on Recent Advances in the Assessment of the Health Effects of Environmental Pollution, Paris, France, June 1974. Luxembourg, Commission of the European Communities, pp 2097-2107.

Ott WR, Rodes CE, Drago RJ, Williams C, & Burmann FJ (1986) Automated data-logging personal exposure monitors for carbon monoxide. J Air Pollut Control Assoc, **36**: 883-887.

Ott W, Thomas J, Mage D, & Wallace L (1988) Validation of the simulation of human activity and pollutant exposure (SHAPE) model using paired days from the Denver, CO, carbon monoxide field study. Atmos Environ, **22**: 2101-2113.

Ott W, Switzer P, & Willits N (1994) Carbon monoxide exposures inside an automobile travelling on an urban arterial highway. Air Waste, **44**: 1010-1018.

Otto DA, Benigus VA, & Prah JD (1979) Carbon monoxide and human time discrimination: failure to replicate Beard-Wertheim experiments. Aviat Space Environ Med, **50**: 40-43.

Pace N, Strajman E, & Walker EL (1950) Acceleration of carbon monoxide elimination in man by high pressure oxygen. Science, **111**: 652-654.

Pankow D & Ponsold W (1972) Leucine aminopeptidase activity in plasma of normal and carbon monoxide poisoned rats. Arch Toxicol, **29**: 279-285.

Pankow D & Ponsold W (1974) [The combined effects of carbon monoxide and other biologically active detrimental factors on the organism.] Z Gesamte Hyg Grenzgeb, **20**: 561-571 (in German).

Pankow D & Ponsold W (1984) Effect of carbon monoxide exposure on heart cytochrome *c* oxidase activity of rats. Biomed Biochim Acta, **43**: 1185-1189.

Pankow D, Ponsold W, & Fritz H (1974) Combined effects of carbon monoxide and ethanol on the activities of leucine aminopeptidase and glutamic-pyruvic transaminase in the plasma of rats. Arch Toxicol, **32**: 331-340.

Pantazopoulou A, Katsouyanni K, Kourea-Kremastinou J, & Trichopoulos D (1995) Short-term effects of air pollution on hospital emergency outpatient visits and admissions in the greater Athens, Greece area. Environ Res, **69**(1): 31-36.

Parving H-H (1972) The effect of hypoxia and carbon monoxide exposure on plasma volume and capillary permeability to albumin. Scand J Clin Lab Invest, **30**: 49-56.

Pauling L (1960) The nature of the chemical bond and the structure of molecules and crystals: an introduction to modern structural chemistry, 3rd ed. Ithaca, Cornell University Press, pp 194-195.

Paulozzi LJ, Spengler RF, Vogt RL, & Carney JK (1993) A survey of carbon monoxide and nitrogen dioxide in indoor ice arenas in Vermont. J Environ Health, **56**(5): 23-25.

Paulson OB, Parving H-H, Olesen J, & Skinhoj E (1973) Influence of carbon monoxide and of hemodilution on cerebral blood flow and blood gases in man. J Appl Physiol, **35**: 111-116.

Peeters LLH, Sheldon RE, Jones MD Jr, Makowski EL, & Meschia G (1979) Blood flow to fetal organs as a function of arterial oxygen content. Am J Obstet Gynecol, **135**: 637-646.

Penn A (1993) Determination of the atherogenic potential of inhaled carbon monoxide. Cambridge, Massachusetts, Health Effects Institute (Research Report No. 57).

Penn A, Butler J, Snyder C, & Albert RE (1983) Cigarette smoke and carbon monoxide do not have equivalent effects upon development of arteriosclerotic lesions. Artery, **12**: 17-131.

Penn A, Currie J, & Snyder C (1992) Inhalation of carbon monoxide does not accelerate arteriosclerosis in cockerels. Eur J Pharmacol, **228**: 155-164.

Penney DG (1984) Carbon monoxide-induced cardiac hypertrophy. In: Zak R ed. Growth of the heart in health and disease. New York, Raven Press, pp 337-362.

Penney DG (1988) A review: hemodynamic response to carbon monoxide. Environ Health Perspect, **77**: 121-130.

Penney DG (1993) Acute carbon monoxide poisoning in an animal model: The effects of altered glucose on morbidity and mortality. Toxicology, **80**(2-3): 85-101.

Penney DG & Chen K (1996) NMDA receptor-blocker ketamine protects during acute carbon monoxide poisoning, while calcium channel-blocker verapamil does not. J Appl Toxicol, **16**(4): 297-304.

Penney DG & Formolo J (1993) Carbon monoxide-induced cardiac hypertrophy is not reduced by *alpha* or *beta*-blockade in the rat. Toxicology, **80**(2-3): 173-187.

Penney DG & Weeks TA (1979) Age dependence of cardiac growth in the normal and carbon monoxide-exposed rat. Dev Biol, **71**: 153-162.

Penney D, Dunham E, & Benjamin M (1974a) Chronic carbon monoxide exposure: time course of hemoglobin, heart weight and lactate dehydrogenase isozyme changes. Toxicol Appl Pharmacol, **28**: 493-497.

Penney D, Benjamin M, & Dunham E (1974b) Effect of carbon monoxide on cardiac weight as compared with altitude effects. J Appl Physiol, **37**: 80-84.

Penney DG, Sodt PC, & Cutilletta A (1979) Cardiodynamic changes during prolonged carbon monoxide exposure in the rat. Toxicol Appl Pharmacol, **50**: 213-218.

Penney DG, Baylerian MS, & Fanning KE (1980) Temporary and lasting cardiac effects of pre- and postnatal exposure to carbon monoxide. Toxicol Appl Pharmacol, **53**: 271-278.

Penney DG, Baylerian MS, Thill JE, Fanning CM, & Yedavally S (1982) Postnatal carbon monoxide exposure: immediate and lasting effects in the rat. Am J Physiol, **243**: H328-H339.

Penney DG, Baylerian MS, Thill JE, Yedavally S, & Fanning CM (1983) Cardiac response of the fetal rat to carbon monoxide exposure. Am J Physiol, **244**: H289-H297.

Penney DG, Barthel BG, & Skoney JA (1984) Cardiac compliance and dimensions in carbon monoxide-induced cardiomegaly. Cardiovasc Res, **18**: 270-276.

Penney DG, Davidson SB, Gargulinski RB, & Caldwell-Ayre TM (1988) Heart and lung hypertrophy, changes in blood volume, hematocrit and plasma renin activity in rats chronically exposed to increasing carbon monoxide concentrations. J Appl Toxicol, **8**: 171-178.

Penney DG, Tucker A, & Bambach GA (1992) Heart and lung alterations in neonatal rats exposed to CO or high altitude. J Appl Physiol, **73**: 1713-1719.

Penney DG, Giraldo A, & VanEgmond EM (1993) Chronic carbon monoxide exposure in young rats alter coronary vessel growth. J Toxicol Environ Health, **39**: 207-222.

Permutt S & Farhi L (1969) Tissue hypoxia and carbon monoxide. In: Effects of chronic exposure to low levels of carbon monoxide on human health, behavior, and performance. Washington, DC, National Academy of Sciences, pp 18-24.

Perry RA, Atkinson R, & Pitts JN Jr (1977) Kinetics of the reactions of OH radicals with C_2H_2 and CO. J Chem Phys, **67**: 5577-5584.

Petersen WB & Allen R (1982) Carbon monoxide exposures to Los Angeles area commuters. J Air Pollut Control Assoc, **32**: 826-833.

Petersen GA & Sabersky RH (1975) Measurements of pollutants inside an automobile. J Air Pollut Control Assoc, **25**: 1028-1032.

Peterson JE & Stewart RD (1970) Absorption and elimination of carbon monoxide by inactive young men. Arch Environ Health, **21**: 165-171.

Peterson JE & Stewart RD (1975) Predicting the carboxyhemoglobin levels resulting from carbon monoxide exposures. J Appl Physiol, **39**: 633-638.

Pettyjohn FS, McNeil RJ, Akers LA, & Faber JM (1977) Use of inspiratory minute volumes in evaluation of rotary and fixed wing pilot workload. Fort Rucker, Alabama, US Army Aeromedical Research Laboratory (Report No. 77-9).

Piantadosi CA (1990) Carbon monoxide intoxication. In: Update in intensive care and emergency medicine — Volume 10. Brussels, Erasme University Hospital.

Piantadosi CA, Sylvia AL, Saltzman HA, & Jobsis-Vandervliet FF (1985) Carbon monoxide-cytochrome interactions in the brain of the fluorocarbon-perfused rat. J Appl Physiol, **58**: 665-672.

Piantadosi CA, Sylvia AL, & Jobsis-Vandervliet FF (1987) Differences in brain cytochrome responses to carbon monoxide and cyanide in vivo. J Appl Physiol, **62**: 1277-1284.

Piantadosi CA, Zhang J, & Demchenko IT (1997) Production of hydroxyl radical in the hippocampus after CO hypoxia or hypoxic hypoxia in the rat. Free Radic Biol Med, **22**(4): 725-732.

Pirnay F, Dujardin J, Deroanne R, & Petit JM (1971) Muscular exercise during intoxication by carbon monoxide. J Appl Physiol, **31**: 573-575.

Pitts GC & Pace N (1947) The effect of blood carboxyhemoglobin concentration on hypoxia tolerance. Am J Physiol, **148**: 139-151.

Porter K & Volman DH (1962) Flame ionization detection of carbon monoxide for gas chromatographic analysis. Anal Chem, **34**: 748-749.

Poulton TJ (1987) Medical helicopters: carbon monoxide risk? Aviat Space Environ Med, **58**: 166-168.

Pratt R & Falconer P (1979) Circumpolar measurements of ozone, particles, and carbon monoxide from a commercial airliner. J Geophys Res Oceans Atmos, **C84**: 7876-7882.

Preziosi TJ, Lindenberg R, Levy D, & Christenson M (1970) An experimental investigation in animals of the functional and morphologic effects of single and repeated exposures to high and low concentrations of carbon monoxide. Ann NY Acad Sci, **174**: 369-384.

Prigge E & Hochrainer D (1977) Effects of carbon monoxide inhalation on erythropoiesis and cardiac hypertrophy in fetal rats. Toxicol Appl Pharmacol, **42**: 225-228.

Putz VR, Johnson BL, & Setzer JV (1976) Effects of CO on vigilance performance: effects of low level carbon monoxide on divided attention, pitch discrimination, and the auditory evoked potential. Cincinnati, Ohio, National Institute for Occupational Safety and Health (Report No. NIOSH-77-124).

Putz VR, Johnson BL, & Setzer JV (1979) A comparative study of the effects of carbon monoxide and methylene chloride on human performance. J Environ Pathol Toxicol, **2**: 97-112.

Quackenboss JJ, Kanarek MS, Spengler JD, & Letz R (1982) Personal monitoring for nitrogen dioxide exposure: methodological considerations for a community study. Environ Int, **8**: 249-258.

Quinlan P, Connor M, & Waters M (1985) Environmental evaluation of stress and hypertension in municipal bus drivers. Cincinnati, Ohio, National Institute for Occupational Safety and Health (Publication No. PB86-144763).

Radford EP & Drizd TA (1982) Blood carbon monoxide levels in persons 3-74 years of age: United States, 1976-80. Hyattsville, Maryland, US Department of Health and Human Services, National Center for Health Statistics (DHHS Publication No. (PHS) 82-1250).

Rahn H & Fenn WO (1955) A graphical analysis of the respiratory gas exchange: the O_2-CO_2 diagram. Washington, DC, American Physiological Society, p 37.

Rai VS & Minty PSB (1987) The determination of carboxyhaemoglobin in the presence of sulphaemoglobin. Forensic Sci Int, **33**: 1-6.

Ramsey JM (1967) Carboxyhemoglobinemia in parking garage employees. Arch Environ Health, **15**: 580-583.

Ramsey JM (1972) Carbon monoxide, tissue hypoxia, and sensory psychomotor response in hypoxaemic subjects. Clin Sci, **42**: 619-625.

Ramsey JM (1973) Effects of single exposures of carbon monoxide on sensory and psychomotor response. Am Ind Hyg Assoc J, **34**: 212-216.

Raphael J-C, Elkharrat D, Jars-Guincestre M-C, Chastang C, Chasles V, Vercken J-B, & Gajdos P (1989) Trial of normobaric and hyperbaric oxygen for acute carbon monoxide intoxication. Lancet, **2**(8660): 414-419.

Rasmussen RA & Khalil MAK (1982) Latitudinal distributions of trace gases in and above the boundary layer. Chemosphere, **11**: 227-235.

Rasmussen RA & Went FW (1965) Volatile organic material of plant origin in the atmosphere. Proc Natl Acad Sci (USA), **53**: 215-220.

Raven PB, Drinkwater BL, Horvath SM, Ruhling RO, Gliner JA, Sutton JC, & Bolduan NW (1974a) Age, smoking habits, heat stress, and their interactive effects with carbon monoxide and peroxyacetylnitrate on man's aerobic power. Int J Biometeorol, **18**: 222-232.

Raven PB, Drinkwater BL, Ruhling RO, Bolduan N, Taguchi S, Gliner J, & Horvath SM (1974b) Effect of carbon monoxide and peroxyacetyl nitrate on man's maximal aerobic capacity. J Appl Physiol, **36**: 288-293.

Redmond CK (1975) Comparative cause-specific mortality patterns by work area within the steel industry. Cincinnati, Ohio, National Institute for Occupational Safety and Health (HEW Publication No. (NIOSH) 75-157).

Redmond CK, Emes JJ, Mazumdar S, Magee PC, & Kamon E (1979) Mortality of steelworkers employed in hot jobs. J Environ Pathol Toxicol, **2**: 75-96.

Reichle HG Jr, Beck SM, Haynes RE, Hesketh WD, Holland JA, Hypes WD, Orr HD III, Sherrill RT, Wallio HA, Casas JC, Saylor MS, & Gormsen BB (1982) Carbon monoxide measurements in the troposphere. Science, **218**: 1024-1026.

Reichle HG Jr, Connors VS, Holland JA, Hypes WD, Wallio HA, Casas JC, Gormsen BB, Saylor MS, & Hesketh WD (1986) Middle and upper tropospheric carbon monoxide mixing ratios as measured by a satellite-borne remote sensor during November 1981. J Geophys Res Atmos, **91**: 10865-10887.

Reichle HG Jr, Connors VS, Holland JA, Sherrill RT, Wallio HA, Casas JC, Condon EP, Gormsen BB, & Seiler W (1990) The distribution of middle tropospheric carbon monoxide during early October 1984. J Geophys Res Atmos, **95**: 9845-9856.

Renaud S, Blache D, Dumont E, Thevenon C, & Wissendanger T (1984) Platelet function after cigarette smoking in relation to nicotine and carbon monoxide. Clin Pharmacol Ther, **36**: 389-395.

Repp M (1977) Evaluation of continuous monitors for carbon monoxide in stationary sources. Research Triangle Park, North Carolina, US Environmental Protection Agency, Environmental Science Research Laboratory (EPA-600/2-77-063).

Research Triangle Institute (1990) An investigation of infiltration and indoor air quality: final report. Albany, New York, New York State Energy Research and Development Authority (Report No. NYERDA-90-11, available from NTIS, Springfield, Virginia, USA).

Reynolds JEF & Prasad AB ed. (1982) Martindale: The extra pharmacopoeia, 28th ed. London, The Pharmaceutical Press, p 804.

Rickert WS, Robinson JC, & Collishaw N (1984) Yields of tar, nicotine, and carbon monoxide in the sidestream smoke from 15 brands of Canadian cigarettes. Am J Public Health, **74**: 228-231.

Rinsland CP & Levine JS (1985) Free tropospheric carbon monoxide concentrations in 1950 and 1951 deduced from infrared total column amount measurements. Nature (Lond), **318**: 250-254.

Robbins RC, Borg KM, & Robinson E (1968) Carbon monoxide in the atmosphere. J Air Pollut Control Assoc, **18**: 106-110.

Robertson G & Lebowitz MD (1984) Analysis of relationships between symptoms and environmental factors over time. Environ Res, **33**: 130-143.

Robinson JP (1977) How Americans use time: a social-psychological analysis of everyday behavior. New York, Praeger Publishers.

Robinson E & Robbins RC (1969) Sources, abundance, and fate of gaseous atmospheric pollutants: supplement (SRI Project PR-6755). Menlo Park, California, Stanford Research Institute

Robinson E & Robbins RC (1970) Atmospheric background concentrations of carbon monoxide. In: Coburn RF ed. Biological effects of carbon monoxide. Ann NY Acad Sci, **174**: 89-95.

Robinson NB, Barie PS, Halebian PH, & Shires GT (1985) Distribution of ventilation and perfusion following acute carbon monoxide poisoning. In: 41st Annual Forum on Fundamental Surgical Problems held at the 71st Annual Clinical Congress of the American College of Surgeons, Chicago, October 1985. Surg Forum, **36**: 115-118.

Roche S, Horvath S, Gliner J, Wagner J, & Borgia J (1981) Sustained visual attention and carbon monoxide: elimination of adaptation effects. Hum Factors, **23**: 175-184.

Rockwell TJ & Weir FW (1975) The interactive effects of carbon monoxide and alcohol on driving skills (CRC-APRAC project CAPM-9-69). Columbus, Ohio, The Ohio State University Research Foundation (Available from NTIS, Springfield, Virginia, USA).

Rodgers PA, Vreman HJ, Dennery PA, & Stevenson DK (1994) Sources of carbon monoxide (CO) in biological systems and applications of CO detection technologies. Semin Perinatol, **18**: 2-10.

Rodkey FL & Collison HA (1979) Effects of oxygen and carbon dioxide on carbon monoxide toxicity. J Combust Toxicol, **6**: 208-212.

Rogers WR, Bass RL III, Johnson DE, Kruski AW, McMahan CA, Montiel MM, Mott GE, Wilbur RL, & McGill HC Jr (1980) Atherosclerosis-related responses to cigarette smoking in the baboon. Circulation, **61**: 1188-1193.

Rogers WR, Carey KD, McMahan CA, Montiel MM, Mott GE, Wigodsky HS, & McGill HC Jr (1988) Cigarette smoking, dietary hyperlipidemia, and experimental atherosclerosis in the baboon. Exp Mol Pathol, **48**: 135-151.

Rosenkrantz H, Grant RJ, Fleischman RW, & Baker JR (1986) Marihuana-induced embryotoxicity in the rabbit. Fundam Appl Toxicol, **7**: 236-243.

Rosenman KD (1984) Cardiovascular disease and work place exposures. Arch Environ Health, **39**: 218-224.

Rosenstock L & Cullen MR (1986a) Cardiovascular disease. In: Clinical occupational medicine. Philadelphia, W.B. Saunders, pp 71-80.

Rosenstock L & Cullen MR (1986b) Neurologic disease. In: Clinical occupational medicine. Philadelphia, W.B. Saunders, pp 118-134.

Roth RA Jr & Rubin RJ (1976a) Role of blood flow in carbon monoxide- and hypoxic hypoxia-induced alterations in hexobarbital metabolism in rats. Drug Metab Dispos, **4**: 460-467.

Roth RA Jr & Rubin RJ (1976b) Comparison of the effect of carbon monoxide and of hypoxic hypoxia. I. In vivo metabolism, distribution and action of hexobarbital. J Pharmacol Exp Ther, **199**: 53-60.

Roth RA Jr & Rubin RJ (1976c) Comparison of the effect of carbon monoxide and of hypoxic hypoxia: II. Hexobarbital metabolism in the isolated, perfused rat liver. J Pharmacol Exp Ther, **199**: 61-66.

Roughton FJW (1970) The equilibrium of carbon monoxide with human hemoglobin in whole blood. In: Coburn RF ed. Biological effects of carbon monoxide. Ann NY Acad Sci, **174**: 177-188.

Roughton FJW & Darling RC (1944) The effect of carbon monoxide on the oxyhemoglobin dissociation curve. Am J Physiol, **141**: 17-31.

Roughton FJW & Root WS (1945) The estimation of small amounts of carbon monoxide in air. J Biol Chem, **160**: 135-148.

Roy TM, Mendieta JM, Ossorio MA, & Walker JF (1989) Perceptions and utilization of hyperbaric oxygen therapy for carbon monoxide poisoning in an academic setting. J Ky Med Assoc, **87**: 223-226.

Rummo N & Sarlanis K (1974) The effect of carbon monoxide on several measures of vigilance in a simulated driving task. J Saf Res, **6**: 126-130.

Saldiva PH, Lichtenfels AJ, Paiva PS, Barone IA, Martins MA, Massad E, Pereira JC, Xavier VP, Singer JM, & Bohm GM (1994) Association between air pollution and mortality due to respiratory diseases in children in Sao Paulo, Brazil: A preliminary report. Environ Res, **65**(2): 218-225.

Saldiva PH, Pope CA III, Schwartz J, Dockery DW, Lichtenfels AJ, Salge JM, Barone I, & Bohm GM (1995) Air pollution and mortality in elderly people: a time-series study in Sao Paulo, Brazil. Arch Environ Health, **50**(2): 159-163.

Salinas M & Vega J (1995) The effect of outdoor air pollution on mortality risk: an ecological study from Santiago, Chile. World Health Stat Q, **48**(2): 118-125.

Salvatore S (1974) Performance decrement caused by mild carbon monoxide levels on two visual functions. J Saf Res, **6**: 131-134.

Sammons JH & Coleman RL (1974) Firefighters' occupational exposure to carbon monoxide. J Occup Med, **16**(8): 543-546.

Sandler DP, Comstock GW, Helsing KJ, & Shore DL (1989) Deaths from all causes in non-smokers who lived with smokers. Am J Public Health, **79**(2): 163-167.

Santiago TV & Edelman NH (1976) Mechanism of the ventilatory response to carbon monoxide. J Clin Invest, **57**: 977-986.

Savolainen H, Kurppa K, Tenhunen R, & Kivisto H (1980) Biochemical effects of carbon monoxide poisoning in rat brain with special reference to blood carboxyhemoglobin and cerebral cytochrome oxidase activity. Neurosci Lett, **19**: 319-323.

Sawday Y, Takahashi M, Ohashi N, Fusamoto H, Maemura K, Kobayashi H, Yoshiora T, & Sugimoto T (1980) Computerized tomography as an indication of long-term outcome after acute carbon monoxide poisoning. Lancet, **1**: 783-784.

Sawicki CA & Gibson QH (1979) A photochemical method for rapid and precise determination of carbon monoxide levels in blood. Anal Biochem, **94**: 440-449.

Schaad G, Kleinhanz G, & Piekarski C (1983) [Influence of carbon monoxide in breath on psychophysical competence.] Wehrmed Monatsschr, **10**: 423-430 (in German).

Schnakenberg GH (1975) Gas detection instrumentation...what's new and what's to come. Coal Age, **80**: 84-92.

Schnakenberg GH Jr (1976) Improvements in coal mine gas detection instrumentation. Pap Symp Underground Min, **2**: 206-216.

Schoendorf KC & Kiely JL (1992) Relationship of sudden infant death syndrome to maternal smoking during and after pregnancy. Pediatrics, **90**(6): 905-908.

Schoenfisch WH, Hoop KA, & Struelens BS (1980) Carbon monoxide absorption through the oral and nasal mucosae of cynomolgus monkeys. Arch Environ Health, **35**: 152-154.

Scholander PF & Roughton FJW (1943) Micro gasometric estimation of the blood gases: II. Carbon monoxide. J Biol Chem, **148**: 551-563.

Schrot J & Thomas JR (1986) Multiple schedule performance changes during carbon monoxide exposure. Neurobehav Toxicol Teratol, **8**: 225-230.

Schrot J, Thomas JR, & Robertson RF (1984) Temporal changes in repeated acquisition behavior after carbon monoxide exposure. Neurobehav Toxicol Teratol, **6**: 23-28.

Schulte JH (1963) Effects of mild carbon monoxide intoxication. Arch Environ Health, **7**: 524-530.

Schwab M, Colome SD, Spengler JD, Ryan PB, & Billick IH (1990) Activity patterns applied to pollutant exposure assessment: data from a personal monitoring study in Los Angeles. Toxicol Ind Health, **6**: 517-532.

Schwartz J & Morris R (1995) Air pollution and hospital admissions for cardiovascular disease in Detroit, Michigan. Am J Epidemiol, **141**(1): 23-35.

Schwetz BA, Smith FA, Leong BKJ, & Staples RE (1979) Teratogenic potential of inhaled carbon monoxide in mice and rabbits. Teratology, **19**: 385-391.

Scragg R, Stewart AW, Mitchell EA, Ford RP, & Thompson JM (1995) Public health policy on bed sharing and smoking in the sudden infant death syndrome. NZ J Med, **108**(1001): 218-222.

Seiler W (1974) The cycle of atmospheric CO. Tellus, **26**: 116-135.

Seiler W & Conrad R (1987) Contribution of tropical ecosystems to the global budget of trace gases especially CH_4, H_2, CO and N_2O. In: Dickinson RE ed. The geophysiology of Amazonia: vegetation and climate. New York, John Wiley & Sons, Inc.

Seiler W & Fishman J (1981) The distribution of carbon monoxide and ozone in the free troposphere. J Geophys Res Oceans Atmos, **C86**: 7255-7265.

Seiler W & Giehl H (1977) Influence of plants on the atmospheric carbon monoxide. Geophys Res Lett, **4**: 329-332.

Seiler W & Junge C (1969) Decrease of carbon monoxide mixing ratio above the polar tropopause. Tellus, **21**: 447-449.

Seiler W & Junge C (1970) Carbon monoxide in the atmosphere. J Geophys Res, **75**: 2217-2226.

Seiler W & Schmidt U (1974) Dissolved nonconservative gases in seawater. In: Goldberg ED ed. The sea — Volume 5: Marine chemistry. New York, John Wiley and Sons, Inc., pp 219-243.

Seiler W & Warneck P (1972) Decrease of the carbon monoxide mixing ratio at the tropopause. J Geophys Res, **77**: 3204-3214.

Seiler W, Giehl H, & Bunse G (1978) The influence of plants on atmospheric carbon monoxide and dinitrogen oxide. Pure Appl Geophys, **116**: 439-451.

Seiler W, Giehl H, Brunke E-G, & Halliday E (1984) The seasonality of CO abundance in the southern hemisphere. Tellus, **B36**: 219-231.

Sekiya S, Sato S, Yamaguchi H, & Harumi K (1983) Effects of carbon monoxide inhalation on myocardial infarct size following experimental coronary artery ligation. Jpn Heart J, **24**: 407-416.

Selvakumar S, Sharan M, & Singh MP (1993) A mathematical model for the elimination of carbon monoxide in humans. J Theor Biol, **162**: 321-336.

Seppanen A (1977) Physical work capacity in relation to carbon monoxide inhalation and tobacco smoking. Ann Clin Res, **9**: 269-274.

Seppanen A, Hakkinen V, & Tenkku M (1977) Effect of gradually increasing carboxyhaemoglobin saturation on visual perception and psychomotor performance of smoking and nonsmoking subjects. Ann Clin Res, **9**: 314-319.

Sexton K & Ryan PB (1988) Assessment of human exposure to air pollution: methods, measurements, and models. In: Watson AY, Bates RR, & Kennedy D ed. Air pollution, the automobile, and public health. Washington, DC, National Academy Press, pp 207-238.

Sexton K, Spengler JD, & Treitman RD (1984) Personal exposure to respirable particles: a case study in Waterbury, Vermont. Atmos Environ, 18: 1385-1398.

Sheppard D, Distefano S, Morse L, & Becker C (1986) Acute effects of routine firefighting on lung function. Am J Ind Med, 9: 333-340.

Sheps DS, Adams KF Jr, Bromberg PA, Goldstein GM, O'Neil JJ, Horstman D, & Koch G (1987) Lack of effect of low levels of carboxyhemoglobin on cardiovascular function in patients with ischemic heart disease. Arch Environ Health, 42: 108-116.

Sheps DS, Herbst MC, Hinderliter AL, Adams KF, Ekelund LG, O'Neil JJ, Goldstein GM, Bromberg PA, Dalton JL, Ballenger MN, Davis SM, & Koch GG (1990) Production of arrhythmias by elevated carboxyhemoglobin in patients with coronary artery disease. Ann Intern Med, 113: 343-351.

Sheps DS, Herbst MC, Hinderliter AL, Adams KF, Ekelund LG, O'Neil JJ, Goldstein GM, Bromberg PA, Ballenger M, Davis SM, & Koch G (1991) Effects of 4 percent and 6 percent carboxyhemoglobin on arrhythmia production in patients with coronary artery disease. Cambridge, Massachusetts, Health Effects Institute (Research Report No. 41).

Shikiya D, Liu C, Kahn M, Juarros J, & Barcikowski W (1989) In-vehicle air toxics characterization study in the South Coast Air Basin. El Monte, California, South Coast Air Quality Management District, Office of Planning and Rules.

Shinomiya K, Orimoto C, & Shinomiya T (1994) [Experimental exposure to carbon monoxide in rats (I) — Relation between the degree of carboxyhemoglobin saturation and the amount of carbon monoxide in the organ tissues of rats.] Jpn J Legal Med, 48: 19-25 (in Japanese).

Sie BKT, Simonaitis R, & Heicklen J (1976) The reaction of OH with CO. Int J Chem Kinet, 8: 85-98.

Sies H (1977) Oxygen gradients during hypoxic steady states in liver: urate oxidase and cytochrome oxidase as intracellular O_2 indicators. Hoppe Seylers Z Physiol Chem, 358: 1021-1032.

Sies H & Brauser B (1970) Interaction of mixed function oxidase with its substrates and associated redox transitions of cytochrome P-450 and pyridine nucleotides in perfused rat liver. Eur J Biochem, 15: 531-540.

Simonaitis R & Heicklen J (1972) Kinetics and mechanism of the reaction of O(3P) with carbon monoxide. J Chem Phys, 56: 2004-2011.

Singh J (1986) Early behavioral alterations in mice following prenatal carbon monoxide exposure. Neurotoxicology, 7: 475-481.

Singh J & Scott LH (1984) Threshold for carbon monoxide induced fetotoxicity. Teratology, 30: 253-257.

Singh MP, Sharan M, & Selvakumar S (1991) A mathematical model for the computation of carboxyhemoglobin in human blood as a function of exposure time. Philos Trans R Soc Lond Biol Sci, B334: 135-147.

Singh J, Smith CB, & Moore Cheatu L (1992) Additivity of protein deficiency and carbon monoxide on placental carboxyhemoglobin in mice. Am J Obstet Gynecol, **167**: 843-846.

Sisovic A & Fugas M (1985) Indoor concentrations of carbon monoxide in selected urban microenvironments. Environ Monit Assess, **5**: 199-204.

Sjostrand T (1948a) A method for the determination of carboxyhaemoglobin concentrations by analysis of the alveolar air. Acta Physiol Scand, **16**: 201-210.

Sjostrand T (1948b) Brain volume, diameter of the blood-vessels in the pia mater, and intracranial pressure in acute carbon monoxide poisoning. Acta Physiol Scand, **15**: 351-361.

Small KA, Radford EP, Frazier JM, Rodkey FL, & Collison HA (1971) A rapid method for simultaneous measurement of carboxy- and methemoglobin in blood. J Appl Physiol, **31**: 154-160.

Smith J & Brandon S (1973) Morbidity from acute carbon monoxide poisoning at three years follow-up. Br Med J, **1**: 318-321.

Smith EE & Crowell JW (1967) Role of an increased hematocrit in altitude acclimatization. Aerosp Med, **38**: 39-43.

Smith JR & Landaw SA (1978a) Smokers' polycythemia. N Engl J Med, **298**: 6-10.

Smith JR & Landaw SA (1978b) Smokers' polycythemia [letter to the editor]. N Engl J Med, **298**: 973.

Smith F & Nelson AC Jr (1973) Guidelines for development of a quality assurance program: reference method for the continuous measurement of carbon monoxide in the atmosphere. Research Triangle Park, North Carolina, US Environmental Protection Agency, Quality Assurance and Environmental Monitoring Laboratory (EPA-R4-73-028a).

Smith RG, Bryan RJ, Feldstein M, Locke DC, & Warner PO (1975) Tentative method for constant pressure volumetric gas analysis for O_2, CO_2, CO, N_2, hydrocarbons (ORSAT). Health Lab Sci, **12**: 177-181.

Smith MV, Hazucha MJ, & Benignus VA (1994) Effect of regional circulation patterns on observed COHb level. J Appl Physiol, **77**: 1659-1665.

Snella M-C & Rylander R (1979) Alteration in local and systemic immune capacity after exposure to bursts of CO. Environ Res, **20**: 74-79.

Snow TR, Vanoli E, De Ferrari G, Stramba-Badiale M, & Dickey DT (1988) Response of cytochrome a,a_3 to carbon monoxide in canine hearts with prior infarcts. Life Sci, **42**: 927-931.

Sokal JA (1985) The effect of exposure duration on the blood level of glucose pyruvate and lactate in acute carbon monoxide intoxication in man. J Appl Toxicol, **5**: 395-397.

Sokal JA & Kralkowska E (1985) The relationship between exposure duration, carboxyhemoglobin, blood glucose, pyruvate and lactate and the severity of intoxication in 39 cases of acute carbon monoxide poisoning in man. Arch Toxicol, **57**: 196-199.

Sokal JA, Majka J, & Palus J (1984) The content of carbon monoxide in the tissues of rats intoxicated with carbon monoxide in various conditions of acute exposure. Arch Toxicol, **56**: 106-108.

Sokal J, Majka J, & Palus J (1986) Effect of work load on the content of carboxymyoglobin in the heart and skeletal muscles of rats exposed to carbon monoxide. J Hyg Epidemiol Microbiol Immunol, **30**: 57-62.

Solanki DL, McCurdy PR, Cuttitta FF, & Schechter GP (1988) Hemolysis in sickle cell disease as measured by endogenous carbon monoxide production: a preliminary report. Am J Clin Pathol, **89**: 221-225.

Somogyi E, Balogh I, Rubanyi G, Sotonyi P, & Szegedi L (1981) New findings concerning the pathogenesis of acute carbon monoxide (CO) poisoning. Am J Forensic Med Pathol, **2**: 31-39.

Song H, Kuo H, Wei SD, Geun L du S, & Yen F (1984) The COHb levels of 1332 blood donors in Beijing. J Chin Prev Med, **18**: 86-88.

Song H, Wang L, & Wang Z (1990) The effects of carbon monoxide on monoamines in brain of rat pups. China Environ Sci, **10**: 127-130.

Sparrow D, Bosse R, Rosner B, & Weiss ST (1982) The effect of occupational exposure on pulmonary function: a longitudinal evaluation of fire fighters and nonfire fighters. Am Rev Respir Dis, **125**: 319-322.

Spengler JD & Soczek ML (1984) Evidence for improved ambient air quality and the need for personal exposure research. Environ Sci Technol, **18**: 268A-280A.

Spengler JD, Stone KR, & Lilley FW (1978) High carbon monoxide levels measured in enclosed skating rinks. J Air Pollut Control Assoc, **28**: 776-779.

Spengler JD, Treitman RD, Tosteson TD, Mage DT, & Soczek ML (1985) Personal exposures to respirable particulates and implications for air pollution epidemiology. Environ Sci Technol, **19**: 700-707.

Steenland K (1992) Passive smoking and the risk of heart disease. J Am Med Assoc, **267**(1): 94-99.

Steenland K, Thun M, Lally C, & Health C Jr (1996) Environmental tobacco smoke and coronary heart disease in the American Cancer Society CPS-II cohort. Circulation, **94**(4): 622-628.

Stender S, Astrup P, & Kjeldsen K (1977) The effect of carbon monoxide on cholesterol in the aortic wall of rabbits. Atherosclerosis, **28**: 357-367.

Stern FB, Lemen RA, & Curtis RA (1981) Exposure of motor vehicle examiners to carbon monoxide: a historical prospective mortality study. Arch Environ Health, **36**: 59-66.

Stern FB, Halperin WE, Hornung RW, Ringenburg VL, & McCammon CS (1988) Heart disease mortality among bridge and tunnel officers exposed to carbon monoxide. Am J Epidemiol, **128**: 1276-1288.

Stetter JR & Blurton KF (1976) Portable high-temperature catalytic reactor: application to air pollution monitoring instrumentation. Rev Sci Instrum, **47**: 691-694.

Stevens RK & Herget WF (1974) Analytical methods applied to air pollution measurements. Ann Arbor, Michigan, Ann Arbor Science Publishers, Inc.

Stewart RD, Peterson JE, Baretta ED, Bachand RT, Hosko MJ, & Herrmann AA (1970) Experimental human exposure to carbon monoxide. Arch Environ Health, **21**: 154-164.

Stewart RD, Peterson JE, Fisher TN, Hosko MJ, Baretta ED, Dodd HC, & Herrmann AA (1973a) Experimental human exposure to high concentrations of carbon monoxide. Arch Environ Health, **26**: 1-7.

Stewart RD, Newton PE, Hosko MJ, & Peterson JE (1973b) Effect of carbon monoxide on time perception. Arch Environ Health, **27**: 155-160.

Stewart RD, Baretta ED, Platte LR, Stewart EB, Kalbfleisch JH, Van Yserloo B, & Rimm AA (1974) Carboxyhemoglobin levels in American blood donors. J Am Med Assoc, **229**: 1187-1195.

Stewart RD, Newton PE, Hosko MJ, Peterson JE, & Mellender JW (1975) The effect of carbon monoxide on time perception, manual coordination, inspection, and arithmetic. In: Weiss B & Laties VG ed. Behavioral toxicology: Proceedings of the International Conference on Environmental Toxicity, Rochester, NY, June 1972. New York, London, Plenum Press, pp 29-60.

Stewart RD, Hake CL, Wu A, Stewart TA, & Kalbfleisch JH (1976) Carboxyhemoglobin trend in Chicago blood donors, 1970-1974. Arch Environ Health, **31**: 280-286.

Stewart RD, Newton PE, Kaufman J, Forster HV, Klein JP, Keelen MH Jr, Stewart DJ, Wu A, & Hake CL (1978) The effect of a rapid 4% carboxyhemoglobin saturation increase on maximal treadmill exercise. New York, Coordinating Research Council, Inc. (Report No. CRC-APRAC-CAPM-22-75, available from NTIS, Springfield, Virginia, USA).

Stock TH, Kotchmar DJ, Contant CF, Buffler PA, Holguin AH, Gehan BM, & Noel LM (1985) The estimation of personal exposures to air pollutants for a community-based study of health effects in asthmatics — design and results of air monitoring. J Air Pollut Control Assoc, **35**: 266-1273.

Stokes DL, MacIntyre NR, & Nadel JA (1981) Nonlinear increases in diffusing capacity during exercise by seated and supine subjects. J Appl Physiol Respir Environ Exercise Physiol, **51**: 858-863.

Storm JE & Fechter LD (1985a) Alteration in the postnatal ontogeny of cerebellar norepinephrine content following chronic prenatal carbon monoxide. J Neurochem, **45**: 965-969.

Storm JE & Fechter LD (1985b) Prenatal carbon monoxide exposure differentially affects postnatal weight and monoamine concentration of rat brain regions. Toxicol Appl Pharmacol, **81**: 139-146.

Storm JE, Valdes JJ, & Fechter LD (1986) Postnatal alterations in cerebellar GABA content, GABA uptake and morphology following exposure to carbon monoxide early in development. Dev Neurosci, **8**: 251-261.

Stuhl F & Niki H (1972) Pulsed vacuum-UV photochemical study of reactions of OH with H2, D2, and CO using a resonance-fluorescent detection method. J Chem Phys, **57**: 3671-3677.

Stupfel M & Bouley G (1970) Physiological and biochemical effects on rats and mice exposed to small concentrations of carbon monoxide for long periods. Ann NY Acad Sci, **174**: 342-368.

Stupfel M, Mordelet-Dambrine M, & Vauzelle A (1981) COHb formation and acute carbon monoxide intoxication in adult male rats and guinea-pigs. Bull Eur Physiopathol Respir, **17**: 43-51.

Styka PE & Penney DG (1978) Regression of carbon monoxide-induced cardiomegaly. Am J Physiol, **235**: H516-H522.

Sue-matsu M, Kashiwagi S, Sano T, Goda N, Shinoda Y, & Ishimura Y (1994) Carbon monoxide as an endogenous modulator of hepatic vascular perfusion. Biochem Biophys Res Commun, **205**: 1333-1337.

Sulkowski WJ & Bojarski K (1988) Hearing loss due to combined exposure to noise and carbon monoxide — a field study. In: Berglund B, Berglund U, Karlsson J, & Lindvall T ed. Noise as a public health problem: Proceedings of the 5th International Congress on Noise as a Public Health Problem, Stockholm, August 1988 — Volume 1: Abstract guide. Stockholm, Sweden, Swedish Council for Building Research, p 179.

Sultzer DL, Brinkhous KM, Reddick RL, & Griggs TR (1982) Effect of carbon monoxide on atherogenesis in normal pigs and pigs with von Willebrand's disease. Atherosclerosis, **43**: 303-319.

Surgeon General of the United States (1986) The health consequences of involuntary smoking: A report of the Surgeon General. Rockville, Maryland, US Department of Health and Human Services, Public Health Service, Office on Smoking and Health (Publication No. DHHS (CDC)87-8398).

Svendsen KH, Kuller LH, Martin MJ, & Ockene JK (1987) Effects of passive smoking in the multiple risk factor intervention trial. Am J Epidemiol, **126**: 783-795.

Swiecicki W (1973) [The effect of vibration and physical training on carbohydrate metabolism in rats intoxicated with carbon monoxide.] Med Pr, **34**: 399-405 (in Polish).

Swinnerton JW & Lamontagne RA (1974) Carbon monoxide in the south Pacific Ocean. Tellus, **26**: 136-142.

Swinnerton JW, Linnenbom VJ, & Cheek CH (1968) A sensitive gas chromatographic method for determining carbon monoxide in seawater. Limnol Oceanogr, **13**: 193-195.

Swinnerton JW, Linnenbom VJ, & Cheek CH (1969) Distribution of methane and carbon monoxide between the atmosphere and natural waters. Environ Sci Technol, **3**: 836-838.

Swinnerton JW, Lamontagne RA, & Linnenbom VJ (1971) Carbon monoxide in rainwater. Science, **172**: 943-945.

Sylvester JT, Scharf SM, Gilbert RD, Fitzgerald RS, & Traystman RJ (1979) Hypoxic and CO hypoxia in dogs: hemodynamics, carotid reflexes, and catecholamines. Am J Physiol, **236**: H22-H28.

Syvertsen GR & Harris JA (1973) Erythropoietin production in dogs exposed to high altitude and carbon monoxide. Am J Physiol, **225**: 293-299.

Szalai A ed. (1972) The use of time: daily activities of urban and suburban populations in 12 countries. The Hague, The Netherlands, Mouton & Co.

Takano T, Motohashi Y, Miyazaki Y, & Okeda R (1985) Direct effect of carbon monoxide on hexobarbital metabolism in the isolated perfused liver in the absence of hemoglobin. J Toxicol Environ Health, **15**: 847-854.

Tanaka T & Knox WE (1959) The nature and mechanism of the tryptophan pyrrolase (peroxidase-oxidase) reaction of *Pseudomonas* and of rat liver. J Biol Chem, **234**: 1162-1170.

Tesarik K & Krejci M (1974) Chromatographic determination of carbon monoxide below the 1 ppm level. J Chromatogr, **91**: 539-544.

Theodore J, O'Donnell RD, & Back KC (1971) Toxicological evaluation of carbon monoxide in humans and other mammalian species. J Occup Med, **13**: 242-255.

Thom S (1990) Carbon monoxide-mediated brain lipid peroxidation in the rat. J Appl Physiol, **68**: 997-1003.

Thom S (1992) Dehydrogenase conversion to oxidase and lipid peroxidation in brain after carbon monoxide poisoning. J Appl Physiol, **73**(4): 1584-1589.

Thom SR (1993) Functional inhibition of leukocyte B2 integrins by hyperbaric oxygen in carbon monoxide-mediated brain injury in rats. Toxicol Appl Pharmacol, **123**(2): 248-256.

Thom SR & Keim LW (1989) Carbon monoxide poisoning: a review — Epidemiology, pathophysiology, clinical findings, and treatment options including hyperbaric oxygen therapy. J Toxicol Clin Toxicol, **27**: 141-156.

Thomsen HK (1974) Carbon monoxide-induced atherosclerosis in primates: an electron-microscopic study on the coronary arteries of *Macaca irus* monkeys. Atherosclerosis, **20**: 233-240.

Thomsen HK & Kjeldsen K (1975) Aortic intimal injury in rabbits: an evaluation of a threshold limit. Arch Environ Health, **30**: 604-607.

Tikuisis P, Kane DM, McLellan TM, Buick F, & Fairburn SM (1992) Rate of formation of carboxyhemoglobin in exercising humans exposed to carbon monoxide. J Appl Physiol, **72**: 1311-1319.

Topping DL, Fishlock RC, Trimble RP, Storer GB, & Snoswell AM (1981) Carboxyhaemoglobin inhibits the metabolism of ethanol by perfused rat liver. Biochem Int, **3**: 157-163.

Touloumi G, Pocock SJ, Katsouyanni K, & Trichopoulos D (1994) Short-term effects of air pollution on daily mortality in Athens: A time-series analysis. Int J Epidemiol, **23**(5): 957-967.

Touloumi G, Samli E, & Katsouyanni K (1996) Daily mortality and "winter type" air pollution in Athens, Greece: A time series analysis within the APHEA project. J Epidemiol Community Health, **50**(suppl 1): 47-51.

Traynor GW, Allen JR, Apte MG, Dillworth JF, Girman JR, Hollowell CD, & Koonce JF Jr (1982) Indoor air pollution from portable kerosene-fired space heaters, wood-burning stoves, and wood-burning furnaces. In: Proceedings of the Air Pollution Control Association Specialty Conference on Residential Wood and Coal Combustion, Louisville, KY, March 1982. Pittsburgh, Pennsylvania, Air Pollution Control Association, pp 253-263.

Traynor GW, Apte MG, Carruthers AR, Dillworth JF, Grimsrud DT, & Gundel LA (1984) Indoor air pollution to emissions from wood burning stoves. Presented at the 77th Annual Meeting of the Air Pollution Control Association, San Francisco, CA, June 1984. Pittsburgh, Pennsylvania, Air Pollution Control Association (Paper No. 84-33.4).

Traynor GW, Apte MG, Carruthers AR, Dillworth JF, Grimsrud DT, & Gundel LA (1987) Indoor air pollution due to emissions from wood-burning stoves. Environ Sci Technol, **21**: 691-697.

Traystman RJ (1978) Effect of carbon monoxide hypoxia and hypoxic hypoxia on cerebral circulation. In: Otto DA ed. Multidisciplinary perspectives in event-related brain potential research: Proceedings of the 4th International Congress on Event-Related Slow Potentials of the Brain (EPIC IV), Hendersonville, NC, April 1976. Washington, DC, US Environmental Protection Agency, Office of Research and Development, pp 453-457 (EPA-600/9-77-043).

Traystman RJ & Fitzgerald RS (1977) Cerebral circulatory responses to hypoxic hypoxia and carbon monoxide hypoxia in carotid baroreceptor and chemoreceptor denervated dogs. Acta Neurol Scand, **56**(suppl 64): 294-295.

Traystman RJ & Fitzgerald RS (1981) Cerebrovascular response to hypoxia in baroreceptor- and chemoreceptor-denervated dogs. Am J Physiol, **241**: H724-H731.

Traystman RJ, Fitzgerald RS, & Loscutoff SC (1978) Cerebral circulatory responses to arterial hypoxia in normal and chemodenervated dogs. Circ Res, **42**: 649-657.

Trese MT, Krohel GB, & Hepler RS (1980) Ocular effects of chronic carbon monoxide exposure. Ann Ophthalmol, **12**: 536-538.

Tsukamoto H & Matsuda Y (1985) [Relationship between smoking and both the carbon monoxide concentration in expired air and the carboxyhemoglobin content.] Kotsu Igaku, **39**: 367-376 (in Japanese).

Turino GM (1981) Effect of carbon monoxide on the cardiorespiratory system. Carbon monoxide toxicity: physiology and biochemistry. Circulation, **63**: 253A-259A.

Turner DM, Lee PN, Roe FJC, & Gough KJ (1979) Atherogenesis in the White Carneau pigeon: further studies of the role of carbon monoxide and dietary cholesterol. Atherosclerosis, **34**: 407-417.

Turner JAM, McNicol MW, & Sillett RW (1986) Distribution of carboxyhaemoglobin concentrations in smokers and non-smokers. Thorax, **41**: 25-27.

UNEP/WHO (1993) UNEP/WHO GEMS/Air — A global programme for urban air quality monitoring and assessment. Nairobi, United Nations Environment Programme/Geneva, World Health Organization (WHO/PEP 93.7; UNEP/GEMS/93.A.1).

US Bureau of the Census (1982) 1980 Census of population and housing supplementary report: provisional estimates of social, economic, and housing characteristics: states and selected standard metropolitan statistical areas. Washington, DC, US Department of Commerce, Bureau of the Census (Report No. PHC 80-S1-1).

US Centers for Disease Control (1982) Carbon monoxide intoxication — a preventable environmental health hazard. Morb Mortal Wkly Rep, **31**: 529-531.

US Department of Commerce (1987) Statistical abstract of the United States 1988, 108th ed. Washington, DC, US Department of Commerce, Bureau of the Census.

US Department of Energy (1982) Estimates of U.S. wood energy consumption from 1949 to 1981. Washington, DC, Energy Information Administration, Office of Coal, Nuclear, Electricity, and Alternate Fuels (Publication No. DOE/EIA-0341, available from NTIS, Springfield, Virginia, USA).

US Department of Energy (1984) Estimates of U.S. wood energy consumption 1980-1983. Washington, DC, Energy Information Administration, Office of Coal, Nuclear, Electric and

Alternate Fuels (Publication No. DOE/EIA-0341[83], available from NTIS, Springfield, Virginia, USA).

US Department of Energy (1988a) Petroleum marketing monthly. Washington, DC, Energy Information Administration (Publication No. DOE/EIA-0380[88/06]).

US Department of Energy (1988b) Coal distribution January-December. Washington, DC, Energy Information Administration (Publication No. DOE/EIA-25[88/4Q]).

US Department of Energy (1988c) Electric power annual. Washington, DC, Energy Information Administration (Publication No. DOE/EIA-0348[87]).

US Department of Health and Human Services (1990) Vital and health statistics — Current estimates from the National Health Interview Survey, 1989. Hyattsville, Maryland, Public Health Service, National Center for Health Statistics (Series 10: Data from the National Health Survey, No. 176) (DHHS Publication No. (PHS) 90-1504).

US Department of Transportation (1988) Highway statistics. Washington, DC, Federal Highway Administration.

US EPA (1979a) Control techniques for carbon monoxide emissions. Research Triangle Park, North Carolina, US Environmental Protection Agency, Office of Air Quality Planning and Standards, Emission Standards and Engineering Division (EPA-450/3-79-006).

US EPA (1979b) Air quality criteria for carbon monoxide. Research Triangle Park, North Carolina, US Environmental Protection Agency, Office of Health and Environmental Assessment, Environmental Criteria and Assessment Office (EPA-600/8-79-022).

US EPA (1983) Controlling emissions from light-duty motor vehicles at higher elevations: a report to Congress. Ann Arbor, Michigan, US Environmental Protection Agency, Office of Mobile Source Air Pollution Control (EPA-460/3-83-001).

US EPA (1985) Compilation of air pollutant emission factors, Volume I: Stationary point and area sources; Volume II: Mobile sources, 4th ed. Research Triangle Park, North Carolina, US Environmental Protection Agency, Office of Air Quality Planning and Standards (EPA Reports Nos AP-42-ED-4-VOL-1 and AP-42-ED-4-VOL-2).

US EPA (1989a) User's guide to MOBILE4 (Mobile Source Emission Factor Model). Ann Arbor, Michigan, US Environmental Protection Agency, Office of Mobile Sources (EPA Report No. EPA-AA-TEB-89-01).

US EPA (1989b) Adjustment of MOBILE4 idle CO emission factors to non-standard operating conditions. Ann Arbor, Michigan, US Environmental Protection Agency, Office of Air and Radiation, Office of Mobile Sources (Unpublished document).

US EPA (1991a) National air quality and emissions trends report, 1990. Research Triangle Park, North Carolina, US Environmental Protection Agency, Office of Air Quality Planning and Standards (EPA/450/4-91/023).

US EPA (1991b) National air pollutant emission estimates 1940-1990. Research Triangle Park, North Carolina, US Environmental Protection Agency, Office of Air Quality Planning and Standards (EPA/450/4-91/028).

US EPA (1991c) National air pollutant emission estimates 1940-1989. Research Triangle Park, North Carolina, US Environmental Protection Agency, Office of Air Quality Planning and Standards (EPA-450/4-91-004).

US EPA (1991d) Air quality criteria for carbon monoxide. Research Triangle Park, North Carolina, US Environmental Protection Agency, National Centre for Environmental Assessment (EPA/600/8-90/045F).

US Forest Service (1988) Wildfire statistics. Washington, DC, US Department of Agriculture, State and Private Forestry.

Van Hoesen KB, Camporesi EM, Moon RE, Hage ML, & Piantadosi CA (1989) Should hyperbaric oxygen be used to treat the pregnant patient for acute carbon monoxide poisoning? A case report and literature review. J Am Med Assoc, **261**: 1039-1043.

Van Netten C, Brubaker RL, Mackenzie CJG, & Godolphin WJ (1987) Blood lead and carboxy-hemoglobin levels in chainsaw operators. Environ Res, **43**: 244-250.

Vanoli E, De Ferrari GM, Stramba-Badiale M, Farber JP, & Schwartz PJ (1989) Carbon monoxide and lethal arrhythmias in conscious dogs with a healed myocardial infarction. Am Heart J, **117**: 348-357.

Van Wijnen JH, Verhoeff AP, Jans HWA, & van Bruggen M (1995) The exposure of cyclists, car drivers and pedestrians to traffic-related air pollutants. Int Arch Occup Environ Health, **67**: 187-193.

Venning H, Roberton D, & Milner AD (1982) Carbon monoxide poisoning in an infant. Br Med J, **284**: 651.

Verdin A (1973) Gas analysis instrumentation. New York, John Wiley & Sons, Inc.

Verhoeff AP, van der Velde HCM, Boleij JSM, Lebret E, & Brunekreef B (1983) Detecting indoor CO exposure by measuring CO in exhaled breath. Int Arch Occup Environ Health, **53**: 167-173.

Verhoeff AP, Hoek G, Schwartz J, & van Wijnen JH (1996) Air pollution and daily mortality in Amsterdam. Epidemiology, **7**(3): 225-230.

Verma A, Hirsch DJ, Glatt CE, Ronnett GV, & Snyder SH (1993) Carbon monoxide — A putative neural messenger. Science, **259**: 381-384.

Verrier RL, Mills AK, & Skornik WA (1990) Acute effects of carbon monoxide on cardiac electrical stability. Cambridge, Massachusetts, Health Effects Institute (Research Report No. 35).

Vincent SR, Das S, & Maines MD (1994) Brain heme oxygenase isoenzymes and nitric oxide synthase are co-localized in select neurons. Neuroscience, **63**: 223-231.

Virtamo M & Tossavainen A (1976) Carbon monoxide in foundry air. Scand J Work Environ Health, **2**(suppl 1): 37-41.

Vogel JA & Gleser MA (1972) Effect of carbon monoxide on oxygen transport during exercise. J Appl Physiol, **32**: 234-239.

Vollmer EP, King BG, Birren JE, & Fisher MB (1946) The effects of carbon monoxide on three types of performance, at simulated altitudes of 10,000 and 15,000 feet. J Exp Psychol, **36**: 244-251.

Volz A, Ehhalt DH, & Derwent RG (1981) Seasonal and latitudinal variation of ^{14}CO and the tropospheric concentration of OH radicals. J Geophys Res Oceans Atmos, **C86**: 5163-5171.

Von Post-Lingen M-L (1964) The significance of exposure to small concentrations of carbon monoxide: results of an experimental study on healthy persons. Proc R Soc Med, **57**: 1021-1029.

Von Restorff W & Hebisch S (1988) Dark adaptation of the eye during carbon monoxide exposure in smokers and nonsmokers. Aviat Space Environ Med, **59**: 928-931.

Vreman HJ, Kwong LK, & Stevenson DK (1984) Carbon monoxide in blood: an improved microliter blood-sample collection system, with rapid analysis by gas chromatography. Clin Chem, **30**: 1382-1386.

Wade WA III, Cote WA, & Yocom JE (1975) A study of indoor air quality. J Air Pollut Control Assoc, **25**: 933-939.

Wagner JA, Horvath SM, & Dahms TE (1975) Carbon monoxide elimination. Respir Physiol, **23**: 41-47.

Wagner JA, Horvath SM, Andrew GM, Cottle WH, & Bedi JF (1978) Hypoxia, smoking history, and exercise. Aviat Space Environ Med, **49**: 785-791.

Wald N, Howard S, Smith PG, & Kjeldsen K (1973) Association between atherosclerotic diseases and carboxyhaemoglobin levels in tobacco smokers. Br Med J, **1**(865): 761-765.

Wald NJ, Idle M, Boreham J, & Bailey A (1981) Carbon monoxide in breath in relation to smoking and carboxyhaemoglobin levels. Thorax, **36**: 366-369.

Wallace LA (1983) Carbon monoxide in air and breath of employees in an underground office. J Air Pollut Control Assoc, **33**: 678-682.

Wallace LA & Ziegenfus RC (1985) Comparison of carboxyhemoglobin concentrations in adult nonsmokers with ambient carbon monoxide levels. J. Air Pollut Control Assoc, **35**: 944-949.

Wallace ND, Davis GL, Rutledge RB, & Kahn A (1974) Smoking and carboxyhemoglobin in the St. Louis metropolitan population: theoretical and empirical considerations. Arch Environ Health, **29**: 136-142.

Wallace LA, Thomas J, & Mage DT (1984) Comparison of end-tidal breath CO estimates of COHb with estimates based on exposure profiles of individuals in the Denver and Washington, DC area. Research Triangle Park, North Carolina, US Environmental Protection Agency, Environmental Monitoring Systems Laboratory (EPA-600/D-84-194).

Wallace LA, Pellizzari ED, Hartwell TD, Sparacino CM, Sheldon LS, & Zelon HS (1985) Results from the first three seasons of the TEAM Study: personal exposures, indoor-outdoor relationships, and breath levels of toxic air pollutants measured for 355 persons in New Jersey. Presented at the 78th Annual Meeting of the Air Pollution Control Association, Detroit, MI, June 1985. Pittsburgh, Pennsylvania, Air Pollution Control Association (Paper No. 85-31.6).

Wallace L, Thomas J, Mage D, & Ott W (1988) Comparison of breath CO, CO exposure, and Coburn model predictions in the U.S. EPA Washington-Denver (CO) study. Atmos Environ, **22**: 2183-2193.

Walsh MP & Nussbaum BD (1978) Who's responsible for emissions after 50,000 miles? Automot Eng, **86**: 32-35.

Ward TV & Zwick HH (1975) Gas cell correlation spectrometer: GASPEC. Appl Opt, **14**: 2896-2904.

Warneck P (1988) Chemistry of the natural atmosphere. New York, London, Academic Press.

Wazawa H, Yamamoto K, Yamamoto Y, Matsumoto H, & Fukui Y (1996) Elimination of carbon monoxide from the body: an experimental study on the rabbit. Nippon Hoigaku Zasshi, **50**: 258-262.

Weber A (1984) Annoyance and irritation by passive smoking. Prev Med, **13**: 618-625.

Weber A, Jermini C, & Grandjean E (1976) Irritating effects on man of air pollution due to cigarette smoke. Am J Public Health, **66**: 672-676.

Weber A, Fischer T, & Grandjean E (1979a) Passive smoking in experimental and field conditions. Environ Res, **20**: 205-216.

Weber A, Fischer T, & Grandjean E (1979b) Passive smoking: irritating effects of the total smoke and the gas phase. Int Arch Occup Environ Health, **43**: 183-193.

Webster WS, Clarkson TB, & Lofland HB (1970) Carbon monoxide-aggravated atherosclerosis in the squirrel monkey. Exp Mol Pathol, **13**: 36-50.

Weir FW, Rockwell TH, Mehta MM, Attwood DA, Johnson DF, Herrin GD, Anglen DM, & Safford RR (1973) An investigation of the effects of carbon monoxide on humans in the driving task: final report (Contracts Nos 68-02-0329 and CRC-APRAC; Project CAPM-9-69). Columbus, Ohio, The Ohio State University Research Foundation (Publication PB-224646, available from NTIS, Springfield, Virginia, USA).

Weiser PC, Morrill CG, Dickey DW, Kurt TL, & Cropp GJA (1978) Effects of low-level carbon monoxide exposure on the adaptation of healthy young men to aerobic work at an altitude of 1,610 meters. In: Folinsbee LJ, Wagner JA, Borgia JF, Drinkwater BL, Gliner JA, & Bedi JF ed. Environmental stress: individual human adaptations. New York, London, Academic Press, pp 101-110.

Weiss HR & Cohen JA (1974) Effects of low levels of carbon monoxide on rat brain and muscle tissue P_{O_2}. Environ Physiol Biochem, **4**: 31-39.

Weissbecker L, Carpenter RD, Luchsinger PC, & Osdene TS (1969) In vitro alveolar macrophage viability: effect of gases. Arch Environ Health, **18**: 756-759.

Wells AJ (1996) Passive smoking as a cause of heart disease. J Am Coll Cardiol, **24**(2): 546-554.

Went FW (1960) Organic matter in the atmosphere, and its possible relation to petroleum formation. Proc Natl Acad Sci (USA), **46**: 212-221.

Went FW (1966) On the nature of Aitken condensation nuclei. Tellus, **18**: 549-556.

Werner B & Lindahl J (1980) Endogenous carbon monoxide production after bicycle exercise in health subjects and in patients with hereditary spherocytosis. Scand J Clin Lab Invest, **40**: 319-324.

Wharton DC & Gibson QH (1976) Cytochrome oxidase from *Pseudomonas aeruginosa*: IV. Reaction with oxygen and carbon monoxide. Biochim Biophys Acta, **430**: 445-453.

White RE & Coon MJ (1980) Oxygen activation by cytochrome P-450. Annu Rev Biochem, **49**: 315-356.

Whitmore RW, Jones SM, & Rosenzweig MS (1984) Final sampling report for the study of personal CO exposure. Research Triangle Park, North Carolina, US Environmental Protection Agency, Environmental Monitoring Systems Laboratory (EPA-600/4-84-034).

WHO (1987) Air quality guidelines for Europe. Copenhagen, World Health Organization, Regional Office for Europe (WHO Regional Publications, European Series, No. 23).

WHO (1997a) Healthy cities — Air management information system. Geneva, World Health Organization (CD- ROM).

WHO (1997b) The world health report 1997. Geneva, World Health Organization.

WHO/UNEP (World Health Organization/United Nations Environment Programme) (1992) Urban air pollution in megacities of the world. Oxford, Blackwell.

Wietlisbach V, Pope CA III, & Ackermann-Liebrich U (1996) Air pollution and daily mortality in three Swiss urban areas. Soz Präventivmed, **41**(2): 107-115.

Wigfield DC, Hollebone BR, MacKeen JE, & Selwin JC (1981) Assessment of the methods available for the determination of carbon monoxide in blood. J Anal Toxicol, **5**: 122-125.

Wilkniss PE, Lamontagne RA, Larson RE, Swinnerton JW, Dickson CR, & Thompson T (1973) Atmospheric trace gases in the southern hemisphere. Nature (Lond), **245**: 45-47.

Wilks SS (1959) Carbon monoxide in green plants. Science, **129**: 964-966.

Wilks SS, Tomashefski JF, & Clark RT Jr (1959) Physiological effects of chronic exposure to carbon monoxide. J Appl Physiol, **14**: 305-310.

Wilson DA, Hanley DF, Feldman MA, & Traystman RJ (1987) Influence of chemoreceptors on neurohypophyseal blood flow during hypoxic hypoxia. Circ Res, **61**(suppl II): II/94-II/101.

Winneke G (1974) Behavioral effects of methylene chloride and carbon monoxide as assessed by sensory and psychomotor performance. In: Xintaras C, Johnson BL, & de Groot I ed. Behavioral toxicology: early detection of occupational hazards — Proceedings of a workshop, Cincinnati, OH, June 1973. Cincinnati, Ohio, National Institute for Occupational Safety and Health, pp 130-144 (DHEW Publication No. (NIOSH) 74-126).

Winston JM & Roberts RJ (1975) Influence of carbon monoxide, hypoxic hypoxia or potassium cyanide pretreatment on acute carbon monoxide and hypoxic hypoxia lethality. J Pharmacol Exp Ther, **193**: 713-719.

Winter PM & Miller JN (1976) Carbon monoxide poisoning. J Am Med Assoc, **236**: 1503.

Wittenberg BA & Wittenberg JB (1985) Oxygen pressure gradients in isolated cardiac myocytes. J Biol Chem, **260**: 6548-6554.

Wittenberg BA & Wittenberg JB (1987) Myoglobin-mediated oxygen delivery to mitochondria of isolated cardiac myocytes. Proc Natl Acad Sci (USA), **84**: 7503-7507.

Wittenberg BA & Wittenberg JB (1993) Effects of carbon monoxide on isolated heart muscles cells. Cambridge, Massachusetts, Health Effects Institute, pp 1-12 (Research Report No. 62).

Wittenberg BA, Wittenberg JB, & Caldwell PRB (1975) Role of myoglobin in the oxygen supply to red skeletal muscle. J Biol Chem, **250**: 9038-9043.

Wohlrab H & Ogunmola GB (1971) Carbon monoxide binding studies of cytochrome a_3 hemes in intact rat liver mitochondria. Biochemistry, **10**: 1103-1106.

Woodman G, Wintoniuk DM, Taylor RG, & Clarke SW (1987) Time course of end-expired carbon monoxide concentration is important in studies of cigarette smoking. Clin Sci, **73**: 553-555.

WMO (1986) Carbon monoxide (CO). In: Atmospheric ozone 1985: Assessment of our understanding of the processes controlling its present distribution and change. Geneva, World Meteorological Organization, vol 1, pp 100-106 (Global Ozone Research and Monitoring Project, Report No. 16).

Wouters EJM, de Jong PA, Cornelissen PJH, Kurver PHJ, van Oel WC, & van Woensel CLM (1987) Smoking and low birth weight: absence of influence by carbon monoxide? Eur J Obstet Gynecol Reprod Biol, **25**: 35-41.

Wright GR & Shephard RJ (1978a) Brake reaction time — effects of age, sex, and carbon monoxide. Arch Environ Health, **33**: 141-150.

Wright GR & Shephard RJ (1978b) Carbon monoxide exposure and auditory duration discrimination. Arch Environ Health, **33**: 226-235.

Wright G, Randell P, & Shephard RJ (1973) Carbon monoxide and driving skills. Arch Environ Health, **27**: 349-354.

Wright GR, Jewczyk S, Onrot J, Tomlinson P, & Shephard RJ (1975) Carbon monoxide in the urban atmosphere: hazards to the pedestrian and the street-worker. Arch Environ Health, **30**: 123-129.

Wu WX (1992) Factors influencing carboxyhemoglobin kinetics in inhalation lung injury. Chung Hua Nei Ko Tsa Chih, **31**: 689-691, 730.

Wyman J, Bishop G, Richey B, Spokane R, & Gill S (1982) Examination of Haldane's first law for the partition of CO and O_2 to hemoglobin A_0. Biopolymers, **21**: 1735-1747.

Yamate G (1974) Emissions inventory from forest wildfires, forest managed burns, and agricultural burns. Research Triangle Park, North Carolina, US Environmental Protection Agency, Office of Air Quality Planning and Standards (EPA-450/3-74-062).

Yang L, Zhang W, He H, & Zhang G (1988) Experimental studies on combined effects of high temperature and carbon monoxide. J Tongji Med Univ, **8**: 60-65.

Yocom JE (1982) Indoor-outdoor air quality relationships: a critical review. J Air Pollut Control Assoc, **32**: 500-520.

Yocom JE, Clink WL, & Cote WA (1971) Indoor/outdoor air quality relationships. J Air Pollut Control Assoc, **21**: 251-259.

Young LJ & Caughey WS (1986) Mitochondrial oxygenation of carbon monoxide. Biochem J, **239**: 225-227.

Young SH & Stone HL (1976) Effect of a reduction in arterial oxygen content (carbon monoxide) on coronary flow. Aviat Space Environ Med, **47**: 142-146.

Young LJ, Choc MG, & Caughey WS (1979) Role of oxygen and cytochrome *c* oxidase in the detoxification of CO by oxidation to CO_2. In: Caughey WS & Caughey H ed. Biochemical and clinical aspects of oxygen: Proceedings of a symposium, Fort Collins, CO, September 1975. New York, London, Academic, pp 355-361.

Young JS, Upchurch MB, Kaufman MJ, & Fechter LD (1987) Carbon monoxide exposure potentiates high-frequency auditory threshold shifts induced by noise. Hear Res, **26**: 37-43.

Zebro T, Wright EA, Littleton RJ, & Prentice AID (1983) Bone changes in mice after prolonged continuous exposure to a high concentration of carbon monoxide. Exp Pathol, **24**: 51-67.

Zimmerman PR, Chatfield RB, Fishman J, Crutzen PJ, & Hanst PL (1978) Estimates on the production of CO and H_2 from the oxidation of hydrocarbon emissions from vegetation. Geophys Res Lett, **5**: 679-682.

Zinkham WH, Houtchens RA, & Caughey WS (1980) Carboxyhemoglobin levels in an unstable hemoglobin disorder (Hb Zuerich): effect on phenotypic expression. Science, **209**: 406-408.

Ziskind RA, Rogozen MB, Carlin T, & Drago R (1981) Carbon monoxide intrusion into sustained-use vehicles. Environ Int, **5**: 109-123.

Ziskind RA, Fite K, & Mage DT (1982) Pilot field study: carbon monoxide exposure monitoring in the general population. Environ Int, **8**: 283-293.

Zorn H (1972) The partial oxygen pressure in the brain and liver at subtoxic concentrations of carbon monoxide. Staub-Reinhalt Luft, **32**: 24-29.

Zwart A & van Kampen EJ (1985) Dyshaemoglobin, especially carboxyhaemoglobin, levels in hospitalized patients. Clin Chem, **31**: 945.

Zwart A, Buursma A, Oeseburg B, & Zijlstra WG (1981a) Determination of hemoglobin derivatives with the IL 282 CO-oximeter as compared with a manual spectrophotometric five-wavelength method. Clin Chem, **27**: 1903-1907.

Zwart A, Buursma A, van Kampen EJ, Oeseburg B, van der Ploeg PHW, & Zijlstra WG (1981b) A multi-wavelength spectrophotometric method for the simultaneous determination of five haemoglobin derivatives. J Clin Chem Clin Biochem, **19**: 457-463.

Zwart A, Buursma A, van Kampen EJ, & Zijlstra WG (1984) Multicomponent analysis of hemoglobin derivatives with a reversed optics spectrophotometer. Clin Chem, **30**: 373-379.

Zwart A, van Kampen EJ, & Zijlstra WG (1986) Results of routine determination of clinically significant hemoglobin derivatives by multicomponent analysis. Clin Chem, **32**: 972-978.

RESUME ET CONCLUSIONS

Le monoxyde de carbone appelé aussi oxyde de carbone (CO) est un gaz incolore et inodore qui peut être toxique pour l'Homme. Il résulte de la combustion incomplète de substances contenant du carbone et se forme également dans l'organisme humain à la faveur de processus naturels ou encore par suite de la biotransformation des halométhanes. Lors d'une exposition externe à des teneurs en monoxyde de carbone supérieures à la normale, des effets subtils peuvent commencer à se manifester et en cas d'exposition à des concentrations encore plus élevées, la mort peut survenir. Les effets toxiques du monoxyde de carbone sont dus pour une grande part à la formation de carboxyhémoglobine (COHb), qui empêche le transport de l'oxygène par le sang.

1. Chimie et méthodes d'analyse

Il existe, pour doser le monoxyde de carbone dans l'air ambiant, toute une série de possibilités qui vont des méthodes totalement automatisées basées sur l'arbsorptiométrie infrarouge non dispersive ou la chromatographie en phase gazeuse à des techniques manuelles semiquantitatives utilisant des tubes détecteurs. Comme la formation de carboxyhémoglobine dans l'organisme humain dépend d'un grand nombre de facteurs, et notamment du caractère variable de la teneur de l'air en monoxyde de carbone, il est préférable de mesurer la concentration de carboxyhémoglobine que de chercher à la calculer. Il existe un certain nombre de méthodes relativement simples pour doser le monoxyde de carbone en analysant le sang ou l'air alvéolaire en équilibre avec le sang. Certaines de ces méthodes ont pu être validées grâce à une études comparatives minutieuses.

2. Sources et concentrations de monoxyde de carbone dans l'environnement

Le monoxyde de carbone est présent dans la troposphère à l'état de traces qui trouvent leur origine dans des processus naturels et dans certaines activités humaines. Etant donné que les végétaux sont capables de métaboliser le monoxyde de carbone et d'en produire, on considère qu'à l'état de traces, ce gaz est un constituant normal de l'environnement naturel. La concentration de monoxyde de carbone dans l'air aux abords des agglomérations et des zones industrielles peut être sensiblement supérieure à la teneur naturelle normale, mais on n'a encore jamais fait état d'effets nocifs sur les plantes ou les microoganismes qui seraient imputables aux teneurs que l'on mesure actuellement dans ces conditions. Il n'en reste pas moins que la présence de monoxyde de carbone à ces concentrations peut être dommageable à la santé humaine, selon les valeurs qu'elles atteignent sur les lieux de travail ou les zones de résidence et en fonction également de la réceptivité des sujets exposés aux effets nocifs potentiels.

En examinant les données de qualité de l'air fournies par les stations de contrôle fixes, on constate une tendance au déclin de la teneur en monoxyde de carbone qui traduit l'efficacité des systèmes antipollution dont sont munis les véhicules récents. Aux Etats-Unis, les émissions dues aux véhicules à moteur qui circulent sur les autoroutes représentent environ 50 % du total, les moyens de transports circulant hors des autoroutes en représentant 13 %. Parmi les autres sources émettant du monoxyde de carbone on peut citer l'utilisation d'autres combustibles que les carburants automobiles, par exemple dans les chaudières (12 %), divers processus industriels (8 %), l'élimination des déchets solides (3 %) et diverses autres sources (14 %).

A l'intérieur des bâtiments, la concentration en monoxyde de carbone dépend de la concentration dans l'air extérieur, de la présence de sources internes, de la ventilation et du brassage de l'air dans chaque pièce et d'une pièce à l'autre. Dans les habitation où il n'en existe pas d'autres sources, la concentration de monoxyde de carbone dans l'air est à peu près la même que la concentration extérieure moyenne. Les concentrations les plus élevées se rencontrent en présence de sources intérieures de combustion, notamment dans les garages fermés, les stations service et les restaurants, par exemple. C'est dans l'air des maisons d'habitation, des églises et des établissements de soins que la concentration en monoxyde de carbone est la plus faible. On a montré que le tabagisme passif résultant de l'exposition à la fumée de cigarette accroît l'exposition des non fumeurs d'environ 1,7 mg/m^3 en moyenne (1,5 ppm) et que l'utilisation d'une cuisinière à gaz augmente la teneur de l'air en CO d'environ 2,9 mg/m^3 (2,5 ppm). Parmi les autres sources de monoxyde de carbone dans l'air intérieur des habitations on peut citer les cheminées, les chauffe-eau ainsi que les poêles à bois ou à charbon.

3. Distribution et transformation dans l'environnement

Il ressort des données récentes sur l'évolution des concentrations en monoxyde de carbone à l'échelle de la planète que celles-ci sont en diminution depuis une dizaine d'années. Le niveau de fond est de l'ordre de 60 à 140 µg/m³ (50 à 120 parties par milliard). Les valeurs sont plus fortes dans l'hémisphère nord que dans l'hémisphère sud. Le niveau de fond moyen fluctue également selon la saison. Les valeurs sont plus fortes durant les mois d'hiver et plus faibles en été. Environ 60 % du monoxyde de carbone présent dans la troposphère hors des agglomérations urbaines trouvent leur origine dans des activités humaines, soit de façon directe par suite de processus de combustion, soit de façon indirecte par suite de l'oxydation d'hydrocarbures et notamment du méthane lors d'activités agricoles, de l'enfouissement de déchets etc. Dans l'atmosphère, le monoxyde de carbone peut intervenir dans des réactions susceptibles d'aboutir à la formation d'ozone au niveau de la troposphère. D'autres réactions peuvent conduire à réduire fortement la concentration en radicaux hydroxyles, qui jouent un rôle clé dans les cycles planétaires d'élimination de nombreux autres gaz résultant de l'activité humaine et qui sont également présents dans l'atmosphère à l'état de traces. Il pourrait s'ensuivre une modification de la chimie atmosphérique et en fin de compte, un changement du climat général de la planète.

4. Exposition de la population au monoxyde de carbone

Au cours de ses activités quotidiennes habituelles, tout un chacun se trouve en présence de monoxyde de carbone dans divers microenvironnements - en se déplaçant à bord d'un véhicule à moteur, en exerçant son activité professionnelle, en se rendant sur des sites urbains où diverses substances sont brûlées, en faisant la cuisine ou en se chauffant au gaz, au charbon ou au bois ou encore en fumant ou en étant exposé à la fumée des autres. En règle générale, c'est à proximité d'un véhicule en marche ou dans certains microenvironnements domestiques que l'exposition au monoxyde de carbone est la plus importante pour la majorité des gens.

Grâce à des moniteurs élecrochimiques personnels (MEP) que leur miniaturisation permet de porter sur soi, il est possible de mesurer les concentrations de monoxyde de carbone qu'un individu est susceptible de rencontrer lorsqu'il se déplace à travers un grand nombre de microenvironnements extérieurs et intérieurs que l'on ne peut surveiller à l'aide de station de mesure fixes. Les résultats des mesures sur le terrain ainsi que les études de modélisation indiquent qu'il n'y a pas de corrélation directe entre l'exposition individuelle déterminée par les MEP et la concentration de monoxyde de carbone mesurée en utilisant uniquement des stations fixes. Cette observation s'explique par la mobilité des personnes et également par les variations spatio-temporelles de la teneur en CO. Les études de grande envergure portant sur l'exposition humaine au monoxyde de carbone ne mettent pas en évidence de corrélation entre les chiffres des moniteurs personnels et ceux des stations de mesure fixes les plus proches, mais elles révèlent que l'exposition individuelle cumulée est plus faible les jours où la concentration ambiante de CO mesurée par les stations fixes est basse et qu'elle est plus élevée lorsque cette concentration est en augmentation. Ces études soulignent la nécessité de compléter les mesures des stations fixes par celles des moniteurs personnels si l'on veut évaluer l'exposition humaine totale. Les données fournies par ces études sur le terrain peuvent être utilisées pour établir et expérimenter des modèles d'exposition humaine rendant compte de l'évolution de l'exposition au CO en fonction du temps et de l'activité.

L'étude des diverses situations conduisant à une exposition humaine au monoxyde de carbone indique que sur certains lieux de travail ou dans certaines habitations où les appareils à combustion sont défectueux ou dont la ventilation n'est pas suffisante, montre que l'exposition peut dépasser 110 mg de monoxyde de carbone par m³ (100 ppm), ce qui conduit souvent à des taux de carboxyhémoglobine de 10 % ou davantage en cas d'exposition continue. En revanche, la population générale est beaucoup moins souvent exposée à de telles concentrations dans les conditions ambiantes habituelles. Les concentrations auxquelles la population générale est exposée pendant des périodes plus ou moins longues se situent plus fréquemment dans la fourchette de 29 à 57 mg de monoxyde de carbone par m³ (25-50 ppm); dans ces circonstances, l'activité physique est généralement réduite et le taux résultant de carboxyhémoglobine ne dépasse pas 1-2 % chez les non fumeurs. Ces valeurs sont comparables à la norme physiologique pour des non fumeurs, qui est de 0,3 à 0,7 %. Chez les fumeurs en revanche, le taux de base de carboxyhémoglobine est de 4 % en moyenne, avec un intervalle de variation qui est habituellement de 3-8 %, et il traduit l'absorption du CO présent dans la fumée inhalée.

Les études consacrées à l'exposition humaine ont montré que les gaz d'échappement des véhicules à moteur sont la source la plus fréquente des concentrations élevées de CO. Elles montrent en particulier qu'à l'intérieur d'un véhicule à moteur, la concentration moyenne maximale en monoxyde de carbone (en moyenne 10-29 mg/m^3, soit 9-25 ppm) est la plus forte de tous les microenvironnements. En outre, pour les habitants des banlieues, la concentration varie dans des proportions extrêmes, et l'exposition peut atteindre dans certains cas plus de 40 mg/m^3 (35 ppm).

Les lieux de travail constituent également un environnement important en ce qui concerne l'exposition au monoxyde de carbone. En général, si l'on excepte les trajets pour se rendre au travail et en revenir, l'exposition pendant l'activité professionnelle est supérieure à ce qu'elle peut être pendant les période de loisirs. Il arrive qu'exposition professionnelle et non professionnelle se juxtaposent, ce qui peut conduire à une concentration encore plus élevée de monoxyde de carbone dans le sang. Mais plus encore, certaines activités professionnelles, de par leur nature même, peuvent comporter un risque accru de forte exposition au CO (par exemple, la conduite et l'entretien de véhicules ou le gardiennage de parkings). Parmi les professions exposées au monoxyde de carbone présent dans les gaz d'échappement de véhicules à moteur, on peut citer les mécaniciens automobiles, les gardiens de parkings et les employés de stations service, les chauffeurs de bus, de camions ou de taxis et enfin les employés des entrepôts. Certains processus industriels peuvent également exposer les ouvriers à du monoxyde de carbone produit directement ou en tant que sous-produit d'autres activités : production d'acier, fours à coke, production de noir de carbone et raffinage du pétrole, par exemple. Les pompiers, les cuisiniers et les ouvriers du bâtiment peuvent également être exposés à de fortes concentrations de monoxyde de carbone. L'exposition au monoxyde de carbone dans l'industrie est la plus forte qui ait été observée au niveau individuel lors des enquêtes sur le terrain.

5. Toxicocinétique et modes d'action du monoxyde de carbone

Le monoxyde de carbone est absorbé au niveau des poumons et la concentration de carboxyhémoglobine à un instant quelconque dépend de plusieurs facteurs. Lorsqu'il y a équilibre avec l'air ambiant, la teneur du sang en carboxyhémoglobine dépend essentielle- ment de la concentration respective du monoxyde de carbone et de l'oxygène inspirés. En revanche, s'il n'y a pas équilibre, la concentration de la carboxyhémoglobine va également dépendre de la durée de l'exposition, de la ventilation pulmonaire et de la quantité de carboxyhémoglobine initialement présente avant l'inhalation d'air contaminé. En plus de sa réaction sur l'hémoglobine, le monoxyde de carbone se combine à la myoglobine, aux cytochromes et aux métalloenzymes telles que la cytochrome-c-oxydase et le cytochrome P-450. On ignore quelle peut être l'incidence de ces réactions sur la santé humaine mais il est probable qu'aux concentrations ambiantes, elles sont moins importantes que la fixation du gaz sur l'hémoglobine.

Les échanges de monoxyde de carbone entre l'air que nous respirons et notre organisme sont sous la dépendance de processus physiques (transferts de masse et diffusion) et physiologiques (par ex. la ventilation alvéolaire et le débit cardiaque). Le monoxyde de carbone passe facilement des poumons au courant sanguin. L'étape finale de ce processus comporte une compétition entre le monoxyde de carbone et l'oxygène pour la fixation sur l'hémoglobine à l'intérieur des hématies, qui conduit respectivement à la formation de carboxyhémoglobine et d'oxyhémoglobine (O_2Hb). La fixation du monoxyde de carbone sur l'hémoglobine pour donner de la carboxyhémoglobine et réduire ainsi la capacité du sang à transporter l'oxygène, se révèle être le principal mécanisme à la base des effets toxiques d'une exposition à de faibles concentration de ce gaz. On ne connaît pas totalement le détail des mécanismes par lesquels la formation de carboxyhémoglobine exerce ses effets toxiques, mais il est probable qu'ils comportent l'apparition d'un état d'hypoxie dans nombre des tissus qui composent les divers organes. On a invoqué d'autres modes d'action ainsi que des mécanismes secondaires pour expliquer la toxicité du monoxyde de carbone (à côté de la formation de carboxyhémoglobine), mais il n'a pu être prouvé que ces mécanismes étaient capables d'intervenir à des niveaux d'exposition relativement faibles (concentrations proches des valeurs ambiantes).

On admet donc, à l'heure actuelle, que le taux sanguin de carboxyhémoglobine constitue un marqueur utile pour l'estimation de la charge interne en monoxyde de carbone due 1) à la formation endogène de cette molécule et 2) à l'exposition au monoxyde de carbone d'origine exogène produit par des sources externes. Le taux de carboxyhémoglobine susceptible de résulter de diverses modalités d'exposition (concernant par exemple la concentration, la durée etc...) à des sources externes peut être calculé assez bien au moyen de l'équation de Coburn-Forster-Kane, appelée aussi équation CFK.

L'exposition au monoxyde de carbone a donc cette caractéristique unique d'être évaluable au moyen du taux de carboxyhémoglobine, qui se révèle donc être un marqueur biologique intéressant de la dose reçu par le sujet exposé. La quantité de carboxyhémoglobine formée dépend de la concentration du gaz et de la durée d'exposition, de l'activité physique exercée lors de l'exposition (qui accroît le volume d'air inhalé par unité de temps), de la température ambiante, de l'état de santé du sujet et de son métabolisme propre. La formation de carboxyhémoglobine est un processus réversible; cependant, en raison de la force de la liaison du monoxyde de carbone à la molécule d'hémoglobine, la demi-vie d'élimination est assez longue et va de 2 à 6,5 h selon la concentration initiale de carboxyhémoglobine et de la fréquence respiratoire du sujet. Il s'ensuit que la carboxyhémoglobine peut s'accumuler et que des concentrations même relativement faibles de monoxyde de carbone sont susceptibles d'en produire de fortes concentrations dans le sang.

On peut déterminer le taux sanguin de carboxyhémoglobine soit directement par analyse du sang, soit indirectement en dosant le monoxyde de carbone dans l'air expiré. Cette dernière méthode a l'avantage de la facilité, de la rapidité, de la précision et d'une meilleure coopération du sujet que dans le cas d'une analyse sanguine. Toutefois, dans le cas d'une exposition à une faible concentration de monoxyde de carbone présente dans l'environnement, l'exactitude de la méthode et la validité de la relation de Haldane entre la concentration dans l'air expiré et la concentration sanguine restent incertaines.

Etant donné la difficulté à obtenir des mesures de carboxyhémoglobine dans la population exposée, on a mis au point des modèles mathématiques qui permettent de déterminer le taux de carboxyhémoglobine résultant d'une exposition à du monoxyde de carbone dans diverses circonstances. Le modèle le mieux adapté aux diverses conditions est l'équation de Coburn-Forster-Kane. La solution linéaire est utile pour l'étude des données de pollution de l'air conduisant à des taux relativement faibles de carboxyhémoglobine, les solutions non linéaires ayant une bonne valeur prédictive même pour les situations où l'exposition au monoxyde de carbone est intense. Les deux modèles de régression peuvent également avoir leur utilité, mais seulement si les conditions d'applications sont très proches de celles dans lesquelles les paramètres ont été établis.

Même si la toxicité du monoxyde de carbone s'explique principalement, en cas d'exposition peu intense, par une hypoxie tissulaire due à la fixation de ce gaz sur l'hémoglobine, il est difficile de bien expliquer certains aspects physiologiques de l'exposition au monoxyde de carbone par une diminution de la pression partielle d'oxygène intracellulaire due à la présence de carboxyhémoglobine. C'est pourquoi de nombreux travaux de recherche sont consacrés à des mécanismes secondaires susceptibles de rendre compte de la toxicité du monoxyde de carbone qui résulte de sa fixation intracellulaire. Il est abondamment attesté que le monoxyde de carbone se combine à divers composés intracellulaires tant *in vivo* qu'*in vitro*. On ignore encore si la fixation intracellulaire du monoxyde de carbone en présence d'hémoglobine est suffisante pour provoquer des troubles organiques aigus ou des effets à long terme. Comme on ne dispose pratiquement pas de techniques sensibles qui puissent nous permettre d'étudier la fixation du monoxyde de carbone dans les conditions physiologiques, diverses méthodes ont été proposées qui abordent le problème de manière indirecte et nombre d'études ont abouti à des résultats négatifs.

A l'heure actuelle, on pense que les protéines intracellulaires qui ont le plus de chances d'être fonctionnellement inhibées par la fixation de monoxyde de carbone lorsque la carboxyhémoglobine atteint un taux significatif sont la myoglobine, qui est principalement présente dans le myocarde et les muscles squelettiques, ainsi que la cytochrome-oxydase. On ignore encore quelle peut être la portée physiologique de la fixation de monoxyde de carbone par la myoglobine, mais il est vrai que si la concentration de carboxy-myoglobine atteint une valeur suffisante, elle peut limiter la fixation de l'oxygène par le muscle pendant l'exercice physique. On dispose de données qui suggèrent que le monoxyde de carbone se combine effectivement à la cytochrome-oxydase dans le tissu cardiaque et cérébral, mais cette combinaison n'est sans doute pas très importante lorsque le taux de carboxyhémoglobine est faible.

6. Effets toxiques de l'exposition au monoxyde de carbone

Du point de vue sanitaire, l'importance de la pollution de l'air par le monoxyde de carbone tient en grande partie au fait que ce gaz se combine énergiquement avec la molécule d'hémoglobine pour former de la carboxyhémoglobine qui réduit la capacité du sang à transporter l'oxygène. La carboxyhémoglobine perturbe également la dissociation intratissulaire de l'oxyhémoglobine, de sorte que cela réduit encore l'apport d'oxygène aux tissus. L'hémoglobine humaine a environ 240 fois plus d'affinité pour le monoxyde de carbone que pour l'oxygène et la proportion relative de carboxyhémoglobine et d'oxyhémoglobine qui se forment dans le sang dépend en grande partie de la pression partielle respective du monoxyde de carbone et de l'oxygène.

Les effets potentiels d'une exposition au monoxyde de carbone ont suscité de nombreuses études tant chez l'Homme que chez l'animal. Divers protocoles expérimentaux ont permis d'obtenir une somme de données sur la toxicité du monoxyde de carbone, sur ses effets directs au niveau du sang et des autres tissus, ainsi que sur la manifestation de ces effets sous forme de perturbations fonctionnelles au niveau des divers organes. Cependant, nombreuses sont les études sur l'animal qui ont été effectuées avec des concentrations extrêmement élevées de monoxyde de carbone (c'est-à-dire ne correspondant nullement à celles qui sont présentes dans l'air ambiant). Il est vrai malgré tout que, même si les graves effets résultant d'une exposition à des concentrations aussi fortes ne sont guère en rapport avec ceux que peut produire l'exposition aux concentrations rencontrées dans l'air ambiant, ces études fournissent néanmoins des renseignements intéressants sur les effets potentiels d'une exposition accidentelle au monoxyde de carbone, en particulier d'une exposition à l'intérieur d'un local.

6.1 Effets cardiovasculaires

On a montré sans ambiguïté qu'à partir d'un taux de carboxyhémoglobine de 5,0 %, il y avait réduction de la fixation d'oxygène et diminution consécutive de la capacité physique dans des conditions d'effort maximal chez de jeunes adultes en bonne santé. En outre, il ressort de plusieurs études que la capacité physique commence déjà à diminuer un peu pour des taux de carboxyhémoglobine n'excédant pas 2,3 à 4,3 %. Ces effets peuvent avoir des conséquences pour la santé de la population, en ce sens qu'ils impliquent un risque de diminution de l'aptitude à exercer certaines activités professionnelles ou récréatives tant soit peu exigeantes sur le plan physique, lorsque l'exposition au monoxyde de carbone est d'une intensité suffisante.

Ce sont cependant certains effets cardiovasculaires qui peuvent être plus à craindre en cas d'exposition à des concentrations ambiantes de monoxyde de carbone plus caractéristiques (notamment l'aggravation d'un angor au cours d'une activité physique) chez une proportion plus faible mais néanmoins non négligeable de la population. Les sujets souffrant d'angine de poitrine chronique sont considérés actuellement comme le groupe le plus réceptif aux effets d'une exposition au monoxyde de carbone, d'après les signes d'aggravation de l'angor constatés chez des patients présentant un taux de carboxyhémoglobine de 2,9 à 4,5 %. Il faudra encore établir de manière plus concluante les relations dose-réponse chez les malades souffrant de cardiopathies coronariennes, mais on ne peut cependant exclure la possibilité que ces effets cardiovasculaires se produisent à des taux de carboxyhémoglobine inférieurs à 2,9 %. C'est pourquoi le présent document comporte une évaluation des études nouvellement publiées sur la question afin de déterminer si le monoxyde de carbone est susceptible d'aggraver un angor à des taux de carboxyhémoglobine situés entre 2 et 6 %.

Cinq études essentielles ont été consacrées à la possibilité d'aggravation d'une ischémie myocardique par une exposition au monoxyde de carbone pendant l'exercice physique chez des malades coronariens. Une première étude avait montré que la durée de l'exercice physique était sensiblement réduite par l'apparition d'une douleur thoracique (angor) chez les sujets souffrant d'angine de poitrine pour des taux post-exposition de carboxyhémoglobine n'excédant pas 2,9 %, c'est-à-dire supérieurs de 1,6 % au taux de base.

Une étude multicentrique de grande envergure a mis en évidence des effets cardiovasculaires chez des malades présentant un angor d'effort reproductible pour des taux post-exposition de carboxyhémoglobine de 3,2 %, c'est-à-dire supérieurs de 2,0 % à la valeur de base. D'autres auteurs ont également observé des effets analogues chez des patients souffrant d'une cardiopathie coronarienne obstructive et mis en évidence une ischémie d'effort à des taux post-exposition de carboxyhémoglobine respectivement égaux à 4,1 et 5,9 %, c'est-à-dire respectivement supérieurs de 2,2 et de 4,2 % au taux de base. Une autre étude, portant sur des sujets souffrant d'angine de poitrine, a mis en évidence des effets cardiovasculaires à un taux de carboxyhémoglobine de 3 %, c'est-à-dire supérieur de 1,5 % à la valeur de base. La valeur la plus faible sans effet nocif observable (LOAEL) chez des malades présentant une ischémie d'effort se situe donc quelque part entre 3 et 4 % de carboxyhémoglobine, c'est-à-dire un taux supérieur de 1,5 à 2,2 % à la valeur de base. On n'a pas étudié les effets du monoxyde de carbone sur les épisodes d'ischémie asymptomatiques, qui représentent en fait la majorité des cas chez ces malades.

Les effets nocifs d'une exposition à de faibles concentrations de monoxyde de carbone chez des malades souffrant de cardiopathie ischémique sont difficiles à prévoir au sein d'une population de cardiaques. On a montré qu'une exposition suffisante pour produire un taux de carboxyhémoglobine de 6 % au moins, augmentait sensiblement le nombre et la complexité des arythmies d'effort en cas de coronaropathie et d'ectopie. Cette observation, si on la rapproche des résultats d'études de morbidité et de mortalité ou d'études épidémiologiques relatives aux ouvriers travaillant dans des tunnels qui sont systématiquement exposés au monoxyde de carbone émis par les véhicules automobiles, incite à penser, sans certitude absolue toutefois, que l'exposition au monoxyde de carbone peut comporter un risque accru de mort subite chez les malades souffrant de coronaropathie.

Des travaux antérieurs consacrés aux effets cardiovasculaires du monoxyde de carbone ont mis en évidence ce qui semble être une relation linéaire entre le taux sanguin de carboxyhémoglobine et une diminution des performances physiques chez les sujets humains mesurées par la consommation maximale d'oxygène. Cette diminution est systématiquement observée pour des taux sanguins de carboxyhémoglobine d'environ 5 %, chez de jeunes sujets non fumeurs en bonne santé. Certaines études ont même mis en évidence une réduction de la durée maximale des exercices physiques courts pour des taux de carboxyhémoglobine n'excédant pas 2,3 à 4,3 %; toutefois cette réduction est si faible qu'elle intéresse davantage les athlètes que les gens ordinaires vaquant à leurs activités quotidiennes.

D'après des considérations théoriques et des études expérimentales sur l'animal, le monoxyde de carbone est susceptible d'exercer des effets nocifs sur le système cardiovasculaire en fonction des conditions d'exposition. Toutefois, les résultats de ces travaux ne sont pas concluants, même s'il révèlent des troubles du rythme et de la conduction chez les animaux en bonne santé comme chez ceux qui présentent des problèmes cardiaques. La concentration la plus faible à laquelle ces troubles ont été observés varie selon les modalités d'exposition et l'espèce étudiée. L'expérimentation animale indique également que l'inhalation de monoxyde de carbone peut accroître la concentration d'hémoglobine et l'hématocrite, ce qui traduit probablement une compensation de la réduction du transport d'oxygène provoquée par ce gaz. A forte concentration de monoxyde de carbone, une augmentation excessive du taux d'hémoglobine et de l'hématocrite peut imposer au coeur une charge supplémentaire et gêner l'apport de sang aux tissus.

Un certain nombre de données contradictoires indiquent qu'une exposition au monoxyde de carbone peut favoriser l'apparition d'athérosclérose chez les animaux de laboratoire, mais la plupart des études ne mettent en évidence aucun effet mesurable. De même, seules quelques études évoquent la possibilité d'une modification du métabolisme des lipides susceptible d'accélérer l'athérosclérose. Cet effet doit être subtil, à tout le moins. Enfin, le monoxyde de carbone agit probablement sur les plaquettes, mais davantage en en inhibant l'agrégation qu'en la favorisant. D'une façon générale, il n'existe guère de données indiquant qu'il puisse se produire dans la population des effets athérogènes consécutifs à une exposition au monoxyde de carbone aux concentrations couramment rencontrées dans l'air ambiant.

6.2   Effets pulmonaires aigus

Il n'est guère probable que le monoxyde de carbone exerce des effets directs sur le tissu pulmonaire sauf en cas d'exposition à des concentrations extrêmement élevées entraînant une intoxication.

L'étude des effets du monoxyde de carbone sur la fonction pulmonaire humaine se heurte à un certain nombre de difficultés, à savoir l'absence d'informations suffisantes sur l'exposition, le petit nombre de sujets étudiés et la durée limitée de l'exposition. Une exposition professionnelle ou accidentelle à des produits de combustion ou de pyrolyse, notamment à l'intérieur d'un local, peut conduire à une chute brutale de la capacité fonctionnelle pulmonaire si le taux de carboxyhémoglobine est élevé. Toutefois, il est difficile de distinguer les effets potentiels du monoxyde de carbone de ceux qui peuvent être dus à des irritants des voies respiratoires présents dans la fumée et les gaz d'échappement. Les études communautaires sur les effets du monoxyde de carbone présent dans l'air ambiant n'ont pas permis de mettre en évidence de relation significative entre ce gaz et certains symptômes ou maladies affectant la fonction pulmonaire.

.3   *Effets sur le système vasculaire cérébral et le comportement*

Il n'existe pas de données fiables qui mettent en évidence une altération des fonctions neurocomportementales chez des jeunes adultes en bonne santé, à des taux de carboxyhémoglobine inférieurs à 5 %. Les résultats des études effectuées en présence de taux égaux ou supérieurs à 5 % sont ambigus. Une grande partie des recherches effectuées sur des sujets dont le taux sanguin de carboxyhémoglobine était de 5 %, n'ont révélé aucun effet, même en considérant des problèmes comportementaux analogues à ceux qui avaient été mis en évidence dans les études portant sur des taux plus élevés de carboxyhémoglobine. Toutefois, il est possible que les chercheurs qui n'avaient pu mettre en évidence d'altérations neurocomportementales à des taux de carboxyhémoglobine égaux ou supérieurs à 5 %, aient utilisé des tests dont la sensibilité était insuffisante pour permettre de déceler les effets subtils du monoxyde de carbone. On peut donc dire, en se fondant sur les données expérimentales, qu'un taux de carboxyhémoglobine supérieur ou égal à 5 % est susceptible d'altérer les fonctions neurocomportementales. Cependant, on me peut affirmer qu'un taux inférieur à 5 % serait sans effet. Il est vrai, toutefois, que l'on n'a étudié que de jeunes adultes en bonne santé en les soumettant à des tests de sensibilité démontrable et dans des conditions où le taux de carboxyhémoglobine était supérieur ou égal à 5 %. Le problème des groupes de population exposés à un risque particulier d'effets neurocomportementaux par suite d'une exposition au monoxyde de carbone reste donc entier.

A noter tout spécialement le cas des personnes qui prennent des médicaments ayant un effet dépresseur primaire ou secondaire et qui risquerait de potentialiser l'altération des fonctions neurocomportementales due à une exposition au monoxyde de carbone. Les autres groupes de population qui peuvent courir un risque accru d'altération neurocomportementale sont les personnes âgées et les malades, mais ce risque n'a pas été évalué.

Dans les conditions normales, le cerveau est capable d'accroître son irrigation ou de capter davantage d'oxygène pour compenser une hypoxie oxycarbonée. La réaction globale du système vasculaire cérébral est analogue chez le foetus, le nouveau-né et l'animal adulte. Toutefois, il semble que plusieurs mécanismes concourent à augmenter le débit sanguin et il est possible qu'interviennent certains facteurs neurologiques et métaboliques, ainsi que la courbe de dissociation de l'oxyhémoglobine, le taux d'oxygène tissulaire et même un effet histotoxique du monoxyde de carbone. La question se pose de savoir si ces mécanismes compensatoires vont continuer à fonctionner dans diverses situations où l'intégrité du système vasculaire cérébral est compromise (par ex. en cas d'accident vasculaire cérébral, de lésion intracrânienne, d'athérosclérose, d'hypertension etc.). La probabilité de lésions ou d'affections de ce genre augmente en tout cas avec l'âge. Il se peut aussi que la sensibilité à la carboxyhémoglobine et les mécanismes compensatoires varient selon les individus.

Ce sont les comportements exigeant une attention ou un effort soutenus qui sont les plus sensibles à une perturbation par la carboxyhémoglobine.Il existe un ensemble d'études sur des sujets humains qui ont été consacrées à la coordination oeil-main (suivi compensatoire), au repérage d'événements rares (vigilance) et à la réalisation de tâches en continu ; ce sont ces études qui rendent compte des effets de la carboxyhémoglobine aux taux n'excédant pas 5 % de la manière la plus cohérente et la plus justifiable. Il est vrai cependant que ces effets du monoxyde de carbone sous faible concentration sont vraiment minimes et quelque peu discutables. Il n'empêche qu'un défaut de coordination et de vigilance pendant l'accomplissement en continu de tâches difficiles, comme lorsqu'on travaille sur des machines ou que l'on conduit des véhicules de transport public, pourrait avoir de sérieuses conséquences.

.4 Effets nocifs sur le développement

Des études sur plusieurs espèces d'animaux de laboratoire tendent fortement à montrer que l'exposition de femelles gravides à du monoxyde de carbone à la concentration de 170-230 mg/m^3 (150-200 ppm), ce qui donne lieu à un taux de carboxyhémoglobine de 15-25 %, entraîne une diminution du poids de naissance, une cardiomégalie, un retard sur le plan comportemental et l'abolition des fonctions cognitives. Des travaux isolés donnent à penser que certains de ces effets pourraient se manifester à des concentrations ne dépassant pas 69-74 mg/m^3 (60-65 ppm; taux de carboxyhémoglobine approximativement égal à 6-11 %) maintenues pendant toute la durée de la gestation. On peut s'inquiéter du fait que des études aient mis en évidence une relation entre l'exposition humaine à du monoxyde de carbone provenant de sources environnementales ou de la fumée de cigarettes et le faible poids de naissance de certains nouveau-nés et cela, du fait de la possibilité de troubles du développement ; il faut dire toutefois que nombre de ces travaux ne prennent pas en compte toutes les sources de monoxyde de carbone. Les données selon lesquelles il existerait un lien entre l'exposition au monoxyde de carbone présent dans l'environnement et la mort subite du nouveau-né ne sont pas convaincantes.

.5 Autres effets généraux

Les études effectuées sur des animaux de laboratoire incitent à penser que la métabolisation enzymatique de composés xénobiotiques pourrait être affectée par une exposition au monoxyde de carbone. La plupart des auteurs de ces études ont toutefois conclu qu'à de faibles taux de carboxyhémoglobine (inférieurs ou égaux à 15 %) les effets étaient attribuables en totalité à l'hypoxie tissulaire due à l'accroissement du taux de carboxyhémoglobine, car ils n'étaient pas plus marqués que ceux qui résultent d'une hypoxie proprement dite. Lorsque l'exposition est plus intense, c'est-à-dire pour un taux de carboxyhémoglobine compris entre 15 et 20 %, le monoxyde de carbone peut inhiber directement l'activité des oxydases à fonction mixte. L'altération du métabolisme des composés xénobiotiques constatée dans les cas d'exposition au monoxyde de carbone pourrait avoir des conséquences importantes pour les sujets qui suivent un traitement médicamenteux.

L'inhalation de monoxyde de carbone sous forte concentration, conduisant à l'apparition dans le sang d'un taux de carboxyhémoglobine supérieur à 10-15 %, produit un certain nombre d'autres effets généraux chez l'animal de laboratoire, ainsi que chez les sujets humains souffrant d'une intoxication oxycarbonée aiguë. Les tissus qui sont le siège d'un intense métabolisme de l'oxygène, comme ceux du coeur, du cerveau, du foie, du rein et des muscles pourraient être particulièrement sensibles à l'intoxication oxycarbonée. On ne connaît pas très bien les effets de fortes concentrations de monoxyde de carbone sur les autres tissus, aussi sont-ils plus incertains. Il est rendu compte, dans la littérature, d'effets sur le foie, le rein, les os et l'immunocompétence du poumon et de la rate. On s'accorde généralement à penser que les lésions tissulaires graves qui se produisent lors d'une intoxication oxycarbonée aiguë ont une ou plusieurs des causes suivantes : 1) ischémie résultant de la formation de carboxyhémoglobine, 2) inhibition de la libération d'oxygène par l'oxyhémoglobine, 3) inhibition de la fonction du cytochrome cellulaire (par ex. des cytochrome-oxydases) et 4) acidose métabolique.

On possède quelques éléments de preuve relativement ténus d'un effet éventuel du monoxyde de carbone sur l'activité fibrinolytique et encore, uniquement en cas d'exposition à de fortes concentrations. De même, certaines données incitent à penser que des effets périnatals sont également possibles (par ex. réduction du poids de naissance, ralentissement du développement postnatal, mort subite du nouveau-né) en cas d'exposition au monoxyde de carbone, mais on manque d'éléments d'appréciation pour pouvoir confirmer qualitativement l'existence d'une telle corrélation chez l'Homme ou établir une relation dose-effet appropriée.

6.6   *Adaptation*

Les seules preuves que l'on possède de la compensation à court ou à long terme d'un taux élevé de carboxyhémoglobine, ou d'une adaptation à cet état de choses, sont des preuves indirectes. Les données obtenues sur des animaux de laboratoire indiquent qu'un accroissement du taux de carboxyhémoglobine produit des réactions physiologiques qui tendent à contrebalancer les autres effets nocifs de l'exposition au monoxyde de carbone. Ces réactions sont les suivantes : 1) augmentation du débit sanguin au niveau des coronaires, 2) augmentation du débit sanguin cérébral, 3) augmentation du taux d'hémoglobine par activation de l'hématopoïèse et 4) augmentation de la consommation d'oxygène par les muscles.

S'agissant du débit sanguin et de la consommation d'oxygène, les réactions compensatoires à court terme peuvent n'être que partielles ou même totalement absentes chez certains sujets. par exemple, on sait, d'après les études sur les animaux de laboratoire, que le débit sanguin au niveau des coronaires s'accroît à mesure qu'augmente le taux de carboxyhémoglobine et les études cliniques sur des sujets humains montrent que chez un patient souffrant de cardiopathie ischémique, une réaction se produit aux taux les plus faibles (6 % ou moins). On peut en déduire que chez certains cardiaques, les mécanismes de compensation à court terme ne jouent plus pleinement leur rôle.

Il apparaît, à la lumière des études neurocomportementales, que l'effet du monoxyde de carbone ne se produit pas systématiquement chez tous les sujets, ni même au cours d'une même étude et on n'a pas mis en évidence de relation dose-effet, c'est-à-dire une intensification avec l'augmentation du taux de carboxyhémoglobine. On peut conclure de ces données que des mécanismes de compensation tels que l'augmentation du débit sanguin, pourraient ne se déclencher qu'à partir d'un certain seuil ou comporter une sorte d'hystérésis. Faute de preuve physiologique directe chez l'animal, ou de préférence, chez l'Homme, cette notion reste ne peut que rester à l'état d'hypothèse.

Le mécanisme d'adaptation à long terme, à supposer qu'on puisse le mettre en évidence chez l'Homme, consisterait semble-t-il en en une augmentation du taux d'hémoglobine consécutive à une activation de l'hématopoïèse. Cette modification de la production d'hémoglobine a été observée à plusieurs reprises chez les animaux de laboratoire, mais aucun travail récent n'indique ou ne suggère qu'il en résulte ou résulterait un avantage adaptatif quelconque. De plus, même si l'accroissement du taux d'hémoglobine signe un mécanisme adaptatif, on ne l'a pas mis en évidence pour de faibles taux de carboxy-hémoglobine.

7. **Exposition au monoxyde de carbone associée à l'altitude, à la prise de médicaments ou de diverses substances chimiques ou encore à une exposition à d'autres polluants atmosphériques ou facteurs environnementaux**

7.1 *Effets d'une altitude élevée*

Il existe de nombreuses études dans lesquelles sont comparés les effets de l'inhalation de monoxyde de carbone et ceux que produit le fait de se trouver à une altitude élevée, mais relativement peu s'intéressent aux effets combinés de ce gaz et de l'altitude. Un certain nombre de données incitent à penser que les deux types d'hypoxie seraient, pour le moins, additifs. Toutefois ces données ont été obtenues dans des conditions où la concentration de monoxyde de carbone était trop élevée pour pouvoir être prise en considération d'un point de vue réglementaire.

S'agissant des effets à long terme du monoxyde de carbone inhalé à haute altitude, les études sont encore moins nombreuses. Les études disponibles indiquent qu'il ne se produit pas grand chose au dessous de 110 mg/m³ (100 ppm) et pour des altitudes inférieures à 4570 m. Il est cependant possible que le foetus soit particulièrement sensible aux effets du monoxyde de carbone en altitude ; c'est particulièrement vrai lorsque la mère est fumeuse et expose l'enfant qu'elle porte à de fortes concentrations de CO.

7.2 *Interaction entre le monoxyde de carbone et les substances ou médicaments divers*

On ne possède guère d'informations directes sur une potentialisation éventuelle de la toxicité du CO en cas de prise de médicaments ou de toxicomanie ; certaines données montrent toutefois que ce peut être un sujet de préoccupation. On possède certains indices d'une action pouvant aller dans le même sens ou en sens inverse ; autrement dit que la toxicité du monoxyde du carbone peut être potentialisée par la prise de certains produits ou que les effets, toxiques ou autres, d'un produit peuvent être atténués par une exposition au CO. Presque toutes les données publiées sur cette question concernent la consommation d'alcool.

La consommation et l'abus de substances psychoactives et d'alcool sont un aspect universel de la vie sociale. En raison de l'effet du monoxyde de carbone sur les fonctions cérébrales, on peut s'attendre à ce qu'il y ait des interactions entre ce gaz et les substances psychoactives. Malheureusement, peu de travaux de recherche systématiques ont été consacrés à cette question. En outre, parmi les quelques travaux effectués, seule une faible proportion se base sur une modélisation des effets probables de ces associations de substances. Dans ces conditions, il est bien souvent impossible de déterminer si les effets combinés de la prise de diverses substances et d'une exposition au CO sont additifs ou non. Il importe de se rendre compte que l'additivité des effets de plusieurs substances peut avoir de l'importance sur le plan clinique, même si le sujet n'a pas conscience de ce genre de risque. Les principales données relatives à des interactions potentiellement importantes avec le monoxyde de carbone proviennent d'études pratiquées sur l'Homme et l'animal, en présence d'une imbibition alcoolique, études qui montrent que les effets sont au moins additifs. Comme les cas d'exposition au CO en présence d'une imbibition alcoolique sont probablement fréquents, ces résultats prennent un relief accru.

.3 Exposition combinée au monoxyde de carbone et à d'autres polluants atmosphériques et facteurs environnementaux

Une grande partie des données relatives aux effets combinés du monoxyde de carbone et d'autres polluants présents dans l'air ambiant provient de l'expérimentation animale. Seules quelques études ont porté sur des sujets humains. Les premiers travaux sur des sujets humains en bonne santé, qui concernaient des polluants atmosphériques courants comme le monoxyde de carbone, le dioxyde d'azote, l'ozone ou le nitrate de peroxyacétyle, n'ont pas mis en évidence d'interactions entre ces composés en cas d'exposition combinée. Lors d'études en laboratoire, aucune interaction n'a été observée après exposition à du monoxyde de carbone et à des polluants atmosphériques courants comme le dioxyde d'azote et le dioxyde de soufre. Cependant, on a tout de même constaté un effet additif après exposition combinée à de fortes concentrations de monoxyde de carbone et à du dioxyde d'azote, un effet synergistique étant observé après exposition combinée au monoxyde de carbone et à l'ozone.

Des interactions toxicologiques entre des produits de combustion, principalement du monoxyde et du dioxyde de carbone ainsi que du cyanure d'hydrogène, présents à des concentrations caractéristiques de celles qui sont produites par des incendies en plein air ou dans un local, ont provoqué des effets synergistiques après exposition au monoxyde et au dioxyde de carbone, et des effets additifs en présence de cyanure d'hydrogène. On a également observé des effets additifs lorsque du monoxyde de carbone, du cyanure d'hydrogène et de l'oxygène à faible concentration étaient simultanément présents ; l'adjonction de dioxyde de carbone à cette association a produit un effet synergistique.

Enfin, d'autres études incitent à penser que certains facteurs environnementaux comme un stress thermique ou le bruit peuvent jouer un rôle important dans les effets toxiques lorsqu'ils s'ajoutent à une exposition à du monoxyde de carbone. Parmi les effets qui ont été ainsi décrits, celui qui pourrait être le plus important dans les cas courants d'exposition humaine consiste dans une diminution des performances physique lorsqu'un stress thermique s'ajoute à une exposition au monoxyde de carbone à la concentration de 57 mg/m^3 (50 ppm).

7.4   Fumée de tabac

Tout en étant une source de monoxyde de carbone pour les fumeurs et les non fumeurs, la fumée de tabac contient également diverses substances chimiques avec lesquelles ce gaz pourrait interagir. Les données dont on dispose incitent fortement à penser qu'une exposition momentanée ou chronique au CO contenu dans la fumée de tabac est de nature à affecter le système cardiopulmonaire, mais il est vrai que les interactions potentielles entre le monoxyde de carbone et d'autres composés présents dans la fumée sont autant de facteurs de confusion. En outre, on ne peut affirmer que les effets d'une augmentation du taux de carboxyhémoglobine par suite d'une exposition au CO présent dans l'environnement s'ajoutent à ceux d'une élévation chronique de ce taux, car il peut se produire une certaine adaptation physiologique.

8. Sous-groupes de population courant un risque d'exposition au monoxyde de carbone

La plupart des données dont on dispose au sujet des effets toxiques du monoxyde de carbone portent sur deux groupes de population bien définis : les adultes jeunes en bonne santé et les malades souffrant d'une cardiopathie coronarienne. D'après les effets qui ont été observés, les malades présentant une ischémie d'effort reproductible sont le groupe de population pour lequel on a le mieux établi qu'ils couraient un risque accru de ressentir des effets inquiétants (réduction de la durée de l'exercice physique par suite de l'exacerbation des symptômes cardiovasculaires) en cas d'exposition à des concentration de CO correspondant aux valeurs ambiantes ou quasi-ambiantes et produisant des taux minimaux de carboxy-hémoglobine de l'ordre de 3 %. Il existe un autre groupe - plus restreint et constitué de sujets en bonne santé - qui est également réceptif aux effets du CO sous la forme d'une diminution de la durée de l'exercice physique, mais cet effet ne s'observe que pendant un exercice physique de courte durée dans des condition d'effort maximal. Cette réduction de l'aptitude à soutenir un effort prolongé pour des personnes en bonne santé devrait donc concerner davantage les athlètes engagés dans des compétitions que des gens ordinaires vaquant à leurs occupations quotidiennes.

On peut toutefois supposer, en se basant sur les résultats cliniques, des considérations théoriques et les données fournies par l'expérimentation animale, que d'autres segments de la population pourraient courir un risque en cas d'exposition au monoxyde de carbone. On peut les regrouper par sexe, par âge (par ex. foetus, nourrissons, personnes âgées), par types génétiques (par ex. porteurs d'hémoglobine anormale), par maladies préexistantes - connues ou non - qui réduisent l'apport d'oxygène aux organes vitaux, ou encore par type de médicaments consommés, de toxicomanie ou en fonction de la qualité de l'environnement (par ex. présence d'autres polluants atmosphériques ou vie à haute altitude). Malheureusement, pour la plupart d'entre eux, on ne possède guère de données expérimentales qui permettraient de définir les risques encourus en cas d'exposition à des concentrations de monoxyde de carbone correspondant aux valeurs ambiantes ou quasi-ambiantes.

9. Intoxication oxycarbonée

La majeure partie de ce document porte sur les effets que des concentrations relativement faibles de monoxyde de carbone peuvent produire chez des sujets humains à des taux de carboxyhémoglobine égaux ou sensiblement égaux aux valeurs les plus faibles qui soient décelables par les techniques biomédicales actuelles. Pourtant, les effets toxiques de ce polluant vont d'anomalies cardiovasculaires ou neurocomportementales subtiles pour de faibles concentrations dans l'air ambiant, à la perte de conscience et à la mort en cas d'exposition instantanée à une forte concentration de monoxyde de carbone. Dans ce dernier cas, la morbidité et la mortalité résultantes peuvent constituer un problème de santé publique non négligeable.

Le monoxyde de carbone est responsable d'une forte proportion des intoxications accidentelles et parfois mortelles qui surviennent chaque année dans le monde. Il peut exister, à l'intérieur de locaux comme en plein air, des conditions qui font qu'une petite fraction de la population se trouve exposée à des concentrations dangereuses de ce gaz. A l'extérieur, c'est aux carrefours où la circulation est intense, à proximité des tuyaux d'échappement des moteurs à combustion interne ou de certaines sources industrielles ou encore dans des parkings et des tunnels mal ventilés que la concentration de monoxyde de carbone est la plus élevée. A l'intérieur, les valeurs les plus fortes s'observent dans les habitations dont les poêles ou autres appareils à combustion fonctionnent mal ou ont une évacuation insuffisante ou dont le système d'aération comporte des refoulements.

Les symptômes d'une intoxication oxycarbonée aiguë ne sont pas en très bonne corrélation avec le taux de carboxyhémoglobine mesuré à l'arrivée à l'hôpital. Un taux de carboxyhémoglobine inférieur à 10 % est généralement asymptomatique. Lorsque la saturation en carboxyhémoglobine atteint une valeur de 10 à 30 %, les symptômes de l'intoxication oxycarbonée peuvent apparaître; ils consistent en céphalées, faiblesse, nausées, confusion, désorientation et troubles visuels. Dans les cas d'exposition chronique aboutissant à des taux de carboxyhémoglobine compris entre 30 et 50 %, on constate les symptômes suivants : dyspnée d'effort, accroissement de la fréquence respiratoire et cardiaque et syncope. Lorsque le taux de carboxyhémoglobine dépasse 50 %, il peut y avoir coma, convulsions et finalement arrêt cardiaque et respiratoire.

L'intoxication oxycarbonée s'accompagne fréquemment de complications (mort immédiate, atteinte du myocarde, hypotension, arythmie, oedème pulmonaire). L'effet le plus insidieux est peut-être l'apparition tardive (dans les 1 à 3 semaines) de séquelles neuropsychiatriques, avec leurs conséquences sur le plan neurocomportemental, notamment chez les enfants. Une intoxication oxycarbonée pendant la grossesse comporte un risque élevé pour la mère, par le fait qu'elle accroît le risque de complications à court terme, de même que pour le foetus par suite du risque de mort foetale, de troubles du développement et de lésions dues à l'anoxie cérébrale. Par ailleurs, ce n'est pas la gravité des troubles présentés par la mère qui permet de juger de celle de l'intoxication foetale.

Les intoxications oxycarbonées sont fréquentes, leurs conséquences sont graves et elles peuvent notamment entraîner une mort immédiate, mais aussi des complications et des séquelles souvent négligées. Il convient donc d'encourager les efforts en vue de les éviter, notamment par l'éducation du public et du corps médical.

10. Valeurs-guides recommandées par l'OMS

Les valeurs-guides suivantes (les chiffres en ppm sont arrondis) et durées d'exposition en moyenne pondérée par rapport au temps ont été établies de manière à ce que le taux de 2,5 % de carboxyhémoglobine ne soit pas dépassé, même si le sujet se livre à une activité physique légère ou modérée.

100 mg/m^3 (87 ppm) pendant 15 min
60 mg/m^3 (52 ppm) pendant 30 min
30 mg/m^3 (26 ppm) pendant 1h
10 mg/m^3 (9 ppm) pendant 8 h

RESUMEN Y CONCLUSIONES

El monóxido de carbono (CO) es un gas incoloro e inodoro que puede ser tóxico para el ser humano. Se produce cuando se queman combustibles con carbono de forma incompleta y también mediante procesos naturales o por la biotransformación de halometanos en el organismo humano. Con la exposición externa a monóxido de carbono adicional pueden comenzar a aparecer efectos ligeros y las concentraciones más elevadas pueden provocar la muerte. Los efectos del monóxido de carbono para la salud se deben fundamentalmente a la formación de carboxihemoglobina (COHb), que reduce la capacidad de transporte de oxígeno de la sangre.

1. Química y métodos analíticos

Los métodos disponibles para la medición del monóxido de carbono en el aire ambiente van desde los sistemas totalmente automatizados, con la técnica de infrarrojos no dispersiva y la cromatografía de gases, hasta los manuales sencillos de tipo semicuantitativo con tubos detectores. Debido a que la formación de carboxihemoglobina en el ser humano depende de numerosos factores, entre ellos la variabilidad de las concentraciones de monóxido de carbono en el aire ambiente, la concentración de carboxihemoglobina se debería medir en lugar de calcularla. Existen varios métodos relativamente sencillos para determinar el monóxido de carbono mediante análisis de la sangre o del aire alveolar, que está en equilibrio con la sangre. Algunos de estos métodos se han validado mediante cuidadosos estudios comparativos.

2. Fuentes y niveles de monóxido de carbono en el medio ambiente

El monóxido de carbono es un constituyente traza de la troposfera que se forma en procesos naturales y en actividades humanas. Teniendo en cuenta que las plantas pueden metabolizar y producir monóxido de carbono, los niveles traza se consideran un constituyente normal del medio ambiente natural. Si bien las concentraciones de monóxido de carbono en el ambiente en las cercanías de las zonas urbanas e industriales puede superar con creces los niveles básicos mundiales, no hay informes de efectos adversos en las plantas o los microorganismos producidos por estos niveles de monóxido de carbono medidos actualmente. Sin embargo, las concentraciones de monóxido de carbono en el medio ambiente pueden ser perjudiciales para la salud y el bienestar humanos, en función de los niveles que alcancen en las zonas de trabajo y de residencia y de la susceptibilidad de las personas expuestas a los efectos potencialmente adversos.

Las tendencias en los datos sobre la calidad del aire obtenidos en estaciones de vigilancia en lugares fijos ponen de manifiesto una disminución general de las concentraciones de monóxido de carbono, gracias a los eficaces sistemas de control de las emisiones de los vehículos más modernos. En los Estados Unidos, las emisiones de los vehículos en las autopistas representan alrededor del 50% del total; las fuentes del transporte fuera de las autopistas contribuyen con un 13%. Los demás tipos de emisiones de monóxido de carbono son otras fuentes de combustión, como las calderas de vapor (12%); los procesos industriales (8%); la eliminación de residuos sólidos (3%); y otras fuentes diversas (14%).

La concentración de monóxido de carbono en los espacios cerrados depende de la presente en el exterior, las fuentes interiores, la infiltración, la ventilación y la mezcla de aire entre las habitaciones y dentro de ellas. En residencias que carecen de fuentes, el promedio de la concentración de monóxido de carbono es prácticamente igual al del exterior. Las concentraciones más altas de monóxido de carbono en los espacios cerrados están relacionadas con fuentes de combustión y se encuentran, por ejemplo, en aparcamientos cerrados, estaciones de servicio y restaurantes. Las concentraciones más bajas de monóxido de carbono en espacios cerrados corresponden a viviendas, iglesias e instalaciones de atención sanitaria. En estudios de exposición se ha puesto de manifiesto que el tabaquismo pasivo está asociado con un aumento de la exposición de los no fumadores de 1,7 mg/m^3 (1,5 ppm) como promedio y el uso de una cocina de gas en el hogar con un aumento de alrededor de 2,9 mg/m^3 (2,5 ppm). Otras fuentes que pueden contribuir al monóxido de carbono en el hogar son el espacio de combustión y los calentadores de agua, así como las cocinas de carbón o de leña.

3. **Distribución y transformación en el medio ambiente**

Hay datos recientes sobre las tendencias mundiales de la concentración de monóxido de carbono en la troposfera que indican una disminución a lo largo del último decenio. Las concentraciones básicas mundiales son del orden de 60-140 $\mu g/m^3$ (50-120 ppmm). Los niveles son más altos en el hemisferio norte que en el hemisferio sur. Las concentraciones básicas medias también tienen una fluctuación estacional. Los niveles son más altos en los meses de invierno y más bajos en los de verano. Alrededor del 60% del monóxido de carbono que se encuentra en la troposfera de las zonas no urbanas es atribuible a actividades humanas, ya sea de forma directa a partir de procesos de combustión como indirecta a través de la oxidación de hidrocarburos y de metano que, a su vez, proceden de las actividades agrícolas, los vertederos y otras fuentes semejantes. Las reacciones atmosféricas en las que interviene el monóxido de carbono pueden producir ozono en la troposfera. En otras reacciones se puede reducir la concentración de radicales hidroxilo, factor fundamental en los ciclos de eliminación mundial de otros muchos gases traza naturales y antropogénicos, contribuyendo posiblemente de esta manera a modificar la química atmosférica y, en último término, al cambio del clima mundial.

4. **Exposición de la población al monóxido de carbono**

Durante las actividades cotidianas normales, la población entra en contacto con el monóxido de carbono en diversos microambientes - al viajar en vehículos de motor, en el lugar de trabajo, al visitar zonas urbanas asociadas con fuentes de combustión o al cocinar y calentarse con fuego de gas, carbón o leña- así como con el humo del tabaco. En general, las exposiciones más importantes al monóxido de carbono para la mayoría de las personas se producen en el vehículo y en microambientes internos.

El perfeccionamiento de pequeños sensores electroquímicos portátiles de control de la exposición personal ha permitido medir las concentraciones de monóxido de carbono que encuentran las personas cuando se desplazan a través de numerosos microambientes internos y externos diversos que no se pueden vigilar con estaciones situadas en un lugar fijo. Los resultados tanto de la vigilancia de la exposición sobre el terreno como de los estudios de creación de modelos ponen de manifiesto que no hay una correlación directa entre la exposición personal individual medida con los sensores de control y las concentraciones de monóxido de carbono determinadas solamente por las estaciones fijas. Esta observación se debe a la movilidad de las personas y a la variabilidad espacial y temporal de las concentraciones de monóxido de carbono. Aunque no permiten establecer una correlación entre la exposición obtenida con los sensores de control personales individuales y la concentración simultánea marcada por la estación fija más cercana, los estudios sobre el terreno en gran escala de la exposición humana al monóxido de carbono indican que la exposición total de las personas es más baja en los días en que las estaciones fijas miden concentraciones más bajas de monóxido de carbono en el ambiente y más altas los días en que la concentración en el ambiente es más alta. Estos estudios señalan que cuando se ha de evaluar la exposición humana total es necesario realizar mediciones personales del monóxido de carbono, a fin de acumular datos de la vigilancia del medio ambiente en lugares fijos. Los datos obtenidos en estos estudios sobre el terreno se pueden utilizar para crear y probar modelos de exposición humana que representen pautas de tiempo y de actividad que se sepa que influyen en la exposición al monóxido de carbono.

La evaluación de las situaciones de exposición humana al monóxido de carbono indica que la exposición ocupacional en algunos puestos de trabajo o en el hogar con aparatos de combustión defectuosos o poco ventilados puede ser superior a 110 mg de monóxido de carbono/m^3 (100 ppm), produciendo con frecuencia, si la exposición es continuada, concentraciones de carboxihemoglobina del 10% o más. En cambio, es mucho menos frecuente que el público general expuesto a los niveles del medio ambiente se encuentre con niveles de exposición tan altos. En la población general es más frecuente la exposición a menos de 29-57 mg de monóxido de carbono/m^3 (25-50 ppm) durante períodos prolongados; en los niveles de ejercicio bajos que son normales en tales circunstancias, la concentración de carboxihemoglobina resultante entre los no fumadores suele ser del 1%-2%. Estos niveles se pueden comparar con los fisiológicos normales de los no fumadores, estimados en el 0,3%-0,7% de carboxihemoglobina. Sin embargo, el promedio de las concentraciones básicas de carboxihemoglobina en los fumadores es del 4%, con una gama normal del 3%-8%, que pone de manifiesto la absorción de monóxido de carbono del humo inhalado.

En los estudios de exposición humana se ha comprobado que los gases de escape de los vehículos de motor son la fuente principal de las elevadas concentraciones de monóxido de carbono que se dan normalmente. Estos estudios indican que el interior de los vehículos de motor tiene como promedio la concentración de monóxido de carbono más alta de todos los microambientes (un promedio de 10-29 mg/m^3 [9-25 ppm]). Además, se ha puesto de manifiesto que la exposición durante el desplazamiento diario al trabajo es enormemente variable, respirando algunos viajeros concentraciones de monóxido de carbono superiores a 40 mg/m^3 (35 ppm).

El puesto de trabajo es otro entorno importante de exposición al monóxido de carbono. En general, si se exceptúa la exposición durante el viaje diario de ida y vuelta al trabajo, la exposición en el entorno laboral es superior a la de los períodos en que no se trabaja. Las exposiciones ocupacional y no ocupacional pueden superponerse y dar lugar a una concentración más alta de monóxido de carbono en la sangre. Un aspecto todavía más importante es que el carácter de determinadas ocupaciones lleva consigo un riesgo mayor de exposición a concentraciones elevadas de monóxido de carbono (por ejemplo, las ocupaciones directamente relacionadas con la conducción, el mantenimiento o el aparcamiento de vehículos). Entre los grupos profesionales expuestos al monóxido de carbono procedente de los gases de escape figuran los mecánicos de automóviles; los empleados de aparcamientos y de gasolineras; los conductores de autobuses, camiones o taxis; los policías; y los trabajadores de almacenes. Determinados procesos industriales pueden provocar la exposición de los trabajadores al monóxido de carbono producido directamente o como subproducto; entre ellos cabe mencionar la producción de acero, los hornos de coque, la producción de negro de humo y el refinado del petróleo. Los bomberos, los cocineros y los trabajadores de la construcción pueden verse también expuestos en el trabajo a concentraciones altas de monóxido de carbono. La exposición ocupacional en industrias o en lugares de producción de monóxido de carbono es de las exposiciones individuales más altas observadas en los estudios de vigilancia sobre el terreno.

5. Toxicocinética y mecanismo de acción del monóxido de carbono

El monóxido de carbono se absorbe a través de los pulmones y la concentración de carboxihemoglobina en la sangre dependerá en todo momento de varios factores. Cuando exista un equilibrio con el aire ambiente, el contenido de carboxihemoglobina de la sangre dependerá fundamentalmente de las concentraciones de monóxido de carbono y de oxígeno inspirados. Sin embargo, si no se ha alcanzado el equilibrio, la concentración de carboxihemoglobina dependerá también de la duración de la exposición, de la ventilación pulmonar y de la concentración de carboxihemoglobina originalmente presente antes de la inhalación del aire contaminado. Además de su reacción con la hemoglobina, el monóxido de carbono se combina con la mioglobina, los citocromos y las enzimas metálicas, como la citocromo c oxidasa y el citocromo P-450. No se conoce completamente la influencia de estas reacciones en la salud, aunque probablemente sea menos importante con los niveles de exposición del medio ambiente que la que tiene la reacción del gas con la hemoglobina.

El intercambio de monóxido de carbono entre el aire que respiramos y el organismo humano está controlado por procesos tanto físicos (por ejemplo, transporte y difusión masivos) como fisiológicos (por ejemplo, ventilación alveolar y rendimiento cardíaco). El monóxido de carbono pasa fácilmente de los pulmones a la corriente sanguínea. La fase final en este proceso consiste en la unión competitiva del monóxido de carbono y el oxígeno a la hemoglobina en los glóbulos rojos, formando carboxihemoglobina y oxihemoglobina (O_2Hb), respectivamente. La unión del monóxido de carbono a la hemoglobina, que produce carboxihemoglobina y reduce la capacidad de transporte de oxígeno de la sangre, parece ser el principal mecanismo de acción que desencadena la inducción de los efectos tóxicos de la exposición a concentraciones bajas de monóxido de carbono. No se conocen totalmente los mecanismos precisos de inducción de los efectos tóxicos mediante la formación de carboxihemoglobina, pero probablemente se deban a la inducción de un estado hipóxico en muchos tejidos de distintos órganos. Se han propuesto como hipótesis mecanismos alternativos o secundarios de toxicidad inducida por el monóxido de carbono (además de la carboxihemoglobina), pero no se ha demostrado en ninguno de ellos que funcione con niveles de exposición al monóxido de carbono relativamente bajos (casi ambientales). Así pues, actualmente se acepta que la concentración de carboxihemoglobina en la sangre representa un marcador fisiológico útil para estimar la carga interna de monóxido de carbono debida a la contribución combinada de: 1) el monóxido de carbono de origen endógeno y 2) el monóxido de carbono de origen exógeno procedente de la exposición a fuentes externas de monóxido de carbono. La concentración de carboxihemoglobina que probablemente se derivará de modalidades concretas (concentraciones, duración, etc.) de exposición externa al monóxido de carbono se puede calcular razonablemente bien utilizando la ecuación de Coburn-Foster-Kane (CFK).

Por consiguiente, una característica única de la exposición al monóxido de carbono es que el nivel de carboxihemoglobina en la sangre representa un marcador biológico útil de la dosis que ha recibido la persona. La cantidad de carboxihemoglobina que se forma depende de la concentración y duración de la exposición al monóxido de carbono, del ejercicio (que aumenta la cantidad de aire inhalado por unidad de tiempo), de la temperatura ambiente, del estado de salud y del metabolismo específico de la persona expuesta. La formación de carboxihemoglobina es un proceso reversible; sin embargo, debido a la fuerte unión del monóxido de carbono a la hemoglobina, el periodo de semieliminación es bastante largo, oscilando entre 2 y 6,5 horas, en función de la concentración inicial de carboxihemoglobina y de la tasa de ventilación de las personas. Esto podría llevar a la acumulación de carboxihemoglobina, y bastarían concentraciones relativamente bajas de monóxido de carbono para producir niveles considerables de carboxihemoglobina en la sangre.

La concentración de carboxihemoglobina en la sangre se puede determinar de manera directa mediante el análisis de la sangre o de modo indirecto midiendo la concentración de monóxido de carbono en el aire expirado. La medición del aire expirado tiene las ventajas de la facilidad, la rapidez, la precisión y una mayor aceptación que la medición de la carboxihemoglobina en la sangre. Sin embargo, la precisión del procedimiento de medición de la respiración y la validez de la relación de Haldane entre la respiración y la sangre siguen sin estar claras para la exposición a concentraciones bajas de monóxido de carbono en el medio ambiente.

Teniendo en cuenta que no se puede disponer fácilmente de mediciones de la carboxihemoglobina en la población expuesta, se han elaborado modelos matemáticos para pronosticar las concentraciones de carboxihemoglobina a partir de exposiciones conocidas al monóxido de carbono en diversas circunstancias. El modelo mejor conocido para el pronóstico de la carboxihemoglobina sigue siendo la ecuación de Coburn, Forster y Kane. La solución lineal es útil para examinar los datos de contaminación del aire que producen niveles relativamente bajos de carboxihemoglobina, mientras que la solución no lineal ofrece una buena capacidad de pronóstico incluso para la exposición a concentraciones elevadas de monóxido de carbono. Los modelos de regresión podría ser útiles sólo cuando las condiciones de aplicación son muy próximas a las reinantes al calcular los parámetros.

Aunque se considera que la causa principal de la toxicidad del monóxido de carbono a niveles de exposición bajos es la hipoxia de los tejidos debida a la unión del monóxido de carbono a la hemoglobina, ciertos aspectos fisiológicos de la exposición al monóxido de carbono no se pueden explicar bien por la disminución de la presión parcial de oxígeno intracelular a causa de la presencia de carboxihemoglobina. Por consiguiente, numerosas investigaciones se han concentrado en los mecanismos secundarios de la toxicidad del monóxido de carbono relacionada con su absorción intracelular. La unión del monóxido de carbono a muchos compuestos intracelulares está bien documentada tanto *in vitro* como *in vivo;*; sin embargo, no se sabe todavía si la absorción intracelular de monóxido de carbono en presencia de hemoglobina es suficiente o no para provocar una disfunción aguda del órgano o efectos en la salud a largo plazo. La práctica inexistencia de técnicas sensibles capaces de evaluar la fijación del monóxido de carbono intracelular en condiciones fisiológicas ha dado lugar a una serie de enfoques indirectos del problema, así como a numerosos estudios negativos.

Los conocimientos actuales relativos a la fijación del monóxido de carbono intracelular apuntan a que con toda probabilidad son la mioglobina, que se encuentra fundamentalmente en el corazón y en el músculo esquelético, y la citocromo oxidasa las proteínas cuya función queda inhibida por la presencia de concentraciones elevadas de carboxihemoglobina. No se conoce en este momento la importancia fisiológica de la absorción de monóxido de carbono por la mioglobina, pero una concentración suficiente de carboximioglobina podría limitar potencialmente la absorción máxima de oxígeno por el músculo en ejercicio. Aunque hay pruebas que parecen indicar la existencia de una unión considerable de monóxido de carbono a la citocromo oxidada en los tejidos del corazón y del cerebro, no es probable que se produzca la unión de una cantidad significativa de monóxido de carbono en presencia de concentraciones bajas de carboxihemoglobina.

6. Efectos en la salud de la exposición al monóxido de carbono

La importancia para la salud del monóxido de carbono presente en el aire ambiente se debe fundamentalmente al hecho de que se une mediante un enlace fuerte a la molécula de la hemoglobina para formar carboxihemoglobina, que limita la capacidad de transporte de oxígeno de la sangre. La presencia de la carboxihemoglobina altera también la disociación de la oxihemoglobina, de manera que la distribución del oxígeno a los tejidos se reduce ulteriormente. La afinidad de la hemoglobina humana por el monóxido de carbono es alrededor de 240 veces superior a la del oxígeno, y las proporciones de carboxihemoglobina y oxihemoglobina que se forman en la sangre dependen en gran medida de la presión parcial del monóxido de carbono y del oxígeno.

La preocupación acerca de los posibles efectos de la exposición al monóxido de carbono para la salud se han expuesto en amplios estudios tanto con seres humanos como con diversas especies de animales. Mediante una serie de protocolos experimentales se ha obtenido abundante información sobre la toxicidad del monóxido de carbono, sus efectos directos en la sangre y en otros tejidos y las manifestaciones de estos efectos en forma de cambios en el funcionamiento de los órganos. Sin embargo, muchos de los estudios con animales se han realizado con concentraciones de monóxido de carbono extremadamente altas (es decir, niveles que no se encuentran en el aire ambiente). Aunque los efectos graves de la exposición a estas concentraciones elevadas de monóxido de carbono no están directamente relacionados con los problemas que se derivan de la exposición a los niveles normales de monóxido de carbono en el medio ambiente, han proporcionado una información valiosa acerca de los posibles efectos de la exposición accidental al monóxido de carbono, en particular de las exposiciones que se producen en espacios cerrados.

.1   Efectos cardiovasculares

En adultos jóvenes sanos se ha observado claramente una menor absorción de oxígeno y la consiguiente disminución de la capacidad de trabajo en condiciones de ejercicio máximo a partir de una concentración de carboxihemoglobina del 5,0%, y en varios estudios se ha detectado una pequeña disminución en la capacidad de trabajo con concentraciones de carboxihemoglobina de sólo 2,3%-4,3%. Estos efectos pueden tener repercusiones en la salud de la población general, en cuanto a la limitación potencial de ciertas actividades profesionales o recreativas que requieren un esfuerzo físico grande en circunstancias de exposición a concentraciones de monóxido de carbono suficientemente altas.

Sin embargo, con los niveles de exposición al monóxido de carbono más comunes en el ambiente son motivo de mayor preocupación determinados efectos cardiovasculares (es decir, el agravamiento de los síntomas de angina durante el ejercicio) que tienen probabilidad de presentarse en un segmento de la población general más pequeño, pero representativo. Este grupo, formado por los enfermos con angina crónica, está considerado actualmente el de mayor riesgo sensible a los efectos de la exposición al monóxido de carbono, basándose en las pruebas de agravamiento de la angina que se produce en estos enfermos cuando se exponen a concentraciones de carboxihemoglobina del 2,9%-4,5%. La relación dosis-respuesta para los efectos cardiovasculares en pacientes con cardiopatía coronaria todavía no se ha definido de manera concluyente y no se puede descartar por el momento la posibilidad de que puedan producirse tales efectos con concentraciones de carboxihemoglobina inferiores al 2,9 % . Por consiguiente, en el presente documento se examinan nuevos estudios publicados para determinar los efectos del monóxido de carbono en el agravamiento de la angina con niveles de carboxihemoglobina del 2%-6%.

En cinco estudios básicos se ha investigado la posibilidad de que la exposición al monóxido de carbono favorezca la aparición de isquemia miocárdica durante el ejercicio en pacientes con cardiopatía coronaria. En un estudio inicial se comprobó que la duración del ejercicio disminuía considerablemente por la aparición del dolor de pecho (angina) en pacientes con angina de pecho después de la exposición a concentraciones de carboxihemoglobina de apenas un 2,9%, lo que representa un aumento de la concentración de carboxihemoglobina del 1,6% sobre el valor de referencia. Los resultados de un amplio estudio multicéntrico demostraron la aparición de efectos en pacientes con angina inducida por el ejercicio reproducible después de la exposición a concentraciones de carboxihemoglobina del 3,2%, correspondiente a un aumento de la concentración de carboxihemoglobina del 2,0% sobre el valor de referencia. En otros estudios se observaron efectos semejantes en pacientes con cardiopatía coronaria obstructiva y pruebas de isquemia inducida por el ejercicio después de la exposición a concentraciones de carboxihemoglobina del 4,1% y 5,9%, respectivamente, lo que representa un aumento del 2,2% y del 4,2% sobre el valor de referencia. En un estudio con enfermos de angina se observaron efectos con concentraciones del 3% de carboxihemoglobina, lo que representa un aumento del 1,5% con respecto al valor de referencia. Así pues, la concentración más baja con efectos adversos observados en pacientes con isquemia inducida por el ejercicio se sitúa entre el 3% y el 4% de carboxihemoglobina, equivalente a un aumento del 1,5% -2,2% con respecto al valor de referencia. No se han estudiado los efectos en los episodios de isquemia silenciosa, que son la mayoría de los que se dan en este tipo pacientes.

Las consecuencias adversas para la salud de la exposición a concentraciones bajas de monóxido de carbono en pacientes con cardiopatía isquémica son muy difíciles de pronosticar en la población de personas con riesgo a causa de enfermedades cardíacas. Se ha observado que la exposición a concentraciones de monóxido de carbono suficientes para alcanzar un 6% de carboxihemoglobina, pero no para valores inferiores de ésta, aumenta considerablemente el número y la complejidad de las arritmias inducidas por el ejercicio en pacientes con cardiopatía coronaria y ectopia básica. Este resultado, junto con los estudios de series cronológicas de morbilidad y mortalidad relacionadas con el monóxido de carbono y los estudios epidemiológicos con trabajadores de túneles, que normalmente están expuestos a los gases de escape de los automóviles, parece indicar que la exposición al monóxido de carbono puede crear un mayor riesgo de muerte repentina por arritmia en pacientes con cardiopatía coronaria, pero no es una prueba concluyente.

En evaluaciones anteriores de los efectos cardiovasculares del monóxido de carbono se ha indicado que parece haber una relación lineal entre la concentración de carboxihemoglobina en la sangre y la disminución del rendimiento humano con el ejercicio máximo, medido como absorción máxima de oxígeno. El rendimiento durante el ejercicio de las personas jóvenes, sanas, no fumadoras disminuye sistemáticamente con una concentración de carboxihemoglobina en la sangre de alrededor del 5%. En algunos estudios se ha observado incluso una disminución a corto plazo de la duración del ejercicio máximo con niveles de sólo un 2,3%-4,3% de carboxihemoglobina; sin embargo, esa disminución es tan pequeña que sólo suscita preocupación en el caso de los atletas de competición, más que en las personas comunes que realizan las actividades cotidianas.

También hay pruebas basadas tanto en consideraciones teóricas como en estudios experimentales con animales de laboratorio de que el monóxido de carbono puede afectar negativamente al sistema cardiovascular, en función de las condiciones de exposición utilizadas en esos estudios. Aunque se han observado alteraciones en el ritmo y la conducción cardíacos en animales sanos y en otros con insuficiencia cardíaca, los resultados de esos estudios no son concluyentes. El nivel más bajo en el cual se han observado efectos varía en función del régimen de exposición utilizado y de las especies sometidas a prueba. Los resultados obtenidos en estudios realizados con animales indican asimismo que el monóxido de carbono inhalado puede aumentar la concentración de hemoglobina y la razón de hematocrito, probablemente como reacción para compensar el efecto de la reducción de la capacidad de transporte de oxígeno debida al monóxido de carbono. Con concentraciones altas de monóxido de carbono, el aumento excesivo de la hemoglobina y el valor hematocrito puede imponer una carga adicional al corazón y comprometer el flujo sanguíneo de los tejidos.

Hay pruebas contradictorias de que la exposición al monóxido de carbono favorece la aparición de aterosclerosis en animales de laboratorio, y en la mayoría de los estudios no aparecen efectos medibles. Igualmente, la posibilidad de que el monóxido de carbono fomente cambios significativos en el metabolismo de los lípidos que puedan acelerar la aterosclerosis solamente se indica en un pequeño número de estudios. Cualquier efecto de este tipo será como máximo poco perceptible. Por último, es probable que el monóxido de carbono inhiba la agregación de las plaquetas más que fomentarla. En general se dispone de pocos datos que indiquen que probablemente se produciría un efecto aterogénico por la exposición en poblaciones humanas con las concentraciones de monóxido de carbono que normalmente se encuentra en el medio ambiente.

6.2 *Efectos pulmonares agudos*

No es probable que el monóxido de carbono tenga ningún efecto directo en el tejido pulmonar, excepto en concentraciones extremadamente altas asociadas con la intoxicación por monóxido de carbono. Los estudios de los efectos del monóxido de carbono en la función pulmonar realizados con personas se complican por la falta de una información adecuada sobre la exposición, el pequeño número de personas estudiadas y la brevedad de las exposiciones investigadas. La exposición ocupacional o occidental a los productos de combustión y la pirólisis, particularmente en espacios cerrados, pueden producir una disminución aguda de la función pulmonar si las concentraciones de carboxihemoglobina son altas. Sin embargo, es difícil separar los efectos potenciales del monóxido de carbono de los correspondientes a otras sustancias del humo y los gases de escape irritantes del aparato respiratorio. En estudios de poblaciones comunitarias sobre el monóxido de carbono en el aire ambiente no se ha encontrado una relación significativa con la función, la sintomatología y las enfermedades pulmonares.

6.3 *Efectos cerebrovasculares y en el neurocomportamiento*

No se han notificado pruebas fidedignas demostrativas de una reducción de la función del neurocomportamiento en adultos jóvenes sanos con concentraciones de carboxihemoglobina inferiores al 5%. Los resultados de los estudios realizados con una concentración de carboxihemoglobina del 5% o superior son equívocos. En gran parte de las investigaciones realizadas con un 5% de carboxihemoglobina no se observó ningún efecto, incluso cuando intervenían comportamientos semejantes a los afectados en estudios con concentraciones más altas de carboxihemoglobina. Sin embargo, los investigadores que no pudieron encontrar una disminución del neurocomportamiento relacionada con el monóxido de carbono con concentraciones de carboxihemoglobina del 5% o superiores tal vez utilizaran pruebas que no eran suficientemente sensibles para detectar de manera fidedigna pequeños efectos del monóxido de carbono. Así pues, teniendo en cuenta las pruebas empíricas, se puede decir que las concentraciones de carboxihemoglobina superiores o iguales al 5% pueden producir una disminución de la función del neurocomportamiento. Sin embargo, no se puede afirmar con seguridad que los niveles de carboxihemoglobina inferiores al 5% no tengan ningún efecto. Sin embargo, solamente se han estudiado adultos jóvenes sanos utilizando pruebas cuya sensibilidad se puede demostrar y concentraciones de carboxihemoglobina del 5% o superiores. Por consiguiente, no se ha investigado el problema de los grupos con riesgo especial de efectos del monóxido de carbono en el neurocomportamiento.

Un caso especial es el de las personas que están tomando medicamentos con efectos depresores primarios o secundarios, en las que cabría esperar que se agravara la disminución del neuro-comportamiento relacionada con el monóxido de carbono. Otros grupos posiblemente con más riesgo en cuanto a los efectos en el neurocomportamiento inducidos por el monóxido de carbono son los ancianos y los enfermos, pero en estos grupos no se ha evaluado tal riesgo.

En circunstancias normales, el cerebro puede aumentar el flujo sanguíneo o la extracción de oxígeno de los tejidos para compensar la hipoxia provocada por la exposición al monóxido de carbono. La respuesta general del sistema cerebrovascular es semejante en el feto, en el recién nacido y en el animal adulto; sin embargo, todavía se sabe poco acerca del mecanismo que determina el aumento del flujo sanguíneo en el cerebro. En realidad, parece probable que el aumento del flujo sanguíneo se deba a varios mecanismos que actúan simultáneamente, en ellos podrían intervenir aspectos metabólicos y neurales, así como la curva de disociación de la oxihemoglobina, los niveles de oxígeno en los tejidos e incluso un efecto histotóxico del monóxido de carbono. Tampoco se sabe si estos mecanismos compensatorios seguirían funcionando con éxito en una serie de condiciones en las que el cerebro y su sistema vascular se ven comprometidos (es decir, apoplejía, traumatismos craneoencefálicos, arteriosclerosis e hipertensión). El envejecimiento aumenta la probabilidad de estas lesiones y enfermedades. También es posible que existan diferencias individuales con respecto a la sensibilidad a la carboxihemoglobina y los mecanismos compensatorios.

Los comportamientos que requieren una atención constante o un rendimiento duradero son los más sensibles a las alteraciones provocadas por la carboxihemoglobina. El grupo de estudios humanos sobre la coordinación entre las manos y los ojos (seguimiento compensatorio), la detección de manifestaciones no frecuentes (vigilancia) y el rendimiento continuado ofrece las pruebas más convincentes y defendibles de los efectos de la carboxihemoglobina en el comportamiento a concentraciones de apenas un 5%. Sin embargo, estos efectos con concentraciones bajas de exposición al monóxido de carbono han sido muy pequeños y algo controvertidos. No obstante, las consecuencias potenciales de una deficiencia en la coordinación y la vigilancia del rendimiento continuado de tareas críticas en operadores de maquinaria como los vehículos de transporte público pueden ser graves.

6.4 Toxicidad en el desarrollo

Los estudios realizados en varias especies de animales de laboratorio aportan pruebas convincentes de que la exposición materna a 170-230 mg/m^3 (150-200 ppm) de monóxido de carbono, que da lugar a alrededor de un 15%-25% de carboxihemoglobina, produce una reducción del peso al nacer, cardiomegalia, retrasos en la evolución del comportamiento y alteración de la función cognoscitiva. De experimentos aislados parece deducirse que algunos de estos efectos pueden estar presentes con concentraciones de sólo 69-74 mg/m^3 (60-65 ppm; alrededor de un 6%-11% de carboxi-hemoglobina) mantenidas durante toda la gestación. Los estudios en los que se relaciona la exposición humana al monóxido de carbono de fuentes del medio ambiente o del humo de los cigarrillos con una reducción del peso al nacer son motivo de preocupación debido al riesgo de trastornos en el desarrollo; sin embargo, en muchos de estos estudios no se han tenido en cuenta todas las fuentes de monóxido de carbono. Los datos actuales relativos a niños que parecen indicar una vinculación entre la exposición al monóxido de carbono del medio ambiente y el síndrome de muerte súbita del lactante son poco convincentes.

6.5 Otros efectos sistémicos

Diversos estudios con animales de laboratorio parecen indicar que el metabolismo enzimático de los compuestos xenobióticos puede verse afectado por la exposición al monóxido de carbono. Sin embargo, la mayoría de los autores de estos estudios ha llegado a la conclusión de que los efectos en el metabolismo cuando las concentraciones de carboxihemoglobina son bajas (≤15%) pueden atribuirse por entero a la hipoxia de los tejidos producida por el aumento del nivel de carboxihemoglobina, debido a que no son superiores a los efectos producidos por niveles comparables de hipoxia hipóxica. Con niveles mayores de exposición, cuando las concentraciones de carboxihemoglobina superan el 15%-20%, el monóxido de carbono puede inhibir directamente la actividad de las oxidasas de funciones múltiples. La disminución del metabolismo xenobiótico que se observa con la exposición al monóxido de carbono puede ser importante para las personas sometidas a tratamiento con medicamentos.

Se ha señalado que la inhalación de niveles altos de monóxido de carbono que dan lugar a concentraciones de carboxihemoglobina superiores al 10%-15% provocan varios otros efectos sistémicos en animales de laboratorio, así como efectos en las personas afectadas por una intoxicación aguda por monóxido de carbono. Los tejidos con un metabolismo de oxígeno muy activo, como el corazón, el cerebro, el hígado, el riñón y el músculo, pueden ser particularmente sensibles a la intoxicación por monóxido de carbono. Los efectos de las concentraciones elevadas de monóxido de carbono en otros tejidos no son tan bien conocidos, por lo que hay más dudas acerca de ellos. En la bibliografía hay informes de efectos en el hígado, el riñón, los huesos y la capacidad inmunitaria del pulmón y el bazo. Por lo general se está de acuerdo en que los daños graves de los tejidos que se producen durante una intoxicación aguda por monóxido de carbono se deben a uno o varios de los siguientes factores: 1) isquemia debida a la formación de carboxihemoglobina, 2) inhibición de la liberación de oxígeno a partir de la oxihemoglobina, 3) inhibición de la función de los citocromos celulares (por ejemplo las oxidasas citocrómicas) y 4) acidosis metabólica.

Sólo hay datos de un valor relativamente escaso que apuntan a posibles efectos del monóxido de carbono en la actividad fibrinolítica, y esto únicamente con niveles de exposición al monóxido de carbono bastante elevados. Asimismo, si bien hay ciertos datos que parecen indicar también que hay efectos perinatales (por ejemplo, peso reducido al nacer, disminución del ritmo de desarrollo postnatal, síndrome de muerte súbita del lactante) asociados con la exposición al monóxido de carbono, las pruebas son insuficientes para confirmar cuantitativamente dicha asociación en el ser humano o establecer cualquier relación pertinente entre la exposición y el efecto.

6.6 Adaptación

Las únicas pruebas de compensación a corto o largo plazo del aumento de la concentración de carboxihemoglobina en la sangre o de la adaptación a él son indirectas. Los datos de animales de experimentación indican que el aumento de la concentración de carboxihemoglobina produce respuestas fisiológicas que tienden a contrarrestar otros efectos nocivos de la exposición al monóxido de carbono. Dichas respuestas son: 1) aumento del flujo de sangre coronaria, 2) aumento del flujo de sangre cerebral, 3) aumento de la hemoglobina mediante una mayor hematopoyesis y 4) mayor consumo de oxígeno en los músculos.

Las respuestas compensatorias a corto plazo en el flujo sanguíneo o el consumo de oxígeno pueden no ser completas, o incluso pueden no existir en ciertas personas. Por ejemplo, se sabe por estudios con animales de laboratorio que el flujo sanguíneo coronario aumenta con la elevación de la carboxihemoglobina y en estudios clínicos humanos se ha demostrado que las personas con cardiopatía isquémica responden a los niveles más bajos de carboxihemoglobina (6% o menos). La consecuencia es que en algunos casos de trastornos cardiacos se altera el mecanismo compensatorio a corto plazo.

Los estudios realizados sobre neurocomportamiento ponen manifiesto que no se produce una disminución debida al monóxido de carbono de manera constante en todos los casos, o incluso en los mismos estudios, y no se ha demostrado una relación dosis-respuesta con el aumento de los niveles de carboxihemoglobina. De estos datos se deduce que podría haber algún umbral o desfase en el mecanismo compensatorio, por ejemplo un aumento del flujo sanguíneo. Sin pruebas fisiológicas directas en animales de laboratorio, o preferiblemente en el ser humano, este concepto no pasa de ser una hipótesis.

En el caso de que pudiera demostrarse en el ser humano, el mecanismo mediante el cual se produce la adaptación a largo plazo se supone que es un aumento de la concentración de hemoglobina debido a una elevación de la hematopoyesis. Esta alteración de la producción de hemoglobina se ha demostrado repetidas veces en estudios con animales de laboratorio, pero no se han realizado recientemente estudios que indiquen o sugieran que se ha obtenido o puede obtenerse algún beneficio de adaptación. Además, aun en el caso de que el aumento de la hemoglobina sea una característica de la adaptación, no se ha demostrado que se produzca con concentraciones bajas de monóxido de carbono en el medio ambiente.

7. Exposición al monóxido de carbono combinada con la altitud, las drogas y otros contaminantes del aire y factores del medio ambiente

.1 Efectos de la altitud elevada

Aunque hay muchos estudios en los que se comparan y contrastan los efectos de la inhalación de monóxido de carbono con los debidos a la exposición a la altitud, son relativamente pocos los informes sobre los efectos combinados de la inhalación de monóxido de carbono con la altitud. Hay datos que respaldan la posibilidad de que los efectos de estos dos factores de hipoxia sean como mínimo aditivos. Estos datos se obtuvieron con concentraciones de monóxido de carbono demasiado elevadas para ser muy significativas a efectos de reglamentación.

Son aún menos los estudios de los efectos a largo plazo del monóxido de carbono a una altitud elevada. Estos estudios indican que hay pocos cambios a concentraciones de monóxido de carbono por debajo de 110 mg/m^3 (100 ppm) y altitudes inferiores a 4570 m. Sin embargo, el feto puede ser particularmente sensible a los efectos del monóxido de carbono con la altitud; esto es particularmente aplicable con niveles elevados de monóxido de carbono asociados con el hábito de fumar de la madre.

.2 Interacción del monóxido de carbono con las drogas

Hay poca información directa disponible sobre el posible aumento de la toxicidad del monóxido de carbono debido al consumo o el abuso concomitante de drogas; sin embargo, algunos datos indican que es un problema preocupante. Hay algunas pruebas de que pueden producirse interacciones entre los efectos de las drogas y la exposición al monóxido de carbono en ambas direcciones, es decir, el consumo de drogas puede aumentar la toxicidad del monóxido de carbono y la exposición al monóxido de carbono puede alterar los efectos tóxicos o de otro tipo de las drogas. Casi todos los datos publicados disponibles sobre la combinación del monóxido de carbono con drogas se refieren al consumo de alcohol.

El consumo y el abuso de drogas psicoactivas y de alcohol están generalizados en la sociedad. Debido al efecto del monóxido de carbono en la función cerebral, cabe prever interacciones entre el monóxido de carbono y las drogas psicoactivas. Por desgracia, apenas hay investigaciones sistemáticas que se hayan ocupado de esta cuestión. Por otra parte, se han realizado pocas investigaciones en las que se hayan utilizado modelos para los efectos previstos derivados de combinaciones de tratamientos. Así pues, con frecuencia es imposible determinar si los efectos combinados de la exposición a las drogas y al monóxido de carbono son aditivos o tienen características distintas. Es importante reconocer que incluso los efectos aditivos de las combinaciones pueden tener importancia clínica, especialmente cuando la persona no es consciente del peligro combinado. Las pruebas más concluyentes de una interacción potencialmente importante del monóxido de carbono proceden de estudios con alcohol tanto en animales de laboratorio como en el ser humano, en los que se han obtenido por lo menos efectos aditivos. La importancia de esto aumenta por la incidencia muy probable de la combinación del consumo de alcohol con la exposición al monóxido de carbono.

7.3 Combinación de la exposición al monóxido de carbono con otros contaminantes del aire y factores del medio ambiente

Muchos de los datos relativos a los efectos combinados del monóxido de carbono y de otros contaminantes presentes en el aire ambiente se basan en experimentos con animales de laboratorio. Sólo se conoce un pequeño número de estudios en el ser humano. En los primeros estudios realizados en personas sanas sobre contaminantes comunes del aire, como el monóxido de carbono, el dióxido de nitrógeno, el ozono o el nitrato de peroxiacetilo, no se consiguió demostrar ninguna interacción debida a la exposición combinada. En estudios de laboratorio no se observó ninguna interacción después de una exposición combinada a monóxido de carbono y contaminantes comunes del aire ambiente, como el dióxido de nitrógeno o el dióxido de azufre. Sin embargo, se observó un efecto aditivo tras la exposición combinada a concentraciones elevadas de monóxido de carbono y óxido nítrico y un efecto sinérgico después de la exposición combinada a monóxido de carbono y ozono.

Las interacciones toxicológicas de los productos de la combustión, fundamentalmente el monóxido de carbono, el anhídrido carbónico y el ácido cianhídrico, en las concentraciones producidas normalmente por los fuegos interiores y exteriores, han puesto de manifiesto un efecto sinérgico tras la exposición al monóxido de carbono más el anhídrido carbónico y un efecto aditivo con el ácido cianhídrico. También se observaron efectos aditivos cuando se combinaban el monóxido de carbono, el ácido cianhídrico y una concentración baja de oxígeno; la adición de anhídrido carbónico a esta combinación fue sinérgica.

Por último, hay estudios que indican que diversos factores del medio ambiente, como la presión térmica y el ruido, puede ser factores determinantes importantes de efectos para la salud si se combinan con la exposición al monóxido de carbono. De los efectos descritos, el único posiblemente con mayor interés para la exposición humana normal es la mayor disminución del rendimiento del ejercicio que se observa cuando la presión térmica se combina con 57 mg de monóxido de carbono/m^3 (50 ppm).

7.4   Humo del tabaco

Además de ser una fuente de monóxido de carbono tanto para los fumadores como para los no fumadores, el humo del tabaco es también una fuente de otras sustancias químicas que pueden tener una interacción con el monóxido de carbono del medio ambiente. Los datos disponibles parecen demostrar de manera convincente que la exposición aguda y crónica al monóxido de carbono atribuida al humo del tabaco puede afectar al sistema cardiopulmonar, pero la posible interacción del monóxido de carbono con otros productos del humo del tabaco crea confusión en los resultados. Además, no está claro si el aumento adicional de la carboxihemoglobina debido a la exposición en el medio ambiente sería en realidad aditivo con las concentraciones de carboxihemoglobina elevadas crónicamente debido al humo del tabaco, puesto que puede haber cierta adaptación fisiológica.

8.   **Evaluación de subpoblaciones potencialmente con riesgo derivado de la exposición al monóxido de carbono**

La mayor parte de la información sobre los efectos del monóxido de carbono para la salud se refieren a dos grupos de población cuidadosamente definidos: adultos sanos jóvenes y pacientes con cardiopatía coronaria diagnosticada. De acuerdo con los efectos conocidos descritos, los pacientes con isquemia inducida por el ejercicio reproducible parecen ser los mejor definidos como grupo sensible dentro de la población general que corre un riesgo mayor de experimentar efectos preocupantes en la salud (es decir, menor duración del ejercicio debido al agravamiento de los síntomas cardiovasculares) con las concentraciones de exposición al monóxido de carbono del medio ambiente o las próximas a ellas que dan lugar a un descenso de los niveles de carboxihemoglobina al 3%. Un grupo sensible más pequeño de individuos sanos experimenta una duración menor del ejercicio con niveles análogos de exposición al monóxido de carbono, pero sólo durante un ejercicio máximo de corta duración. Por consiguiente, la disminución de la duración del ejercicio en la población sana debería preocupar sobre todo a los atletas de competición, más que a la población normal que lleva cabo las actividades habituales de la vida cotidiana.

Sin embargo, el trabajo tanto clínico como teórico y la investigación experimental en animales de laboratorio pueden llevar a la hipótesis de que algunos otros grupos de población tal vez corran un riesgo probable derivado de la exposición al monóxido de carbono. Los grupos de riesgo probable identificables se pueden clasificar en función de las diferencias por razón de sexo; la edad (por ejemplo, fetos, niños pequeños y ancianos); las variaciones genéticas (es decir, anomalías de la hemoglobina); enfermedades existentes anteriormente, conocidas o desconocidas, ya que reducen la disponibilidad de oxígeno para tejidos críticos; o el uso de medicamentos, drogas de esparcimiento o alteraciones del medio ambiente (por ejemplo, exposición a otros contaminantes del aire o a una altitud elevada). Por desgracia, actualmente se dispone de pocas pruebas empíricas que permitan especificar los efectos para la salud asociados con la exposición al monóxido de carbono del medio ambiente o a una concentración próxima para la mayoría de los grupos con riesgo probable.

9. Intoxicación por monóxido de carbono

El presente documento se ocupa en su mayor parte de las concentraciones relativamente bajas de monóxido de carbono que inducen efectos en ser humano en el margen inferior de la detección de la carboxihemoglobina mediante la tecnología médica actual, o cerca de dicho margen. Ahora bien, los efectos asociados con la exposición a este contaminante van desde los cardiovasculares y de neurocomportamiento más leves a concentraciones bajas en el medio ambiente hasta la inconsciencia y la muerte tras una exposición aguda a concentraciones elevadas de monóxido de carbono. La morbilidad y la mortalidad debidas a estas últimas exposiciones pueden ser motivo de preocupación importante para la salud pública.

Al monóxido de carbono se debe un porcentaje elevado de las intoxicaciones y muertes accidentales notificadas en todo el mundo cada año. Se dan ciertas condiciones tanto del medio ambiente interior como del exterior que provocan la exposición de un pequeño porcentaje de la población a concentraciones peligrosas de monóxido de carbono. En el exterior, las mayores concentraciones de monóxido de carbono se registran cerca de los cruces de calles, con un tráfico intenso, cerca de los gases de escape de los motores de combustión interna y de fuentes industriales y en zonas poco ventiladas, como los aparcamientos cerrados y los túneles. En espacios interiores, las concentraciones de monóxido de carbono alcanzan un nivel máximo en los lugares de trabajo o los hogares que tienen aparatos de combustión defectuosos o mal ventilados o bien con corrientes descendentes o contracorrientes.

Es escasa la correlación entre los síntomas y los signos de la intoxicación aguda por monóxido de carbono y el nivel de carboxihemoglobina medido en el momento de la llegada al hospital. Los niveles de carboxihemoglobina inferiores al 10% no se suelen asociar con síntomas. A concentraciones mayores de carboxihemoglobina, del 10%-30%, pueden producirse síntomas neurológicos debidos a la intoxicación por monóxido de carbono, por ejemplo dolor de cabeza, mareos, debilidad, náuseas, confusión, desorientación y alteraciones visuales. Con una exposición continua, que produce niveles de carboxihemoglobina del 30% al 50%, se observa disnea por esfuerzo, aumento del ritmo del pulso y la respiración y síncope.

Cuando las concentraciones de carboxihemoglobina son superiores al 50%, puede producirse coma, convulsiones y paro cardiopulmonar.

En la intoxicación por monóxido de carbono se producen con frecuencia complicaciones (muerte inmediata, trastornos del miocardio, hipotensión, arritmias, edema pulmonar). Tal vez el efecto más insidioso de la intoxicación por monóxido de carbono sea la aparición retardada de trastornos neuropsiquiátricos en un plazo de una a tres semanas, junto con consecuencias para el neurocomportamiento, especialmente en los niños. La intoxicación por monóxido de carbono durante el embarazo representa un riesgo elevado para la madre, aumentando la tasa de complicaciones a corto plazo, y también para el feto, provocando la muerte fetal, trastornos del desarrollo y lesiones anóxicas cerebrales. Además, la gravedad de la intoxicación fatal no se puede evaluar por la tasa materna.

La intoxicación por monóxido de carbono se produce con frecuencia, tiene consecuencias graves, incluso la muerte inmediata, lleva consigo complicaciones y secuelas tardías y con frecuencia se la pasa por alto. Hay que fomentar los esfuerzos de prevención y de educación pública y médica.

10. Valores indicativos recomendados por la OMS

Los siguientes valores indicativos (valores redondeados en ppm) y los períodos de exposición como promedio ponderado por el tiempo se han determinado de manera que no se supere la concentración del 2,5% de carboxihemoglobina, incluso cuando una persona normal realice un ejercicio ligero o moderado:

100 mg/m^3 (87 ppm) durante 15 min
60 mg/m^3 (52 ppm) durante 30 min
30 mg/m^3 (26 ppm) durante una hora
10 mg/m^3 (9 ppm) durante ocho horas

INDEX

Absorption, 5, 124, 133
Adaptation, 14, 15, 200, 210, 211, 272-274
    at high altitude, 336
    in smokers, 17, 278, 323, 346
Age
    and carboxyhaemoglobin level, 99
    as risk factor, 12, 15, 18, 174, 336, 337
Agricultural burning, 42, 43, 48
Aircraft, 40, 124-126, 132, 133, 326
Air pollutants, 16-18, 240, 241, 248, 249, 255, 275-277, 300-305, 316, 348
    — *see also* individual air pollutants
Alcohol, 15, 16, 237-240, 274, 275
Altitude, 18, 101, 123, 133, 172, 210, 316
    in combination with carbon monoxide exposure, 15, 200, 210, 246-248, 319-327, 335, 336, 342, 343, 348
    effect on carbon monoxide concentrations, 67
    effect on carbon monoxide emissions, 101, 320, 321
    effect on carboxyhaemoglobin formation, 321, 322
    effect on oxygen partial pressure, 124
    and haematocrit ratio, 177
    and pulmonary hypertension, 176
Ambient air monitoring, 2, 3, 25, 71-77, 89, 93, 345
Anaemia, 123, 135, 137, 141, 294, 295, 335, 340, 348
Angina, 8, 9, 126, 280-293, 298, 338
Arrhythmogenic effects, 9, 10, 19, 163-166, 201, 202, 295-298, 308, 339
Atherosclerosis, 10, 12, 179-186, 205, 206
Atmospheric lifetime, 60-62, 67, 69
Auditory effects, 263, 277, 278

Barbiturates, 239 — *see also* Drugs; Psychoactive drugs
Behaviour, 11, 12, 190-200, 209, 217, 218, 235, 259-272, 274, 275, 336 — *see also* Neurobehavioural effects
Binding
    to cytochrome oxidase, 5, 7, 154, 156-160
    to cytochrome P-450, 5, 155, 156
    to haemoglobin, 7, 8, 137, 140, 150
    to myoglobin, 5, 7, 150-155
    of oxygen to haemoglobin, 5, 35, 140, 150, 212, 213
Biological exposure index, 122, 125
Biological monitoring — *see* Measurement, in blood; Measurement, in exhaled breath
Biological tolerance limit, 122

Smoking 70
Occupational Exposures 71, 77
Fixed site vs personal Exp 75
Commuting 77
Home studies well Gas stus 78
Table of CO Concentrations in microenvironments 110
Tobacco, 118

Food additives and contaminants in food, principles for the safety assessment of (No. 70, 1987)
Formaldehyde (No. 89, 1989)
Genetic effects in human populations, guidelines for the study of (No. 46, 1985)
Glyphosate (No. 159, 1994)
Guidance values for human exposure limits (No. 170, 1994)
Heptachlor (No. 38, 1984)
Hexachlorobenzene (No. 195, 1997)
Hexachlorobutadiene (No. 156, 1994)
Alpha- and beta-hexachlorocyclohexanes (No. 123, 1992)
Hexachlorocyclopentadiene (No. 120, 1991)
n-Hexane (No. 122, 1991)
Hydrazine (No. 68, 1987)
Hydrogen sulfide (No. 19, 1981)
Hydroquinone (No. 157, 1994)
Immunotoxicity associated with exposure to chemicals, principles and methods for assessment (No. 180, 1996)
Infancy and early childhood, principles for evaluating health risks from chemicals during (No. 59, 1986)
Isobenzan (No. 129, 1991)
Isophorone (No. 174, 1995)
Kelevan (No. 66, 1986)
Lasers and optical radiation (No. 23, 1982)
Lead (No. 3, 1977)[a]
Lead, inorganic (No. 165, 1995)
Lead – environmental aspects (No. 85, 1989)
Lindane (No. 124, 1991)
Linear alkylbenzene sulfonates and related compounds (No. 169, 1996)
Magnetic fields (No. 69, 1987)
Man-made mineral fibres (No. 77, 1988)
Manganese (No. 17, 1981)
Mercury (No. 1, 1976)[a]
Mercury – environmental aspects (No. 86, 1989)
Mercury, inorganic (No. 118, 1991)
Methanol (No. 196, 1997)
Methomyl (No. 178, 1996)
2-Methoxyethanol, 2-ethoxyethanol, and their acetates (No. 115, 1990)
Methyl bromide (No. 166, 1995)
Methylene chloride (No. 32, 1984, 1st edition) (No. 164, 1996, 2nd edition)
Methyl ethyl ketone (No. 143, 1992)
Methyl isobutyl ketone (No. 117, 1990)
Methylmercury (No. 101, 1990)
Methyl parathion (No. 145, 1992)
Methyl tertiary-butyl ether (No. 206, 1998)
Mirex (No. 44, 1984)
Morpholine (No. 179, 1996)
Mutagenic and carcinogenic chemicals, guide to short-term tests for detecting (No. 51, 1985)

Mycotoxins (No. 11, 1979)
Mycotoxins, selected: ochratoxins, trichothecenes, ergot (No. 105, 1990)
Nephrotoxicity associated with exposure to chemicals, principles and methods for the assessment of (No. 119, 1991)
Neurotoxicity associated with exposure to chemicals, principles and methods for the assessment of (No. 60, 1986)
Nickel (No. 108, 1991)
Nitrates, nitrites, and N-nitroso compounds (No. 5, 1978)[a]
Nitrogen oxides (No. 4, 1977, 1st edition)[a] (No. 188, 1997, 2nd edition)
2-Nitropropane (No. 138, 1992)
Noise (No. 12, 1980)[a]
Organophosphorus insecticides: a general introduction (No. 63, 1986)
Paraquat and diquat (No. 39, 1984)
Pentachlorophenol (No. 71, 1987)
Permethrin (No. 94, 1990)
Pesticide residues in food, principles for the toxicological assessment of (No. 104, 1990)
Petroleum products, selected (No. 20, 1982)
Phenol (No. 161, 1994)
d-Phenothrin (No. 96, 1990)
Phosgene (No. 193, 1997)
Phosphine and selected metal phosphides (No. 73, 1988)
Photochemical oxidants (No. 7, 1978)
Platinum (No. 125, 1991)
Polybrominated biphenyls (No. 152, 1994)
Polybrominated dibenzo-p-dioxins and dibenzofurans (No. 205, 1998)
Polychlorinated biphenyls and terphenyls (No. 2, 1976, 1st edition)[a] (No. 140, 1992, 2nd edition)
Polychlorinated dibenzo-p-dioxins and dibenzofurans (No. 88, 1989)
Polycyclic aromatic hydrocarbons, selected non-heterocyclic (No. 202, 1998)
Progeny, principles for evaluating health risks associated with exposure to chemicals during pregnancy (No. 30, 1984)
1-Propanol (No. 102, 1990)
2-Propanol (No. 103, 1990)
Propachlor (No. 147, 1993)
Propylene oxide (No. 56, 1985)
Pyrrolizidine alkaloids (No. 80, 1988)
Quintozene (No. 41, 1984)
Quality management for chemical safety testing (No. 141, 1992)
Radiofrequency and microwaves (No. 16, 1981)
Radionuclides, selected (No. 25, 1983)
Resmethrins (No. 92, 1989)
Synthetic organic fibres, selected (No. 151, 1993)
Selenium (No. 58, 1986)